Poincaré-Andronov-Melnikov Analysis for Non-Smooth Systems

Poincaré-Andronov-Melnikov Analysis for Non-Smooth Systems

Michal Fečkan
Faculty of Mathematics, Physics and Informatics, Comenius
University in Bratislava, Department of Mathematical Analysis
and Numerical Mathematics, Mlynská dolina, 842 48 Bratislava,
Slovak Republic

Michal Pospíšil
Mathematical Institute of Slovak Academy of Sciences,
Štefánikova 49, 814 73 Bratislava, Slovak Republic

AMSTERDAM · BOSTON · HEIDELBERG · LONDON
NEW YORK · OXFORD · PARIS · SAN DIEGO
SAN FRANCISCO · SINGAPORE · SYDNEY · TOKYO
Academic Press is an imprint of Elsevier

Academic Press is an imprint of Elsevier
32 Jamestown Road, London NW1 7BY, UK
525 B Street, Suite 1800, San Diego, CA 92101-4495, USA
50 Hampshire Street, 5th Floor, Cambridge, MA 02139, USA
The Boulevard, Langford Lane, Kidlington, Oxford OX5 1GB, UK

Library of Congress Cataloging-in-Publication Data
A catalog record for this book is available from the Library of Congress

British Library Cataloguing in Publication Data
A catalogue record for this book is available from the British Library

ISBN: 978-0-12-804294-6

For information on all Academic Press Publications
visit our website at http://www.elsevier.com/

  Working together
to grow libraries in
developing countries

www.elsevier.com • www.bookaid.org

Publisher: Nikki Levi
Acquisition Editor: Graham Nisbet
Editorial Project Manager: Susan Ikeda
Production Project Manager: Poulouse Joseph
Designer: Mark Rogers

Typeset by SPi Global, India

To our beloved families

CONTENTS

ACKNOWLEDGMENT

Partial support of Grants VEGA-MS 1/0071/14 and VEGA-SAV 2/0153/16, an award from Literárny fond and by the Slovak Research and Development Agency under contract No. APVV-14-0378 are appreciated.

Michal Fečkan and Michal Pospíšil
April 2016

PREFACE

Discontinuous systems describe many real processes characterized by instantaneous changes, such as electrical switching or impacts of a bouncing ball. This is the reason why many papers and books have appeared on this topic in the last few years. This book is a contribution to this direction; namely, it is devoted to the study of bifurcations of periodic solutions for general n-dimensional discontinuous systems. First, we study these systems under assumptions of transversal intersections with discontinuity/switching boundaries and sufficient conditions are derived for the persistence of single periodic solutions under nonautonomous perturbations from single solutions; or under autonomous perturbations from non-degenerate families of solutions; or from isolated solutions. Furthermore, bifurcations of periodic sliding solutions are studied from sliding periodic solutions of unperturbed discontinuous equations. Then bifurcations of forced periodic solutions are investigated for impact systems from single periodic solutions of unperturbed impact equations. We also study weakly coupled discontinuous systems. In addition, local asymptotic properties of derived perturbed periodic solutions are investigated for all studied problems. The relationship between non-smooth systems and their continuous approximations is investigated as well. Many examples of discontinuous ordinary differential equations and impact systems are given to illustrate the theoretical results. To achieve our results, we mostly use the so-called discontinuous Poincaré mapping, which maps a point to its position after one period of the periodic solution. This approach is rather technical. On the other hand, by this method we can get results for general dimensions of spatial variables and parameters as well as asymptotic results such as stability, instability and hyperbolicity of solutions. Moreover, we explain how this approach can be modified for differential inclusions. These are the aims of this book and make it unique, since no one else in any book has ever before studied bifurcations of periodic solutions in discontinuous systems in such general settings. Therefore, our results in this book are original.

Some parts of this book are related to our previous works. But we are substantially improving these results, give more details in the proofs and present more examples. Needless to say, this book contains brand new parts. So the aim of this book is to collect and improve our previous results, as well as to continue with new results. Numerical computations described by figures are given with the help of the computational software *Mathematica*.

This book is intended for post-graduate students, mathematicians, physicists and theoretically inclined engineers studying either oscillations of nonlinear discontinuous mechanical systems or electrical circuits by applying the modern theory of bifurcation methods in dynamical systems.

Michal Fečkan and Michal Pospíšil
Bratislava, Slovakia
April 2016

ABOUT THE AUTHORS

Michal Fečkan is Professor of Mathematics at the Department of Mathematical Analysis and Numerical Mathematics on the Faculty of Mathematics, Physics and Informatics at the Comenius University in Bratislava, Slovak Republic. He obtained his Ph.D. (mathematics) from the Mathematical Institute of Slovak Academy of Sciences in Bratislava, Slovak Republic. He is interested in nonlinear functional analysis, bifurcation theory and dynamical systems with applications to mechanics and vibrations.

Michal Pospíšil is senior researcher at the Mathematical Institute of Slovak Academy of Sciences in Bratislava, Slovak Republic. He obtained his Ph.D. (applied mathematics) from the Mathematical Institute of Slovak Academy of Sciences in Bratislava, Slovak Republic. He is interested in discontinuous dynamical systems and delayed differential equations.

An introductory example

Let us consider a reflected pendulum sketched in Figure 0.1.

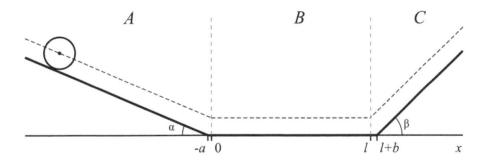

Figure 0.1 Reflected pendulum

In part A, the ball of radius $r > 0$ is moving on an inclined plane due to gravity. Its position with respect to the slanted surface obeys $\ddot{z} = g \sin \alpha$, i.e. the horizontal position satisfies $\ddot{x} = g \sin \alpha \cos \alpha$. When $x = 0$, the ball touches the horizontal segment resulting in the change $\dot{x} \mapsto \lambda_1 \dot{x}$, $0 < \lambda_1 \leq 1$. So we consider the case of a heavy ball, which is the reason why the ball does not jump above the surface (see [1, p. 123] for more details). Next, it moves in a horizontal rectilinear motion with a constant speed, $\ddot{x} = 0$. At $x = l \geq 0$, the ball hits the other slanted plane, and its horizontal speed is changed by $\dot{x} \mapsto \lambda_2 \dot{x}$, $0 < \lambda_2 \leq 1$. In part C, gravity slows down the ball, i.e. $\ddot{x} = -g \sin \beta \cos \beta$. Note that in the limit case $l = 0$, the system is reduced to a single

Poincaré-Andronov-Melnikov Analysis for Non-Smooth Systems.
http://dx.doi.org/10.1016/B978-0-12-804294-6.50001-1

impact at $x = 0$. The distances $a = r \tan \frac{\alpha}{2}$, $b = r \tan \frac{\beta}{2}$ describe positions of slanted segments so the ball with the center at $x = 0$ or $x = l$ touches the horizontal and one of the slanted segments simultaneously. Setting $\lambda_3 = g \sin \alpha \cos \alpha$ and $\lambda_4 = g \sin \beta \cos \beta$, we get the equation

$$
\begin{aligned}
&\dot{x} = y, \quad \dot{y} = \lambda_3 \quad \text{for} \quad x < 0, \\
&x(t_1^+) = x(t_1^-) = 0, \quad y(t_1^+) = \lambda_1 y(t_1^-), \\
&\dot{x} = y, \quad \dot{y} = 0 \quad \text{for} \quad 0 < x < l, \\
&x(t_2^+) = x(t_2^-) = l, \quad y(t_2^+) = \lambda_2 y(t_2^-), \\
&\dot{x} = y, \quad \dot{y} = -\lambda_4 \quad \text{for} \quad l < x,
\end{aligned}
\tag{0.1}
$$

where times t_1 and t_2 are unknown. When $\lambda_1 = \lambda_2 = 1$, then (0.1) is reduced to

$$
\begin{aligned}
&\dot{x} = y, \quad \dot{y} = \lambda_3 \quad \text{for} \quad x < 0, \\
&\dot{x} = y, \quad \dot{y} = 0 \quad \text{for} \quad 0 < x < l, \\
&\dot{x} = y, \quad \dot{y} = -\lambda_4 \quad \text{for} \quad l < x.
\end{aligned}
\tag{0.2}
$$

System (0.2) is piecewise-linear, i.e. the plane (x, y) is divided by the lines $x = 0$ and $x = l$ into three regions

$$
A = (-\infty, 0) \times \mathbb{R}, \quad B = (0, l) \times \mathbb{R}, \quad C = (l, \infty) \times \mathbb{R},
$$

where on each of these regions the vector field of (0.2) is fixed and linear. Of course, (0.2) can be continuous on $\mathbb{R}^2$ only if $\lambda_3 = \lambda_4 = 0$, i.e. $\alpha = \beta = 0$, which we do not consider, since it is trivial. The aim of this book is to study the other kind, the discontinuous/non-smooth systems. Moreover, when either $0 < \lambda_1 < 1$ or $0 < \lambda_2 < 1$, then (0.1) contains (0.2) with additional impact conditions when the ball is passing through $x = 0$ and $x = l$, respectively. We call systems like (0.1) hybrid. Systems of this kind are also studied in Part II.

Now we study in more detail the dynamics of (0.1) and (0.2). It is clear that the function

$$
\frac{y^2}{2} + h(x)
\tag{0.3}
$$

for

$$
h(x) = -\lambda_3 x \frac{1 - \operatorname{sgn} x}{2} + \lambda_4 (x - l) \frac{1 + \operatorname{sgn}(x - l)}{2}
$$

is a first integral of (0.1) and (0.2) in the region $O = A \cup B \cup C$. First we study (0.2). Any of its solutions with $x(0) < 0$ reaches a time $t_0 \in \mathbb{R}$ when $y(t_0) = 0$ and $x(t_0) < 0$. So shifting the time, we can consider (0.2) with initial value conditions $x(0) = \xi < 0$

and $y(0) = 0$. Then the solution satisfies

$$\frac{y^2}{2} + h(x) = -\lambda_3 \xi > 0. \tag{0.4}$$

The contour plot of (0.4) consists of periodic curves as can be seen in Figure 0.2.

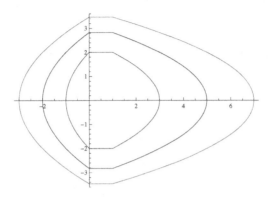

Figure 0.2 Solutions of (0.2) with $l = 1$, $\lambda_3 = 2$ and $\lambda_4 = 1$

Now we switch to the system (0.1) with possibly damping impacts. The general solution is given by

$$x(t) = \xi + \frac{\lambda_3 t^2}{2}, \quad y(t) = \lambda_3 t \quad \text{for} \quad t \in [0, t_1),$$

$$x(t) = \lambda_1 \lambda_3 t_1 (t - t_1), \quad y(t) = \lambda_1 \lambda_3 t_1 \quad \text{for} \quad t \in (t_1, t_2),$$

$$x(t) = l + \lambda_1 \lambda_2 \lambda_3 t_1 (t - t_2) - \lambda_4 \frac{(t - t_2)^2}{2},$$

$$y(t) = \lambda_1 \lambda_2 \lambda_3 t_1 - \lambda_4 (t - t_2) \quad \text{for} \quad t \in (t_2, \bar{t}_2),$$

$$x(t) = l - \lambda_1 \lambda_2^2 \lambda_3 t_1 (t - \bar{t}_2), \quad y(t) = -\lambda_1 \lambda_2^2 \lambda_3 t_1 \quad \text{for} \quad t \in (\bar{t}_2, \bar{t}_1),$$

$$x(t) = -\lambda_1^2 \lambda_2^2 \lambda_3 t_1 (t - \bar{t}_1) + \lambda_3 \frac{(t - \bar{t}_1)^2}{2},$$

$$y(t) = -\lambda_1^2 \lambda_2^2 \lambda_3 t_1 + \lambda_3 (t - \bar{t}_1) \quad \text{for} \quad t \in (\bar{t}_1, T], \tag{0.5}$$

where

$$t_1 = \sqrt{-\frac{2\xi}{\lambda_3}}, \quad t_2 = t_1 + \frac{l}{\lambda_1 \lambda_3 t_1}, \quad \bar{t}_2 = t_2 + \frac{2\lambda_1 \lambda_2 \lambda_3 t_1}{\lambda_4},$$

$$\bar{t}_1 = \bar{t}_2 + \frac{l}{\lambda_1 \lambda_2^2 \lambda_3 t_1}, \quad T = \bar{t}_1 + \lambda_1^2 \lambda_2^2 t_1.$$

Note

$$x(t_1) = x(\bar{t}_1) = 0, \quad x(t_2) = x(\bar{t}_2) = l, \quad x(T) = \lambda_1^4 \lambda_2^4 \xi, \quad y(T) = 0.$$

Taking the section $\Sigma = (-\infty, 0)$, we get the Poincaré mapping $P \colon \Sigma \to \Sigma$ of (0.1) given by

$$P(\xi) = \lambda_1^4 \lambda_2^4 \xi, \quad \xi \in \Sigma. \tag{0.6}$$

This is a generalization of Poincaré mappings for continuous dynamical systems [2–4] to discontinuous/non-smooth ones. Taking $\lambda_1 = \lambda_2 = 1$, we get (0.2) with $P(\xi) = \xi$, $\xi \in \Sigma$, which corresponds to the above observation on the existence of periodic solutions with periods

$$T(\xi) = 2\sqrt{-\frac{2\xi}{\lambda_3}} \left(-\frac{l}{2\xi} + \frac{\lambda_3}{\lambda_4} + 1 \right) \geq T_0 = 4\sqrt{\frac{l}{\lambda_3} + \frac{l}{\lambda_4}}, \tag{0.7}$$

when the equality is achieved at $\xi_0 = -\frac{\lambda_4 l}{2(\lambda_3 + \lambda_4)}$. Furthermore, the function $T(\xi)$ is decreasing on $(-\infty, \xi_0]$ from ∞ to T_0, and increasing on $[\xi_0, 0)$ from T_0 to ∞. Thus the equation

$$T(\xi) = \bar{T}$$

has a solution if and only if $\bar{T} \geq T_0$, and has two different solutions for

$$\bar{T} > T_0.$$

When also $l = 0$ and there is a string force, then (0.2) is the reflection pendulum mentioned in [5] (see also Section II.1.3) with phase portrait in Figure III.1.6.

Now we consider a periodically forced and weakly damped (0.1) of the form

$$\begin{aligned}
\dot{x} &= y, \quad \dot{y} = \lambda_3 - \varepsilon\eta_4 y + \varepsilon\eta_3 \cos\omega t \quad \text{for} \quad x < 0, \\
&x(\tilde{t}_1^+) = x(\tilde{t}_1^-) = 0, \quad y(\tilde{t}_1^+) = (1 + \varepsilon\eta_1)y(\tilde{t}_1^-), \\
\dot{x} &= y, \quad \dot{y} = -\varepsilon\eta_4 y + \varepsilon\eta_3 \cos\omega t \quad \text{for} \quad 0 < x < l, \\
&x(\tilde{t}_2^+) = x(\tilde{t}_2^-) = l, \quad y(\tilde{t}_2^+) = (1 + \varepsilon\eta_2)y(\tilde{t}_2^-), \\
\dot{x} &= y, \quad \dot{y} = -\lambda_4 - \varepsilon\eta_4 y + \varepsilon\eta_3 \cos\omega t \quad \text{for} \quad l < x
\end{aligned} \tag{0.8}$$

where $\tilde{t}_1, \tilde{t}_2$ are hitting times, $\omega > 0$, η_i, $i = 1, 2, 3, 4$ are constants and ε is a small parameter. We are interested in the persistence of periodic solutions of (0.2) to perturbed (0.8). The first step is the resonance condition

$$T(\xi) = \frac{2\pi k}{\omega} \tag{0.9}$$

for some $\xi < 0$ and $k \in \mathbb{N}$. So we start with

$$\frac{2\pi k}{\omega} > T_0. \tag{0.10}$$

Then (0.9) has two different solutions $\xi_1 < \xi_2 < 0$. Our task is to show that (0.8) has a $\bar{T} = \frac{2\pi k}{\omega}$-periodic solution for any $\varepsilon \neq 0$ small near (0.5) with either ξ_1 or ξ_2, and $\lambda_1 = \lambda_2 = 1$. To achieve this, we construct a stroboscopic Poincaré mapping of (0.8). So we consider the initial values $x(0) = \xi < 0$ and $y(0) = 0$. Since (0.8) is nonautonomous, we need to shift the time to consider

$$
\begin{aligned}
\dot{x} = y, \quad \dot{y} &= \lambda_3 - \varepsilon\eta_4 y + \varepsilon\eta_3 \cos\omega(t + \alpha) \quad \text{for} \quad x < 0, \\
x(\tilde{t}_1^+) &= x(\tilde{t}_1^-) = 0, \quad y(\tilde{t}_1^+) = (1 + \varepsilon\eta_1)y(\tilde{t}_1^-), \\
\dot{x} = y, \quad \dot{y} &= -\varepsilon\eta_4 y + \varepsilon\eta_3 \cos\omega(t + \alpha) \quad \text{for} \quad 0 < x < l, \\
x(\tilde{t}_2^+) &= x(\tilde{t}_2^-) = l, \quad y(\tilde{t}_2^+) = (1 + \varepsilon\eta_2)y(\tilde{t}_2^-), \\
\dot{x} = y, \quad \dot{y} &= -\lambda_4 - \varepsilon\eta_4 y + \varepsilon\eta_3 \cos\omega(t + \alpha) \quad \text{for} \quad l < x \\
x(0) &= \xi < 0, \quad y(0) = 0
\end{aligned}
\tag{0.11}
$$

for a parameter $\alpha \in \mathbb{R}$. Then as above we can construct a stroboscopic Poincaré mapping by taking the solution $(x(\cdot), y(\cdot))$ of (0.11) and setting

$$P(\xi, \varepsilon, \alpha) = (x(\bar{T}), y(\bar{T})). \tag{0.12}$$

To get a $\bar{T}$-periodic solution, we need to solve

$$P(\xi, \varepsilon, \alpha) = (\xi, 0). \tag{0.13}$$

Equation (0.13) is a system of two equations with three unknowns. Our aim is to solve $\xi = \xi(\varepsilon)$ and $\alpha = \alpha(\varepsilon)$ for ε small. The analytical procedure is derived in Section II.1.3. A similar problem for forced billiards is studied in Chapter II.2.

We roughly explained above the main idea and the aim of this book: To present a complex method for the persistence of periodic solutions and their local asymptotic properties for periodically perturbed non-smooth, including hybrid, systems in any finite spatial dimension. Of course, we also present more examples. Furthermore, there is a vast literature from those working in this area, such as [5–31]. Some of these results are considered in more detail in this book.

REFERENCE

[1] E. DiBenedetto, *Classical Mechanics: Theory and Mathematical Modeling*, Cornerstones, Birkhäuser 2011.
[2] C. Chicone, *Ordinary Differential Equations with Applications*, Texts in Applied Mathematics 34, Springer 2006.
[3] S. N. Chow, J. K. Hale, *Methods of Bifurcation Theory*, Texts in Applied Mathematics 34, Springer-Verlag 1982.
[4] J. Guckenheimer, P. Holmes, *Nonlinear Oscillations, Dynamical Systems and Bifurcations of Vector*

Fields, Springer-Verlag 1983.

[5] M. Kunze, *Non-smooth Dynamical Systems*, Lecture Notes in Mathematics 1744, Springer 2000.

[6] V. Acary, O. Bonnefon, B. Brogliato, *Nonsmooth Modeling and Simulation for Switched Circuits*, Springer 2011.

[7] Z. Afsharnezhad, M. Karimi Amaleh, Continuation of the periodic orbits for the differential equation with discontinuous right hand side, *J. Dynam. Differential Equations* **23** (2011) 71–92.

[8] M. U. Akhmet, Periodic solutions of strongly nonlinear systems with non classical right-side in the case of a family of generating solutions, *Ukrainian Math. J.* **45** (1993) 215–222.

[9] M. U. Akhmet, D. Arugaslan, Bifurcation of a non-smooth planar limit cycle from a vertex, *Nonlinear Anal.* **71** (2009) 2723–2733.

[10] J. Andres, L. Górniewicz, *Topological Fixed Point Principles for Boundary Value Problems*, Kluwer 2003.

[11] A. A. Andronov, A. A. Vitt, S. E. Khaikin, *Theory of Oscillators*, Pergamon Press 1966.

[12] J. Awrejcewicz, M. M. Holicke, *Smooth and Nonsmooth High Dimensional Chaos and the Melnikov-Type Methods*, World Scientific Publishing Company 2007.

[13] M. di Bernardo, C. J. Budd, A. R. Champneys, P. Kowalczyk, *Piecewise-smooth Dynamical Systems: Theory and Applications*, Applied Mathematical Sciences 163, Springer-Verlag 2008.

[14] B. Brogliato, *Nonsmooth Impact Mechanics*, Lecture Notes in Control and Information Sciences 220, Springer 1996.

[15] L. O. Chua, M. Komuro, T. Matsumoto, The double scroll family, *IEEE Trans. Circuits Syst.* **33** (1986) 1073–1118.

[16] M. Fečkan, M. Pospíšil, On the bifurcation of periodic orbits in discontinuous sytems, *Commun. Math. Anal.* **8** (2010) 87–108.

[17] M. Fečkan, M. Pospíšil, Bifurcation from family of periodic orbits in discontinuous systems, *Differ. Equ. Dyn. Syst.* **20** (2012) 207–234.

[18] M. Fečkan, M. Pospíšil, Bifurcation from single periodic orbit in discontinuous autonomous systems, *Appl. Anal.* **92** (2013) 1085–1100.

[19] M. Fečkan, M. Pospíšil, Bifurcation of periodic orbits in periodically forced impact systems, *Math. Slovaca* **64** (2014) 101–118.

[20] M. Fečkan, M. Pospíšil, Bifurcation of sliding periodic orbits in periodically forced discontinuous systems, *Nonlinear Anal. Real World Appl.* **14** (2013) 150–162.

[21] M. Fečkan, M. Pospíšil, Discretization of dynamical systems with first integrals, *Discrete Cont. Dyn. Syst.* **33** (2013) 3543–3554.

[22] A. Fidlin, *Nonlinear Oscillations in Mechanical Engineering*, Springer 2006.

[23] A. F. Filippov, *Differential Equations with Discontinuous Righthand Sides*, Mathematics and Its Applications 18, Kluwer Academic 1988.

[24] U. Galvanetto, C. Knudsen, Event maps in a stick-slip system, *Nonlinear Dynam.* **13** (1997) 99–115.

[25] A. Kovaleva, The Melnikov criterion of instability for random rocking dynamics of a rigid block with an attached secondary structure, *Nonlinear Anal. Real World Appl.* **11** (2010) 472–479.

[26] M. Kunze, T. Küpper, *Non-smooth dynamical systems: an overview*, Springer 2001 pp. 431–452.

[27] M. Kunze, T. Küpper, Qualitative bifurcation analysis of a non-smooth friction-oscillator model, *Z. Angew. Math. Phys.* **48** (1997) 87–101.

[28] R. I. Leine, H. Nijmeijer, *Dynamics and Bifurcations of Non-smooth Mechanical Systems*, Lecture Notes in Applied and Computational Mechanics 18, Springer-Verlag 2004.

[29] S. Lenci, G. Rega, Heteroclinic bifurcations and optimal control in the nonlinear rocking dynamics of generic and slender rigid blocks, *Internat. J. Bifur. Chaos Appl. Sci. Engrg.* **15** (2005) 1901–1918.

[30] J. Llibre, O. Makarenkov, Asymptotic stability of periodic solutions for nonsmooth differential equations with application to the nonsmooth van der Pol oscillator, *SIAM J. Math. Anal.* **40** (2009) 2478–2495.

[31] W. Xu, J. Feng, H. Rong, Melnikov's method for a general nonlinear vibro-impact oscillator, *Nonlinear Anal.* **71** (2009) 418–426.

PART I

Piecewise-smooth systems of forced ODEs

Introduction

This part is devoted to perturbed piecewise-smooth nonlinear dynamical systems (NDS) under which we understand a differential equation

$$\dot{x} = F(t, x, \chi),$$

where $F(t, x, \chi)$ is a smooth function on $\mathbb{R} \times (\mathbb{R}^n \backslash S) \times \mathbb{R}^m$ periodic in $t \in \mathbb{R}$. Here S denotes the discontinuity set – in this work it is a sufficiently smooth hypersurface of $\mathbb{R}^n$. Moreover, we suppose that $F(t, x, 0) = F(x)$, i.e. function $F(t, x, \chi)$ is independent of t at $\chi_0 = 0$ and the associated autonomous system

$$\dot{x} = F(x)$$

possesses a periodic solution $\gamma(t)$ that hits S. For the case of transverse crossing we prove the persistence of a periodically forced isolated solution and the bifurcation from a non-degenerate family or a single periodic solution under autonomous perturbation (here $F(t, x, \chi) = F(x, \chi)$ for any χ). If $\gamma(t)$ does not cross the boundary, we investigate a periodically forced sliding periodic solution. We also investigate hyperbolicity, stability and instability of persisting solutions. All theoretical results are illustrated by concrete examples different from known works such as [1–8].

CHAPTER I.1

Periodically forced discontinuous systems

I.1.1. Setting of the problem and main results

In this chapter, we investigate the persistence of a periodic orbit in an autonomous discontinuous system under a small nonautonomous perturbation. More precisely, we assume that the unperturbed equation possesses a periodic solution that transversally crosses the discontinuity boundary, and we look for sufficient conditions on the perturbation such that the perturbed equation has a periodic solution which is close to the original one and has the same period.

Now we formulate the problem. Let $\Omega \subset \mathbb{R}^n$ be an open set in $\mathbb{R}^n$ and $h(x)$ be a C^r-function on $\overline{\Omega}$, with $r \geq 2$. We set $\Omega_\pm := \{x \in \Omega \mid \pm h(x) > 0\}$, $\Omega_0 := \{x \in \Omega \mid h(x) = 0\}$. Let $f_\pm \in C_b^r(\overline{\Omega})$, $g \in C_b^r(\overline{\Omega} \times \mathbb{R} \times \mathbb{R} \times \mathbb{R}^p)$ and $h \in C_b^r(\overline{\Omega}, \mathbb{R})$. Furthermore, we suppose that g is T-periodic in $t \in \mathbb{R}$ and 0 is a regular value of h. Let $\varepsilon, \alpha \in \mathbb{R}$ and $\mu \in \mathbb{R}^p, p \geq 1$ be parameters.

Definition I.1.1. We say that a function $x(t)$ is a solution of the equation

$$\dot{x} = f_\pm(x) + \varepsilon g(x, t + \alpha, \varepsilon, \mu), \quad x \in \overline{\Omega}_\pm, \tag{I.1.1}$$

if it is continuous, piecewise C^1, satisfies equation (I.1.1) on $\Omega_\pm$ and, moreover, the following holds: if for some t_0 we have $x(t_0) \in \Omega_0$, then there exists $\rho > 0$ such that for any $t \in (t_0 - \rho, t_0)$ we have $x(t) \in \Omega_\pm$, and for any $t \in (t_0, t_0 + \rho)$ we have $x(t) \in \Omega_\mp$.

We assume (see Figure I.1.1):

H1) For $\varepsilon = 0$ equation (I.1.1) has a T-periodic solution $\gamma(t)$ which has a starting point $x_0 \in \Omega_+$ and consists of three branches

$$\gamma(t) = \begin{cases} \gamma_1(t) & \text{if } t \in [0, t_1], \\ \gamma_2(t) & \text{if } t \in [t_1, t_2], \\ \gamma_3(t) & \text{if } t \in [t_2, T], \end{cases} \tag{I.1.2}$$

where $0 < t_1 < t_2 < T$, $\gamma_1(t) \in \Omega_+$ for $t \in [0, t_1)$, $\gamma_2(t) \in \Omega_-$ for $t \in (t_1, t_2)$, $\gamma_3(t) \in$

Poincaré-Andronov-Melnikov Analysis for Non-Smooth Systems.
http://dx.doi.org/10.1016/B978-0-12-804294-6.50003-5

Ω_+ for $t \in (t_2, T]$, and

$$
\begin{aligned}
x_1 &:= \gamma_1(t_1) = \gamma_2(t_1) \in \Omega_0, \\
x_2 &:= \gamma_2(t_2) = \gamma_3(t_2) \in \Omega_0, \\
x_0 &:= \gamma_3(T) = \gamma_1(0) \in \Omega_+.
\end{aligned} \qquad (I.1.3)
$$

H2) Moreover, we also assume that

$$
Dh(x_1)f_\pm(x_1) < 0 \quad \text{and} \quad Dh(x_2)f_\pm(x_2) > 0.
$$

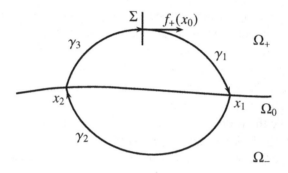

Figure I.1.1 Notation used

Let $x_\pm(\tau, \xi)(t, \varepsilon, \mu, \alpha)$ denote a solution of the initial value problem

$$
\begin{aligned}
\dot{x} &= f_\pm(x) + \varepsilon g(x, t + \alpha, \varepsilon, \mu) \\
x(\tau) &= \xi
\end{aligned} \qquad (I.1.4)_\pm
$$

with a corresponding sign.

As an example we consider a two-position automatic pilot for a ship's controller with periodic forcing [9], which after scaling parameters takes the form

$$
\begin{aligned}
\dot{u} &= v, \\
\dot{v} &= -v - 1 + \varepsilon \cos \omega t
\end{aligned} \qquad (I.1.5)
$$

in the region $\Omega_- = \{(u, v) \in \mathbb{R}^2 \mid u + \beta v > 0\}$, and

$$
\begin{aligned}
\dot{u} &= v, \\
\dot{v} &= -v + 1 + \varepsilon \cos \omega t
\end{aligned} \qquad (I.1.6)
$$

in the region $\Omega_+ = \{(u, v) \in \mathbb{R}^2 \mid u + \beta v < 0\}$ where $\beta, \omega \in \mathbb{R}$ are given constants. The unperturbed system is as follows

$$\dot{u} = v,$$
$$\dot{v} = -v - 1 \tag{I.1.7}$$

in the region Ω_-, and

$$\dot{u} = v,$$
$$\dot{v} = -v + 1 \tag{I.1.8}$$

in the region Ω_+. Certainly, now

$$h(u, v) = -u - \beta v, \quad f_\pm(u, v) = \begin{pmatrix} v \\ -v \pm 1 \end{pmatrix}, \quad g(u, v, t, \varepsilon, \mu) = \begin{pmatrix} 0 \\ \cos \omega t \end{pmatrix}.$$

We refer the reader to [9] for the full analysis of system (I.1.7), (I.1.8). Here we only note that for $\beta < 0$, system (I.1.7), (I.1.8) has a unique stable periodic solution (a limit cycle) given as follows: Let $T > 0$ be the unique solution of $1 - \beta = \frac{T}{4} \coth \frac{T}{4}$, i.e.

$$\beta = 1 - \frac{T}{4} \coth \frac{T}{4}. \tag{I.1.9}$$

Then we take $\Sigma = \{(u, 0) \in \mathbb{R}^2 \mid u < 0\}$ and

$$\gamma_1(t) = \begin{pmatrix} -1 + e^{-t} + t + \frac{T}{4} + \ln \frac{2}{1+e^{\frac{T}{2}}} \\ 1 - e^{-t} \end{pmatrix} \text{ for } t \in [0, t_1], \ t_1 = \ln \frac{1 + e^{\frac{T}{2}}}{2},$$

$$\gamma_2(t) = \begin{pmatrix} 1 - e^{\frac{T}{2}-t} - t + \frac{T}{4} - \ln \frac{2}{1+e^{\frac{T}{2}}} \\ e^{\frac{T}{2}-t} - 1 \end{pmatrix} \text{ for } t \in [t_1, t_2], \ t_2 = t_1 + \frac{T}{2},$$

$$\gamma_3(t) = \begin{pmatrix} -1 + e^{T-t} + t - \frac{3T}{4} + \ln \frac{2}{1+e^{\frac{T}{2}}} \\ 1 - e^{T-t} \end{pmatrix} \text{ for } t \in [t_2, T],$$

$$x_0 = \left(\ln \operatorname{sech} e^{\frac{T}{4}}, 0 \right), \quad x_1 = -x_2 = \left(\frac{T}{4} - \tanh \frac{T}{4}, \tanh \frac{T}{4} \right).$$

This system is studied in more detail in Section I.1.3.

Now we proceed with the study of (I.1.1). Using the implicit function theorem (IFT) [10], we show that there are some trajectories in the neighborhood of $\gamma(t)$, and then we select periodic ones from these.

Lemma I.1.2. *Assume* H1) *and* H2). *Then there exist* $\varepsilon_3, r_3 > 0$ *and a Poincaré mapping*

$$P(\cdot, \varepsilon, \mu, \alpha) \colon B(x_0, r_3) \to \Sigma$$

for all fixed $\varepsilon \in (-\varepsilon_3, \varepsilon_3)$, $\mu \in \mathbb{R}^p$, $\alpha \in \mathbb{R}$ *where* $\Sigma = \{x \in \mathbb{R}^n \mid \langle x - x_0, f_+(x_0) \rangle = 0\}$ *and* $B(x, r)$ *is the ball of radius* r *and center at* x. *Moreover,* P *is* C^r-*smooth in all arguments.*

Proof. We denote $\mathcal{A}(\tau, \xi, t, \varepsilon, \mu, \alpha) = h(x_+(\tau, \xi)(t, \varepsilon, \mu, \alpha))$. Since

$$\mathcal{A}(0, x_0, t_1, 0, \mu, \alpha) = 0, \quad D_t\mathcal{A}(0, x_0, t_1, 0, \mu, \alpha) = Dh(x_1)f_+(x_1) < 0,$$

IFT yields the existence of $\tau_1, r_1, \delta_1, \varepsilon_1 > 0$ and C^r-function

$$t_1(\cdot, \cdot, \cdot, \cdot, \cdot): (-\tau_1, \tau_1) \times B(x_0, r_1) \times (-\varepsilon_1, \varepsilon_1) \times \mathbb{R}^p \times \mathbb{R} \to (t_1 - \delta_1, t_1 + \delta_1)$$

such that $\mathcal{A}(\tau, \xi, t, \varepsilon, \mu, \alpha) = 0$ for $\tau \in (-\tau_1, \tau_1)$, $\xi \in B(x_0, r_1) \subset \Omega_+$, $\varepsilon \in (-\varepsilon_1, \varepsilon_1)$, $\mu \in \mathbb{R}^p$, $\alpha \in \mathbb{R}$ and $t \in (t_1 - \delta_1, t_1 + \delta_1)$ if and only if $t = t_1(\tau, \xi, \varepsilon, \mu, \alpha)$.

Next we set

$$\mathcal{B}(\tau, \xi, t, \varepsilon, \mu, \alpha) = h(x_-(t_1(\tau, \xi, \varepsilon, \mu, \alpha), x_+(\tau, \xi)(t_1(\tau, \xi, \varepsilon, \mu, \alpha), \varepsilon, \mu, \alpha))(t, \varepsilon, \mu, \alpha)).$$

Then

$$\mathcal{B}(0, x_0, t_2, 0, \mu, \alpha) = 0, \quad D_t\mathcal{B}(0, x_0, t_2, 0, \mu, \alpha) = Dh(x_2)f_-(x_2) > 0.$$

Hence IFT implies that there exist $\tau_2, r_2, \delta_2, \varepsilon_2 > 0$ and C^r-function

$$t_2(\cdot, \cdot, \cdot, \cdot, \cdot): (-\tau_2, \tau_2) \times B(x_0, r_2) \times (-\varepsilon_2, \varepsilon_2) \times \mathbb{R}^p \times \mathbb{R} \to (t_2 - \delta_2, t_2 + \delta_2)$$

such that $\mathcal{B}(\tau, \xi, t, \varepsilon, \mu, \alpha) = 0$ for $\tau \in (-\tau_2, \tau_2)$, $\xi \in B(x_0, r_2) \subset \Omega_+$, $\varepsilon \in (-\varepsilon_2, \varepsilon_2)$, $\mu \in \mathbb{R}^p$, $\alpha \in \mathbb{R}$ and $t \in (t_2 - \delta_2, t_2 + \delta_2)$ if and only if $t = t_2(\tau, \xi, \varepsilon, \mu, \alpha)$.

Once more we use IFT on function C defined as

$$C(\tau, \xi, t, \varepsilon, \mu, \alpha) = \langle x_+(t_2(\tau, \xi, \varepsilon, \mu, \alpha), x_-(t_1(\tau, \xi, \varepsilon, \mu, \alpha), x_+(\tau, \xi)$$
$$(t_1(\tau, \xi, \varepsilon, \mu, \alpha), \varepsilon, \mu, \alpha)) \ (t_2(\tau, \xi, \varepsilon, \mu, \alpha), \varepsilon, \mu, \alpha))(t, \varepsilon, \mu, \alpha) - x_0, f_+(x_0)\rangle.$$

Since

$$C(0, x_0, T, 0, \mu, \alpha) = 0, \quad D_tC(0, x_0, T, 0, \mu, \alpha) = \|f_+(x_0)\|^2 > 0,$$

there exist $\tau_3, r_3, \delta_3, \varepsilon_3 > 0$ and C^r-function

$$t_3(\cdot, \cdot, \cdot, \cdot, \cdot): (-\tau_3, \tau_3) \times B(x_0, r_3) \times (-\varepsilon_3, \varepsilon_3) \times \mathbb{R}^p \times \mathbb{R} \to (T - \delta_3, T + \delta_3)$$

such that $C(\tau, \xi, t, \varepsilon, \mu, \alpha) = 0$ for $\tau \in (-\tau_3, \tau_3)$, $\xi \in B(x_0, r_3) \subset \Omega_+$, $\varepsilon \in (-\varepsilon_3, \varepsilon_3)$, $\mu \in \mathbb{R}^p$, $\alpha \in \mathbb{R}$ and $t \in (T - \delta_3, T + \delta_3)$ if and only if $t = t_3(\tau, \xi, \varepsilon, \mu, \alpha)$. Moreover

$$t_1(0, x_0, 0, \mu, \alpha) = t_1, \quad t_2(0, x_0, 0, \mu, \alpha) = t_2, \quad t_3(0, x_0, 0, \mu, \alpha) = T.$$

Now we can define the Poincaré mapping from the statement

$$P(\xi, \varepsilon, \mu, \alpha) = x_+(t_2(0, \xi, \varepsilon, \mu, \alpha), x_-(t_1(0, \xi, \varepsilon, \mu, \alpha), x_+(0, \xi)(t_1(0, \xi, \varepsilon, \mu, \alpha), \varepsilon, \mu, \alpha))$$
$$(t_2(0, \xi, \varepsilon, \mu, \alpha), \varepsilon, \mu, \alpha))(t_3(0, \xi, \varepsilon, \mu, \alpha), \varepsilon, \mu, \alpha).$$

$$(I.1.10)$$

Obviously, P maps $B(x_0, r_3)$ to Σ. $\square$

Our aim is to find T-periodic orbits, which is the reason for solving the following system

$$P(\xi, \varepsilon, \mu, \alpha) = \xi$$
$$t_3(0, \xi, \varepsilon, \mu, \alpha) = T$$

for ξ and ε sufficiently close to x_0 and 0, respectively. This problem can be reduced to one equation

$$F(\xi, \varepsilon, \mu, \alpha) := \xi - \widetilde{P}(\xi, \varepsilon, \mu, \alpha) = 0 \qquad (I.1.11)$$

where

$$\widetilde{P}(\xi, \varepsilon, \mu, \alpha) = x_+(t_2(0, \xi, \varepsilon, \mu, \alpha), x_-(t_1(0, \xi, \varepsilon, \mu, \alpha), x_+(0, \xi)(t_1(0, \xi, \varepsilon, \mu, \alpha), \varepsilon, \mu, \alpha))$$
$$(t_2(0, \xi, \varepsilon, \mu, \alpha), \varepsilon, \mu, \alpha))(T, \varepsilon, \mu, \alpha)$$

$$(I.1.12)$$

is the so-called stroboscopic Poincaré mapping (cf. [2]). It is easy to see that $(\xi, \varepsilon) = (x_0, 0)$ solves equation (I.1.11) for any $\mu \in \mathbb{R}^p, \alpha \in \mathbb{R}$. However, IFT cannot be used here, which is proved in the next lemma (see [11, 12]).

Lemma I.1.3. *Let $\widetilde{P}(\xi, \varepsilon, \mu, \alpha)$ be defined by (I.1.12). Then $\widetilde{P}_\xi(x_0, 0, \mu, \alpha)$ has eigenvalue 1 with corresponding eigenvector $f_+(x_0)$, i.e. $\widetilde{P}_\xi(x_0, 0, \mu, \alpha) f_+(x_0) = f_+(x_0)$, where $\widetilde{P}_\xi$ denotes the partial derivative of $\widetilde{P}$ with respect to ξ.*

Proof. Let V be a sufficiently small neighborhood of 0. Then

$$x_+(0, x_+(0, x_0)(t, 0, \mu, \alpha))(t_1(0, x_+(0, x_0)(t, 0, \mu, \alpha), 0, \mu, \alpha), 0, \mu, \alpha)$$
$$= x_+(0, x_0)(t + t_1(0, x_+(0, x_0)(t, 0, \mu, \alpha), 0, \mu, \alpha), 0, \mu, \alpha) \qquad (I.1.13)$$

for any $t \in V$, where the left-hand side of (I.1.13) is from Ω_0, and the right-hand side is a point of $\gamma(t)$. Thereafter $t + t_1(0, x_+(0, x_0)(t, 0, \mu, \alpha), 0, \mu, \alpha) = t_1$, i.e. it is constant for all $t \in V$. Similarly

$$x_-(t_1(0, x_+(0, x_0)(t, 0, \mu, \alpha), 0, \mu, \alpha), x_+(0, x_+(0, x_0)(t, 0, \mu, \alpha))(t_1(0, x_+(0, x_0)$$
$$(t, 0, \mu, \alpha), 0, \mu, \alpha), 0, \mu, \alpha)(t_2(0, x_+(0, x_0)(t, 0, \mu, \alpha), 0, \mu, \alpha), 0, \mu, \alpha)$$
$$= x_-(t_1(0, x_+(0, x_0)(t, 0, \mu, \alpha), 0, \mu, \alpha), x_+(0, x_0)(t_1, 0, \mu, \alpha))$$
$$(t_2(0, x_+(0, x_0)(t, 0, \mu, \alpha), 0, \mu, \alpha), 0, \mu, \alpha)$$
$$= x_-(t_1 - t, x_+(0, x_0)(t_1, 0, \mu, \alpha))(t_2(0, x_+(0, x_0)(t, 0, \mu, \alpha), 0, \mu, \alpha), 0, \mu, \alpha)$$
$$= x_-(t_1, x_1)(t_2(0, x_+(0, x_0)(t, 0, \mu, \alpha), 0, \mu, \alpha) + t, 0, \mu, \alpha),$$

and we obtain $t + t_2(0, x_+(0, x_0)(t, 0, \mu, \alpha), 0, \mu, \alpha) = t_2$ for all $t \in V$.

With these results we can derive

$$\widetilde{P}(x_+(0, x_0)(t, 0, \mu, \alpha), 0, \mu, \alpha)$$
$$= x_+(t_2(0, x_+(0, x_0)(t, 0, \mu, \alpha), 0, \mu, \alpha), x_-(t_1(0, x_+(0, x_0)(t, 0, \mu, \alpha), 0, \mu, \alpha),$$
$$x_+(0, x_+(0, x_0)(t, 0, \mu, \alpha))(t_1(0, x_+(0, x_0)(t, 0, \mu, \alpha), 0, \mu, \alpha), 0, \mu, \alpha))$$
$$(t_2(0, x_+(0, x_0)(t, 0, \mu, \alpha), 0, \mu, \alpha), 0, \mu, \alpha))(T, 0, \mu, \alpha)$$
$$= x_+(t_2(0, x_+(0, x_0)(t, 0, \mu, \alpha), 0, \mu, \alpha), x_2)(T, 0, \mu, \alpha)$$
$$= x_+(t_2 - t, x_2)(T, 0, \mu, \alpha) = x_+(t_2, x_2)(T + t, 0, \mu, \alpha)$$

and finally

$$\widetilde{P}_\xi(x_0, 0, \mu, \alpha)f(x_0) = D_t\left[\widetilde{P}(x_+(0, x_0)(t, 0, \mu, \alpha), 0, \mu, \alpha)\right]_{t=0}$$
$$= D_t\left[x_+(t_2, x_2)(T + t, 0, \mu, \alpha)\right]_{t=0}$$
$$= f(x_+(t_2, x_2)(T + t, 0, \mu, \alpha))|_{t=0} = f(x_0).$$

$\square$

In the next step we construct the linearization $\widetilde{P}_\xi(x_0, 0, \mu, \alpha)$ which will be important in further work.

Differentiating $(I.1.4)_+$ with respect to ξ at the point $(\tau, \xi, \varepsilon) = (0, x_0, 0)$ we get

$$\dot{x}_{+\xi}(0, x_0)(t, 0, \mu, \alpha) = Df_+(\gamma(t))x_{+\xi}(0, x_0)(t, 0, \mu, \alpha)$$
$$x_{+\xi}(0, x_0)(0, 0, \mu, \alpha) = \mathbb{I}$$

where $\mathbb{I}$ denotes an $n \times n$ identity matrix. Denote by $X_1(t)$ the matrix solution satisfying this linearized equation on $[0, t_1]$, i.e.

$$\dot{X}_1(t) = Df_+(\gamma(t))X_1(t)$$
$$X_1(0) = \mathbb{I}. \tag{I.1.14}$$

So $x_{+\xi}(0, x_0)(t, 0, \mu, \alpha) = X_1(t)$. By differentiation of $(I.1.4)_+$ with respect to τ at the same point we get

$$\dot{x}_{+\tau}(0, x_0)(t, 0, \mu, \alpha) = Df_+(\gamma(t))x_{+\tau}(0, x_0)(t, 0, \mu, \alpha)$$
$$x_{+\tau}(0, x_0)(0, 0, \mu, \alpha) = -f_+(x_+(0, x_0)(0, 0, \mu, \alpha)).$$

Hence

$$x_{+\tau}(0, x_0)(t, 0, \mu, \alpha) = -X_1(t)f_+(x_0)$$

for $t \in [0, t_1]$. Also the derivative of $(I.1.4)_+$ with respect to ε at $(0, x_0, 0)$ will be needed. We obtain the initial value problem

$$\dot{x}_{+\varepsilon}(0, x_0)(t, 0, \mu, \alpha) = Df_+(\gamma(t))x_{+\varepsilon}(0, x_0)(t, 0, \mu, \alpha) + g(\gamma(t), t + \alpha, 0, \mu)$$
$$x_{+\varepsilon}(0, x_0)(0, 0, \mu, \alpha) = 0$$

which, solved by variation of constants, gives the equality

$$x_{+\varepsilon}(0, x_0)(t, 0, \mu, \alpha) = \int_0^t X_1(t)X_1^{-1}(s)g(\gamma(s), s + \alpha, 0, \mu)ds$$

holding on $[0, t_1]$.

The first intersection point on Ω_0 fulfills

$$h(x_+(\tau, \xi)(t_1(\tau, \xi, \varepsilon, \mu, \alpha), \varepsilon, \mu, \alpha)) = 0$$

for all (τ, ξ, ε) sufficiently close to $(0, x_0, 0)$ and $\mu \in \mathbb{R}^p$, $\alpha \in \mathbb{R}$. Thus differentiating the latter identity with respect to ξ, τ and ε at $(\tau, \xi, \varepsilon) = (0, x_0, 0)$ yields

$$Dh(x_1)(X_1(t_1) + f_+(x_1)t_{1\xi}(0, x_0, 0, \mu, \alpha)) = 0$$
$$t_{1\xi}(0, x_0, 0, \mu, \alpha) = -\frac{Dh(x_1)X_1(t_1)}{Dh(x_1)f_+(x_1)},$$

$$Dh(x_1)(-X_1(t_1)f_+(x_0) + f_+(x_1)t_{1\tau}(0, x_0, 0, \mu, \alpha)) = 0$$
$$t_{1\tau}(0, x_0, 0, \mu, \alpha) = \frac{Dh(x_1)X_1(t_1)f_+(x_0)}{Dh(x_1)f_+(x_1)}$$

and

$$Dh(x_1)\left(f_+(x_1)t_{1\varepsilon}(0, x_0, 0, \mu, \alpha) + \int_0^{t_1} X_1(t_1)X_1^{-1}(s)g(\gamma(s), s + \alpha, 0, \mu)ds\right) = 0$$
$$t_{1\varepsilon}(0, x_0, 0, \mu, \alpha) = -\frac{Dh(x_1)\int_0^{t_1} X_1(t_1)X_1^{-1}(s)g(\gamma(s), s + \alpha, 0, \mu)ds}{Dh(x_1)f_+(x_1)},$$

respectively.

Next, differentiating (I.1.4)_ with respect to ξ, τ and ε at the point $(\tau, \xi, \varepsilon) = (t_1, x_1, 0)$ we obtain

$$\dot{x}_{-\xi}(t_1, x_1)(t, 0, \mu, \alpha) = Df_-(\gamma(t))x_{-\xi}(t_1, x_1)(t, 0, \mu, \alpha)$$
$$x_{-\xi}(t_1, x_1)(t_1, 0, \mu, \alpha) = \mathbb{I},$$

$$\dot{x}_{-\tau}(t_1, x_1)(t, 0, \mu, \alpha) = Df_-(\gamma(t)))x_{-\tau}(t_1, x_1)(t, 0, \mu, \alpha)$$
$$x_{-\tau}(t_1, x_1)(t_1, 0, \mu, \alpha) = -f_-(x_-(t_1, x_1)(t_1, 0, \mu, \alpha))$$

and

$$\dot{x}_{-\varepsilon}(t_1, x_1)(t, 0, \mu, \alpha) = Df_-(\gamma(t))x_{-\varepsilon}(t_1, x_1)(t, 0, \mu, \alpha) + g(\gamma(t), t + \alpha, 0, \mu)$$
$$x_{-\varepsilon}(t_1, x_1)(t_1, 0, \mu, \alpha) = 0,$$

respectively, for $t \in [t_1, t_2]$. Using the matrix solution $X_2(t)$ of the first equation satisfying

$$\dot{X}_2(t) = Df_-(\gamma(t))X_2(t)$$
$$X_2(t_1) = \mathbb{I},$$

(I.1.15)

i.e. $x_{-\xi}(t_1, x_1)(t, 0, \mu, \alpha) = X_2(t)$, we can rewrite the other two solutions as

$$x_{-\tau}(t_1, x_1)(t, 0, \mu, \alpha) = -X_2(t)f_-(x_1),$$

$$x_{-\varepsilon}(t_1, x_1)(t, 0, \mu, \alpha) = \int_{t_1}^{t} X_2(t)X_2^{-1}(s)g(\gamma(s), s + \alpha, 0, \mu)ds$$

for $t \in [t_1, t_2]$.

The second intersection point is characterized by

$$h(x_-(t_1(\tau, \xi, \varepsilon, \mu, \alpha), x_+(\tau, \xi)(t_1(\tau, \xi, \varepsilon, \mu, \alpha), \varepsilon, \mu, \alpha))(t_2(\tau, \xi, \varepsilon, \mu, \alpha), \varepsilon, \mu, \alpha)) = 0.$$

From that we derive

$$Dh(x_2)(x_{-\tau}(t_1, x_1)(t_2, 0, \mu, \alpha)t_{1\xi}(0, x_0, 0, \mu, \alpha) + x_{-\xi}(t_1, x_1)(t_2, 0, \mu, \alpha)$$
$$\times [x_{+\xi}(0, x_0)(t_1, 0, \mu, \alpha) + x_{+t}(0, x_0)(t_1, 0, \mu, \alpha)t_{1\xi}(0, x_0, 0, \mu, \alpha)]$$
$$+ x_{-t}(t_1, x_1)(t_2, 0, \mu, \alpha)t_{2\xi}(0, x_0, 0, \mu, \alpha)) = 0$$

$$t_{2\xi}(0, x_0, 0, \mu, \alpha) = -\frac{Dh(x_2)X_2(t_2)S_1X_1(t_1)}{Dh(x_2)f_-(x_2)},$$

$$Dh(x_2)(x_{-\tau}(t_1, x_1)(t_2, 0, \mu, \alpha)t_{1\tau}(0, x_0, 0, \mu, \alpha) + x_{-\xi}(t_1, x_1)(t_2, 0, \mu, \alpha)$$
$$\times [x_{+\tau}(0, x_0)(t_1, 0, \mu, \alpha) + x_{+t}(0, x_0)(t_1, 0, \mu, \alpha)t_{1\tau}(0, x_0, 0, \mu, \alpha)]$$
$$+ x_{-t}(t_1, x_1)(t_2, 0, \mu, \alpha)t_{2\tau}(0, x_0, 0, \mu, \alpha)) = 0$$

$$t_{2\tau}(0, x_0, 0, \mu, \alpha) = \frac{Dh(x_2)X_2(t_2)S_1X_1(t_1)f_+(x_0)}{Dh(x_2)f_-(x_2)}$$

and

$$Dh(x_2)(x_{-\tau}(t_1, x_1)(t_2, 0, \mu, \alpha)t_{1\varepsilon}(0, x_0, 0, \mu, \alpha) + x_{-\xi}(t_1, x_1)(t_2, 0, \mu, \alpha)$$
$$\times [x_{+\varepsilon}(0, x_0)(t_1, 0, \mu, \alpha) + x_{+t}(0, x_0)(t_1, 0, \mu, \alpha)t_{1\varepsilon}(0, x_0, 0, \mu, \alpha)]$$
$$+ x_{-t}(t_1, x_1)(t_2, 0, \mu, \alpha)t_{2\varepsilon}(0, x_0, 0, \mu, \alpha) + x_{-\varepsilon}(t_1, x_1)(t_2, 0, \mu, \alpha)) = 0$$

$$t_{2\varepsilon}(0, x_0, 0, \mu, \alpha) = -\frac{Dh(x_2)}{Dh(x_2)f_-(x_2)} \left(X_2(t_2)S_1 \int_0^{t_1} X_1(t_1)X_1^{-1}(s) \right.$$

$$\left. \times g(\gamma(s), s + \alpha, 0, \mu)ds + \int_{t_1}^{t_2} X_2(t_2)X_2^{-1}(s)g(\gamma(s), s + \alpha, 0, \mu)ds \right),$$

where

$$S_1 = \mathbb{I} + \frac{(f_-(x_1) - f_+(x_1))Dh(x_1)}{Dh(x_1)f_+(x_1)} \tag{I.1.16}$$

is the so-called saltation matrix [11, 13].

Finally, we calculate derivatives of (I.1.4)$_+$ with respect to ξ, τ and ε at $(\tau, \xi, \varepsilon) = (t_2, x_2, 0)$ to obtain

$$\dot{x}_{+\xi}(t_2, x_2)(t, 0, \mu, \alpha) = Df_+(\gamma(t))x_{+\xi}(t_2, x_2)(t, 0, \mu, \alpha)$$
$$x_{+\xi}(t_2, x_2)(t_2, 0, \mu, \alpha) = \mathbb{I},$$

$$\dot{x}_{+\tau}(t_2, x_2)(t, 0, \mu, \alpha) = Df_+(\gamma(t))x_{+\tau}(t_2, x_2)(t, 0, \mu, \alpha)$$
$$x_{+\tau}(t_2, x_2)(t_2, 0, \mu, \alpha) = -f_+(x_+(t_2, x_2)(t_2, 0, \mu, \alpha))$$

and

$$\dot{x}_{+\varepsilon}(t_2, x_2)(t, 0, \mu, \alpha) = Df_+(\gamma(t))x_{+\varepsilon}(t_2, x_2)(t, 0, \mu, \alpha) + g(\gamma(t), t + \alpha, 0, \mu)$$
$$x_{+\varepsilon}(t_2, x_2)(t_2, 0, \mu, \alpha) = 0,$$

respectively, on $[t_2, T]$. The matrix solution $X_3(t)$ for the first equation that for $t \in [t_2, T]$ satisfies

$$\dot{X}_3(t) = Df_+(\gamma(t))X_3(t)$$
$$X_3(t_2) = \mathbb{I}, \tag{I.1.17}$$

i.e. $x_{+\xi}(t_2, x_2)(t, 0, \mu, \alpha) = X_3(t)$, simplifies expressions for the other two solutions:

$$x_{+\tau}(t_2, x_2)(t, 0, \mu, \alpha) = -X_3(t)f_+(x_2),$$

$$x_{+\varepsilon}(t_2, x_2)(t, 0, \mu, \alpha) = \int_{t_2}^t X_3(t)X_3^{-1}(s)g(\gamma(s), s + \alpha, 0, \mu)ds$$

for $t \in [t_2, T]$. Now we can state the following lemma.

Lemma I.1.4. *Let $\widetilde{P}(\xi, \varepsilon, \mu, \alpha)$ be defined by (I.1.12). Then*

$$\widetilde{P}_\xi(x_0, 0, \mu, \alpha) = X_3(T)S_2X_2(t_2)S_1X_1(t_1), \tag{I.1.18}$$

$$\widetilde{P}_\varepsilon(x_0, 0, \mu, \alpha) = \int_0^T A(s)g(\gamma(s), s + \alpha, 0, \mu)ds, \tag{I.1.19}$$

where $\widetilde{P}_\xi$ and $\widetilde{P}_\varepsilon$ denote the partial derivatives of $\widetilde{P}$ with respect to ξ and ε, respectively, $X_1(t)$, $X_2(t)$ and $X_3(t)$ are matrix solutions of corresponding linearized equations (I.1.14), (I.1.15) and (I.1.17), respectively, S_1 is the saltation matrix given by (I.1.16), S_2 is a second saltation matrix given by

$$S_2 = \mathbb{I} + \frac{(f_+(x_2) - f_-(x_2))Dh(x_2)}{Dh(x_2)f_-(x_2)} \tag{I.1.20}$$

and

$$A(t) = \begin{cases} X_3(T)S_2X_2(t_2)S_1X_1(t_1)X_1^{-1}(t) & \text{if } t \in [0, t_1), \\ X_3(T)S_2X_2(t_2)X_2^{-1}(t) & \text{if } t \in [t_1, t_2), \\ X_3(T)X_3^{-1}(t) & \text{if } t \in [t_2, T]. \end{cases} \tag{I.1.21}$$

Proof. Direct differentiation of (I.1.12) and the use of previous results give the statement of the lemma:

$$\widetilde{P}_\xi(x_0, 0, \mu, \alpha) = x_{+\tau}(t_2, x_2)(T, 0, \mu, \alpha)t_{2\xi}(0, x_0, 0, \mu, \alpha) + x_{+\xi}(t_2, x_2)(T, 0, \mu, \alpha)$$
$$\times [x_{-\tau}(t_1, x_1)(t_2, 0, \mu, \alpha)t_{1\xi}(0, x_0, 0, \mu, \alpha) + x_{-\xi}(t_1, x_1)(t_2, 0, \mu, \alpha)$$
$$\times [x_{+\xi}(0, x_0)(t_1, 0, \mu, \alpha) + x_{+t}(0, x_0)(t_1, 0, \mu, \alpha)t_{1\xi}(0, x_0, 0, \mu, \alpha)]$$
$$+ x_{-t}(t_1, x_1)(t_2, 0, \mu, \alpha)t_{2\xi}(0, x_0, 0, \mu, \alpha)]$$

$$= X_3(T)f_+(x_2)\frac{Dh(x_2)X_2(t_2)S_1X_1(t_1)}{Dh(x_2)f_-(x_2)} + X_3(T)\left[X_2(t_2)f_-(x_1)\frac{Dh(x_1)X_1(t_1)}{Dh(x_1)f_+(x_1)}\right.$$
$$+ X_2(t_2)\left[X_1(t_1) - f_+(x_1)\frac{Dh(x_1)X_1(t_1)}{Dh(x_1)f_+(x_1)}\right] - f_-(x_2)\frac{Dh(x_2)X_2(t_2)S_1X_1(t_1)}{Dh(x_2)f_-(x_2)}\right]$$
$$= X_3(T)S_2X_2(t_2)S_1X_1(t_1).$$

Equality (I.1.19) can be shown in the same way.   □

For further work, we recall the following well-known result (cf. [14]).

Lemma I.1.5. *Let $X(t)$ be a fundamental matrix solution of equation $X' = UX$. Then $X(t)^{-1*}$ is a fundamental matrix solution of the adjoint equation*

$$\left(X(t)^{-1*}\right)' = -U^*X(t)^{-1*}.$$

We solve equation (I.1.11) via the Lyapunov-Schmidt reduction. As was already shown in Lemma I.1.3, $\dim \mathcal{N}(\mathbb{I} - \widetilde{P}_\xi(x_0, 0, \mu, \alpha)) \geq 1$. From now on we suppose that
H3) $\dim \mathcal{N}(\mathbb{I} - \widetilde{P}_\xi(x_0, 0, \mu, \alpha)) = 1$,

and therefore codim $\mathcal{R}(\mathbb{I} - \widetilde{P}_\xi(x_0, 0, \mu, \alpha)) = 1$. We denote

$$R_1 = \mathcal{R}(\mathbb{I} - \widetilde{P}_\xi(x_0, 0, \mu, \alpha)), \qquad R_2 = \left[\mathcal{R}(\mathbb{I} - \widetilde{P}_\xi(x_0, 0, \mu, \alpha))\right]^\perp \qquad \text{(I.1.22)}$$

the image of the corresponding operator and its orthogonal complement in $\mathbb{R}^n$. Then two linear projections are considered, $\mathcal{P}: \mathbb{R}^n \to R_2$ and $\mathcal{Q}: \mathbb{R}^n \to R_1$, defined by

$$\mathcal{P}y = \frac{\langle y, \psi \rangle}{\|\psi\|^2}\psi, \qquad \mathcal{Q}y = (\mathbb{I} - \mathcal{P})y = y - \frac{\langle y, \psi \rangle}{\|\psi\|^2}\psi$$

where $\psi \in R_2$ is fixed. We assume that the initial point ξ of the perturbed periodic trajectory is an element of Σ. Equation (I.1.11) for $(\xi, \alpha) \in \Sigma \times \mathbb{R}$ is equivalent to the couple of equations

$$\mathcal{Q}F(\xi, \varepsilon, \mu, \alpha) = 0, \qquad \mathcal{P}F(\xi, \varepsilon, \mu, \alpha) = 0$$

for $(\xi, \alpha) \in \Sigma \times \mathbb{R}$ with parameters $(\varepsilon, \mu) \in \mathbb{R} \times \mathbb{R}^p$. The first one can be solved via IFT which implies the existence of $r_0, \varepsilon_0 > 0$ and a C^r-function

$$\xi: (-\varepsilon_0, \varepsilon_0) \times \mathbb{R}^p \times \mathbb{R} \to B(x_0, r_0) \cap \Sigma$$

such that $\mathcal{Q}F(\xi, \varepsilon, \mu, \alpha) = 0$ for $\varepsilon \in (-\varepsilon_0, \varepsilon_0)$, $\mu \in \mathbb{R}^p$, $\alpha \in \mathbb{R}$ and $\xi \in B(x_0, r_0) \cap \Sigma$ if and only if $\xi = \xi(\varepsilon, \mu, \alpha)$. Moreover $\xi(0, \mu, \alpha) = x_0$.

Then the second equation has the form

$$\langle \xi(\varepsilon, \mu, \alpha) - \widetilde{P}(\xi(\varepsilon, \mu, \alpha), \varepsilon, \mu, \alpha), \psi \rangle = 0. \qquad \text{(I.1.23)}$$

Again, if $\varepsilon = 0$ this equation is satisfied for any $(\mu, \alpha) \in \mathbb{R}^p \times \mathbb{R}$. Differentiation with respect to ε at 0 gives

$$\left\langle \xi_\varepsilon(0, \mu, \alpha) - \widetilde{P}_\xi(x_0, 0, \mu, \alpha)\xi_\varepsilon(0, \mu, \alpha) - \widetilde{P}_\varepsilon(x_0, 0, \mu, \alpha), \psi \right\rangle$$

$$= \left\langle (\mathbb{I} - \widetilde{P}_\xi(x_0, 0, \mu, \alpha))\xi_\varepsilon(0, \mu, \alpha) - \widetilde{P}_\varepsilon(x_0, 0, \mu, \alpha), \psi \right\rangle$$

$$= \left\langle (\mathbb{I} - \widetilde{P}_\xi(x_0, 0, \mu, \alpha))\xi_\varepsilon(0, \mu, \alpha), \psi \right\rangle - \left\langle \widetilde{P}_\varepsilon(x_0, 0, \mu, \alpha), \psi \right\rangle$$

$$= -\left\langle \int_0^T A(s)g(\gamma(s), s + \alpha, 0, \mu)ds, \psi \right\rangle$$

$$= -\int_0^T \langle A(s)g(\gamma(s), s + \alpha, 0, \mu), \psi \rangle ds$$

$$= -\int_0^T \langle g(\gamma(s), s + \alpha, 0, \mu), A^*(s)\psi \rangle ds$$

where

$$
A^*(t) = \begin{cases} X_1^{-1*}(t)X_1^*(t_1)S_1^*X_2^*(t_2)S_2^*X_3^*(T) & \text{if } t \in [0,t_1), \\ X_2^{-1*}(t)X_2^*(t_2)S_2^*X_3^*(T) & \text{if } t \in [t_1,t_2), \\ X_3^{-1*}(t)X_3^*(T) & \text{if } t \in [t_2,T]. \end{cases} \tag{I.1.24}
$$

Note that by Lemma I.1.5, $A^*(t)$ solves the adjoint variational equation

$$
\begin{aligned}
X' &= -Df_+^*(\gamma(t))X & \text{if } 0 < t < t_1, \\
X' &= -Df_-^*(\gamma(t))X & \text{if } t_1 < t < t_2, \\
X' &= -Df_+^*(\gamma(t))X & \text{if } t_2 < t < T
\end{aligned} \tag{I.1.25}
$$

of (I.1.1). Differentiation of the left-hand side of (I.1.23) with respect to ε and α at $\varepsilon = 0$ gives

$$
-\int_0^T \langle D_t g(\gamma(s), s + \alpha, 0, \mu), A^*(s)\psi\rangle ds.
$$

In conclusion, we obtain the next result.

Theorem I.1.6. *Let conditions* H1), H2), H3) *hold,* $\gamma(t)$, R_2 *and* $A^*(t)$ *be defined by* (I.1.2), (I.1.22) *and* (I.1.24), *respectively, and* $\psi \in R_2$ *be arbitrary and fixed. If* $\alpha_0 \in \mathbb{R}$ *is a simple root of function* $M^{\mu_0}(\alpha)$ *given by*

$$
M^{\mu}(\alpha) = \int_0^T \langle g(\gamma(t), t + \alpha, 0, \mu), A^*(t)\psi\rangle dt, \tag{I.1.26}
$$

i.e. $M^{\mu_0}(\alpha_0) = 0$, $DM^{\mu_0}(\alpha_0) \neq 0$ *then there exists a neighborhood* U *of the point* $(0, \mu_0)$ *in* $\mathbb{R} \times \mathbb{R}^p$ *and a* C^{r-1}-*function* $\alpha(\varepsilon, \mu)$, *with* $\alpha(0, \mu_0) = \alpha_0$, *such that equation* (I.1.1) *with* $\alpha = \alpha(\varepsilon, \mu)$ *possesses a unique* T-*periodic piecewise* C^1-*smooth solution for each* $(\varepsilon, \mu) \in U$.

Proof. Let us denote

$$
\mathcal{D}(\varepsilon, \mu, \alpha) = \begin{cases} \frac{1}{\varepsilon}\langle \xi(\varepsilon, \mu, \alpha) - \widetilde{P}(\xi(\varepsilon, \mu, \alpha), \varepsilon, \mu, \alpha), \psi\rangle & \text{for } \varepsilon \neq 0, \\ D_\varepsilon\langle \xi(\varepsilon, \mu, \alpha) - \widetilde{P}(\xi(\varepsilon, \mu, \alpha), \varepsilon, \mu, \alpha), \psi\rangle & \text{for } \varepsilon = 0. \end{cases}
$$

Then $\mathcal{D}$ is C^{r-1}-smooth and the assumptions on M^{μ_0} are fulfilled if and only if

$$
\mathcal{D}(0, \mu_0, \alpha_0) = 0, \qquad D_\alpha\mathcal{D}(0, \mu_0, \alpha_0) \neq 0.
$$

IFT implies the existence of the function $\alpha(\varepsilon, \mu)$ from the statement of the theorem. □

Function $M^{\mu}(\alpha)$ is a Poincaré-Andronov-Melnikov function for system (I.1.1).

Remark I.1.7.

1. If g is discontinuous in x, i.e.

$$g(x, t, \varepsilon, \mu) = \begin{cases} g_+(x, t, \varepsilon, \mu) & \text{if } x \in \Omega_+, \\ g_-(x, t, \varepsilon, \mu) & \text{if } x \in \Omega_-, \end{cases}$$

it is possible to show that Theorem I.1.6 still holds. Of course, g has to be T-periodic in t.

2. It can be shown that in Theorem I.1.6 we can take any other solution of the adjoint variational system consisting of the adjoint variational equation (I.1.25) and corresponding impulsive and boundary conditions (see Lemma I.2.4).

3. Using the preceding calculation (see also (I.2.6)), we get

$$P_\xi(x_0, 0, \mu, \alpha) = (\mathbb{I} - S_{x_0})\widetilde{P}_\xi(x_0, 0, \mu, \alpha),$$

where S_{x_0} is the orthogonal projection onto the 1-dimensional space $[f_+(x_0)]$ defined by

$$S_{x_0} u = \frac{\langle u, f_+(x_0) \rangle f_+(x_0)}{\|f_+(x_0)\|^2}. \tag{I.1.27}$$

I.1.2. Geometric interpretation of assumed conditions

Consider the linearization of the unperturbed problem of (I.1.1) along $\gamma(t)$, given by

$$\dot{x} = Df_\pm(\gamma(t))x. \tag{I.1.28}$$

Then (I.1.28) splits into two unperturbed equations

$$\begin{aligned} \dot{x} &= Df_+(\gamma(t))x && \text{if } t \in [0, t_1] \cup [t_2, T], \\ \dot{x} &= Df_-(\gamma(t))x && \text{if } t \in (t_1, t_2) \end{aligned}$$

with impulsive conditions [11–13]

$$x(t_1+) = S_1 x(t_1-), \qquad x(t_2+) = S_2 x(t_2-)$$

where $x(t\pm) = \lim_{s \to t^\pm} x(s)$. We already know (from (I.1.14), (I.1.15), (I.1.17)) that they have the fundamental matrices $X_1(t)$ resp. $X_3(t)$ and $X_2(t)$ satisfying $X_1(0) = X_2(t_1) = X_3(t_2) = \mathbb{I}$. Consequently, the fundamental matrix solution of the discontinuous variational equation (I.1.28) is given by

$$X(t) = \begin{cases} X_1(t) & \text{if } t \in [0, t_1), \\ X_2(t)S_1 X_1(t_1) & \text{if } t \in [t_1, t_2), \\ X_3(t)S_2 X_2(t_2)S_1 X_1(t_1) & \text{if } t \in [t_2, T]. \end{cases}$$

Then a T-periodic solution of (I.1.28) with an initial point ξ fulfills $\xi = X(T)\xi$ or, equivalently, $(\mathbb{I} - X(T))\xi = 0$. Now one can easily conclude the following result.

Proposition I.1.8. *Condition H3) is equivalent to say that discontinuous variational equation (I.1.28) has a unique T-periodic solution up to a scalar multiple.*

I.1.3. Two-position automatic pilot for ship's controller with periodic forcing

We continue with the study of system (I.1.5), (I.1.6) by applying Theorem I.1.6. In this part we omit the argument μ, e.g. in $\widetilde{P}(x, \varepsilon, \mu, \alpha)$, as the perturbed problem is independent of μ. Since

$$h(\gamma_1(t)) = 1 - e^{-t} - t - \frac{T}{4} + (e^{-t} - 1)\left(1 - \frac{T}{4}\coth\frac{T}{4}\right) - \ln\frac{2}{1 + e^{T/2}},$$

we get

$$\frac{d}{dt}h(\gamma_1(t)) = e^{-t}\frac{T}{4}\coth\frac{T}{4} - 1$$

which is a decreasing function from $\frac{d}{dt}h(\gamma_1(0)) = \frac{T}{4}\coth\frac{T}{4} - 1 = -\beta > 0$ to $\frac{d}{dt}h(\gamma_1(t_1))$ $= \frac{2 - 2e^{\frac{T}{2}} + T}{2(e^{\frac{T}{2}} - 1)} < 0$. But $h(\gamma_1(0)) = h(x_0) = \ln\cosh\frac{T}{4} > 0$ and $h(\gamma_1(t_1)) = h(x_1) = 0$, so $\gamma_1(t) \in \Omega_+$ for $t \in [0, t_1)$. Similarly we have

$$h(\gamma_2(t)) = t - \frac{T}{4} + \left(e^{\frac{T}{2} - t} - 1\right)\frac{T}{4}\coth\frac{T}{4} + \ln\frac{2}{1 + e^{\frac{T}{2}}},$$

so

$$\frac{d}{dt}h(\gamma_2(t)) = 1 - e^{\frac{T}{2} - t}\frac{T}{4}\coth\frac{T}{4}$$

which is an increasing function from $\frac{d}{dt}h(\gamma_2(t_1)) = 1 - \frac{T}{4} - \frac{T}{4}\coth\frac{T}{4} = \beta - \frac{T}{4} < 0$ to $\frac{d}{dt}h(\gamma_2(t_2)) = \frac{2 - 2e^{\frac{T}{2}} + T}{2(1 - e^{\frac{T}{2}})} > 0$. But $h(\gamma_2(t_1)) = h(x_1) = 0$ and $h(\gamma_2(t_2)) = h(x_2) = 0$, so $\gamma_2(t) \in \Omega_-$ for $t \in (t_1, t_2)$. Finally, we have

$$h(\gamma_3(t)) = -t + \frac{3T}{4} + (1 - e^{T-t})\frac{T}{4}\coth\frac{T}{4} - \ln\frac{2}{1 + e^{\frac{T}{2}}},$$

so

$$\frac{d}{dt}h(\gamma_3(t)) = e^{T-t}\frac{T}{4}\coth\frac{T}{4} - 1 \geq \frac{T}{4}\coth\frac{T}{4} - 1 = -\beta > 0.$$

But $h(\gamma_3(t_2)) = h(x_2) = 0$, so $\gamma_3(t) \in \Omega_+$ for $t \in (t_2, T)$. Consequently, H1) is satisfied. Next, we compute

$$Dh(x_1)f_\pm(x_1) \leq -\tanh\frac{T}{4} + \beta\tanh\frac{T}{4} - \beta$$

$$= \frac{T}{4}\coth\frac{T}{4} - \frac{T}{4} - 1 = \frac{d}{dt}h(\gamma_1(t_1)) < 0,$$

$$Dh(x_2)f_\pm(x_2) \geq \tanh\frac{T}{4} - \beta\tanh\frac{T}{4} + \beta$$

$$= \frac{T}{4} + 1 - \frac{T}{4}\coth\frac{T}{4} = \frac{d}{dt}h(\gamma_2(t_2)) > 0,$$

hence H2) is satisfied as well. Furthermore, due to the linearity of (I.1.7), (I.1.8), we easily obtain

$$X_1(t) = \begin{pmatrix} 1 & 1-e^{-t} \\ 0 & e^{-t} \end{pmatrix}, \quad X_2(t) = \begin{pmatrix} 1 & 1-\frac{1}{2}e^{-t}(1+e^{\frac{T}{2}}) \\ 0 & \frac{1}{2}e^{-t}(1+e^{\frac{T}{2}}) \end{pmatrix},$$

$$X_3(t) = \begin{pmatrix} 1 & 1-\frac{1}{2}e^{\frac{T}{2}-t}(1+e^{\frac{T}{2}}) \\ 0 & \frac{1}{2}e^{\frac{T}{2}-t}(1+e^{\frac{T}{2}}) \end{pmatrix},$$

$$S_1 = S_2 = \begin{pmatrix} 1 & 0 \\ -\frac{8}{4+T-T\coth\frac{T}{4}} & -1+\frac{2T}{4+T-T\coth\frac{T}{4}} \end{pmatrix}.$$

Then we derive

$$\mathbb{I} - \widetilde{P}_\xi(x_0, 0, \alpha) = -\begin{pmatrix} \frac{16\sinh^2\frac{T}{4}(T-2\sinh\frac{T}{2})}{(2-2e^{\frac{T}{2}}+T)^2} & 0 \\ \frac{16e^{-T}(-1+e^T)}{(4+T-T\coth\frac{T}{4})^2} & 0 \end{pmatrix}.$$

Thus also H3) is verified, and

$$\psi = \begin{pmatrix} \frac{2}{2-T\operatorname{csch}\frac{T}{2}} \\ 1 \end{pmatrix}.$$

We also derive

$$A(t) = \begin{cases} \begin{pmatrix} \frac{e^{-T}(2+e^{\frac{T}{2}}(T-2))^2}{(2-2e^{\frac{T}{2}}+T)^2} & \frac{(1-e^t)(T-2+2e^{-\frac{T}{2}})^2}{(T+2-2e^{-\frac{T}{2}})^2} \\ \frac{16e^{-T}(e^T-1)}{(4+T-T\coth\frac{T}{4})^2} & e^t + \frac{4(1-e^t)(e^T-2e^{\frac{T}{2}}+2e^{-\frac{T}{2}}-e^{-T})}{(T+2-2e^{-\frac{T}{2}})^2} \end{pmatrix} & \text{if } t \in [0, t_1), \\ \begin{pmatrix} 1 + \frac{-4+4e^{-\frac{T}{2}}}{4+T-T\coth\frac{T}{4}} & \frac{(e^t-e^{\frac{T}{2}})(2+e^{\frac{T}{2}}(-2+T))}{2e^{\frac{3T}{2}}-e^T(2+T)} \\ -\frac{4+4e^{-\frac{T}{2}}}{4+T-T\coth\frac{T}{4}} & \frac{e^{-T}(4e^{\frac{T}{2}}+4e^T-e^t(4+T+T\coth\frac{T}{4}))}{-4-T+T\coth\frac{T}{4}} \end{pmatrix} & \text{if } t \in [t_1, t_2), \\ \begin{pmatrix} 1 & 1-e^{t-T} \\ 0 & e^{t-T} \end{pmatrix} & \text{if } t \in [t_2, T]. \end{cases}$$

We consider the simplest resonance condition

$$\omega = \frac{2\pi}{T}.$$

(I.1.29)

Finally, the Poincaré-Andronov-Melnikov function (I.1.26) is as follows

$$M(\alpha) = \int_0^T \left\langle A(t) \begin{pmatrix} 0 \\ \cos \frac{2\pi}{T}(t+\alpha) \end{pmatrix}, \psi \right\rangle dt$$

$$= \int_0^{t_1} \frac{\cos \frac{2\pi(\alpha+t)}{T}(2 - Te^t \operatorname{csch} \frac{T}{2})}{2 - T \operatorname{csch} \frac{T}{2}} dt + \int_{t_1}^{t_2} \frac{(1 - e^T + e^t T)\cos \frac{2\pi(\alpha+t)}{T}}{-1 + e^T - e^{\frac{T}{2}}T} dt$$

$$+ \int_{t_2}^T \frac{e^{-T} \cos \frac{2\pi(\alpha+t)}{T}(-2e^T + e^t T \operatorname{csch} \frac{T}{2})}{-2 + T \operatorname{csch} \frac{T}{2}} dt$$

$$= a(T) \cos \frac{2\pi\alpha}{T} + b(T) \sin \frac{2\pi\alpha}{T},$$

where

$$a(T) = \frac{1}{\pi(4\pi^2 + T^2)(T - 2\sinh \frac{T}{2})} \left(4\pi T^3 \cos \frac{2\pi \ln \frac{2}{1+e^{\frac{T}{2}}}}{T} \cosh^2 \frac{T}{4} \right.$$

$$\left. -4T \sin \frac{2\pi \ln \frac{2}{1+e^{\frac{T}{2}}}}{T} \left(\pi^2 T \left(1 + \cosh \frac{T}{2}\right) - (4\pi^2 + T^2)\sinh \frac{T}{2} \right) \right)$$

and

$$b(T) = \frac{1}{\pi(4\pi^2 + T^2)(T - 2\sinh \frac{T}{2})} \left(4\pi T^3 \cosh^2 \frac{T}{4} \sin \frac{2\pi \ln \frac{2}{1+e^{\frac{T}{2}}}}{T} \right.$$

$$\left. +4T \cos \frac{2\pi \ln \frac{2}{1+e^{\frac{T}{2}}}}{T} \left(\pi^2 T \left(1 + \cosh \frac{T}{2}\right) - (4\pi^2 + T^2)\sinh \frac{T}{2} \right) \right).$$

Next we derive

$$a(T)^2 + b(T)^2 = \frac{8c(T)T^2 \cosh^2 \frac{T}{4}}{\pi^2(4\pi^2 + T^2)\left(T - 2\sinh \frac{T}{2}\right)^2}$$

for

$$c(T) = -16\pi^2 - (4 - \pi^2)T^2 + (4T^2 + \pi^2(16 + T^2))\cosh \frac{T}{2} - 8\pi^2 T \sinh \frac{T}{2}.$$

Since

$$c(T)' = d(T)T \sinh \frac{T}{4}, \quad d(T) = (4 + \pi^2)T \cosh \frac{T}{4} + 4(4 - \pi^2) \sinh \frac{T}{4},$$

$$c(0) = d(0) = 0, \quad d(T)' = 8 \cosh \frac{T}{4} + \frac{1}{4}(4 + \pi^2)T \sinh \frac{T}{4} > 0$$

for $T > 0$, we get $a(T)^2 + b(T)^2 > 0$ for $T > 0$. So $M(\alpha)$ is not identically zero and it has a simple root on $[0, T]$. Consequently, by Theorem I.1.6 we get the following result.

Theorem I.1.9. *System* (I.1.5), (I.1.6) *under assumptions* (I.1.9) *and* (I.1.29) *has a T-periodic solution for any ε close to* 0.

Moreover, since

$$\widetilde{P}_\xi(x_0, 0, \alpha) = \begin{pmatrix} \frac{e^{-T}(2 + e^{\frac{T}{2}}(T-2))^2}{(2 - 2e^{\frac{T}{2}} + T)^2} & 0 \\ \frac{16e^{-T}(e^T - 1)}{(4 + T - T \coth \frac{T}{4})^2} & 1 \end{pmatrix},$$

for its spectrum we get

$$\sigma(\widetilde{P}_\xi(x_0, 0, \alpha)) = \left\{ \frac{e^{-T}(2 + e^{\frac{T}{2}}(T - 2))^2}{(2 - 2e^{\frac{T}{2}} + T)^2}, 1 \right\}.$$

Using

$$1 - \frac{e^{-T}(2 + e^{\frac{T}{2}}(T - 2))^2}{(2 - 2e^{\frac{T}{2}} + T)^2} = \frac{16 \sinh^2 \frac{T}{4}(2 \sinh \frac{T}{2} - T)}{(2 - 2e^{\frac{T}{2}} + T)^2} > 0,$$

and applying (I.1.27), we see that $\gamma(t)$ is stable, so the perturbed T-periodic orbit is also stable. The stability of $\gamma(t)$ is mentioned in [9] but we verify it above analytically.

I.1.4. Nonlinear planar applications

Here we consider the following piecewise-nonlinear planar problem

$$
\begin{aligned}
\dot{x} &= \omega_1(y - \delta) + \varepsilon g_1(x, y, t + \alpha, \varepsilon, \mu) \\
\dot{y} &= -\omega_1 x + \varepsilon g_2(x, y, t + \alpha, \varepsilon, \mu)
\end{aligned}
\qquad \text{if } y > 0,
$$

$$
\begin{aligned}
\dot{x} &= \eta x + \omega_2(y + \delta) \\
&\quad + \left[x^2 + (y + \delta)^2 \right] \left[-ax - b(y + \delta) \right] + \varepsilon g_1(x, y, t + \alpha, \varepsilon, \mu) \\
\dot{y} &= -\omega_2 x + \eta(y + \delta) \\
&\quad + \left[x^2 + (y + \delta)^2 \right] \left[bx - a(y + \delta) \right] + \varepsilon g_2(x, y, t + \alpha, \varepsilon, \mu)
\end{aligned}
\qquad \text{if } y < 0
$$

$$(\text{I.1.30})_\varepsilon$$

with assumptions

$$
\eta, \delta, \omega_1, \omega_2, \omega, a > 0, \quad b \in \mathbb{R}, \quad \omega_2 - \frac{\eta b}{a} > 0, \quad \frac{\eta}{a} > \delta^2. \tag{I.1.31}
$$

Note the dependence on ε in the notation $(\text{I.1.30})_\varepsilon$. Hence $(\text{I.1.30})_0$ refers to the unperturbed system.

Due to linearity, the first part of $(\text{I.1.30})_0$ can be easily solved, e.g. via a matrix exponential. For the starting point $(x_0, y_0) = \left(0, \delta + \sqrt{\frac{\eta}{a}}\right)$ and $t \in [0, t_1]$, the solution is

$$
\gamma_1(t) = \left(\sqrt{\frac{\eta}{a}} \sin \omega_1 t, \delta + \sqrt{\frac{\eta}{a}} \cos \omega_1 t \right). \tag{I.1.32}
$$

Time t_1 of the first intersection with discontinuity boundary $\Omega_0 = \{(x, y) \in \mathbb{R}^2 \mid y = 0\}$ and the point (x_1, y_1) of this intersection are obtained from the relations $h(\gamma_1(t_1)) = 0$ for $h(x, y) = y$ and $(x_1, y_1) = \gamma_1(t_1)$, respectively:

$$
t_1 = \frac{1}{\omega_1} \arccos\left(-\sqrt{\frac{a}{\eta}} \delta \right), \qquad (x_1, y_1) = \left(\sqrt{\frac{\eta}{a} - \delta^2}, 0 \right).
$$

After transformation $x = r \cos \theta$, $y + \delta = r \sin \theta$ in the second part of $(\text{I.1.30})_0$, we get

$$
\begin{aligned}
\dot{r} &= \eta r - a r^3 \\
\dot{\theta} &= -\omega_2 + b r^2
\end{aligned}
$$

from which one can see that the second part of $(\text{I.1.30})_0$ possesses a stable limit cycle/circle with the center at $(0, -\delta)$ and radius $\sqrt{\frac{\eta}{a}}$, which intersects boundary Ω_0. Now it is obvious that (x_1, y_1) is a point of this cycle and the direction of rotation remains the same as in $\Omega_+ = \{(x, y) \in \mathbb{R}^2 \mid y > 0\}$. Therefore $\gamma_2(t)$ is a part of the circle, given by

$$\gamma_2(t) = (x_1 \cos \omega_3(t - t_1) + \delta \sin \omega_3(t - t_1),$$
$$-\delta - x_1 \sin \omega_3(t - t_1) + \delta \cos \omega_3(t - t_1)) \tag{I.1.33}$$

for $t \in [t_1, t_2]$, where $\omega_3 = \omega_2 - \frac{nb}{a}$. Equation $h(\gamma_2(t_2)) = 0$ together with the symmetry of $\gamma_2(t)$ give the couple of equations

$$x_1 \cos \omega_3(t_2 - t_1) + \delta \sin \omega_3(t_2 - t_1) = -x_1,$$
$$-\delta - x_1 \sin \omega_3(t_2 - t_1) + \delta \cos \omega_3(t_2 - t_1) = 0.$$

From these we obtain

$$t_2 = \frac{1}{\omega_3}\left(\pi + \operatorname{arccot}\frac{-\delta^2 + x_1^2}{2\delta x_1}\right) + t_1.$$

Point (x_2, y_2) is the second intersection point of the limit cycle and Ω_0, i.e.

$$(x_2, y_2) = \gamma_2(t_2) = \left(-\sqrt{\frac{\eta}{a} - \delta^2}, 0\right).$$

Next, solution $\gamma(t)$ continues in Ω_+ following the solution of the first part of (I.1.30)$_0$. Thus we have

$$\gamma_3(t) = (x_2 \cos \omega_1(t - t_2) - \delta \sin \omega_1(t - t_2),$$
$$\delta - x_2 \sin \omega_1(t - t_2) - \delta \cos \omega_1(t - t_2)) \tag{I.1.34}$$

for $t \in [t_2, T]$. Period T obtained from the identity $\gamma_3(T) = (x_0, y_0)$ is

$$T = \frac{1}{\omega_1} \arccos\left(-\sqrt{\frac{a}{\eta}}\delta\right) + t_2.$$

The next theorem is due to Diliberto (cf. [15–17]), and we shall use it to find the fundamental matrix solution of the variational equation.

Theorem I.1.10. *Let $\gamma(t)$ be a solution of the differential equation $\dot{x} = f(x)$, $x \in \mathbb{R}^2$. If $\gamma(0) = p$, $f(p) \neq 0$ then the variational equation along $\gamma(t)$,*

$$\dot{V} = Df(\gamma(t))V,$$

has the fundamental matrix solution $\Phi(t)$ satisfying $\det \Phi(0) = \|f(p)\|^2$, given by

$$\Phi(t) = [f(\gamma(t)), V(t)]$$

where $[\lambda_1, \lambda_2]$ stands for a matrix with columns λ_1 and λ_2, and

$$V(t) = a(t)f(\gamma(t)) + b(t)f^{\perp}(\gamma(t)),$$

$$a(t) = \int_0^t \left[2\kappa(\gamma(s))\|f(\gamma(s))\| + \operatorname{div} f^{\perp}(\gamma(s)) \right] b(s)\,ds,$$

$$b(s) = \frac{\|f(p)\|^2}{\|f(\gamma(t))\|^2} e^{\int_0^t \operatorname{div} f(\gamma(s))ds},$$

$$\operatorname{div} f(x) = \frac{\partial f_1(x)}{\partial x_1} + \frac{\partial f_2(x)}{\partial x_2}, \qquad \operatorname{div} f^{\perp}(x) = -\frac{\partial f_2(x)}{\partial x_1} + \frac{\partial f_1(x)}{\partial x_2},$$

$$\kappa(\gamma(t)) = \frac{1}{\|f(\gamma(t))\|^3} \left[f_1(\gamma(t))\dot{f}_2(\gamma(t)) - f_2(\gamma(t))\dot{f}_1(\gamma(t)) \right].$$

Lemma I.1.11. *Assuming (I.1.31), unperturbed system (I.1.30)$_0$ has fundamental matrices X_1, X_2 and X_3 satisfying (I.1.14), (I.1.15) and (I.1.17), respectively, given by*

$$X_1(t) = \begin{pmatrix} \cos\omega_1 t & \sin\omega_1 t \\ -\sin\omega_1 t & \cos\omega_1 t \end{pmatrix},$$

$$X_2(t) = \frac{a}{\eta}[\lambda_1, \lambda_2], \qquad X_3(t) = X_1(t - t_2)$$

where

$$\lambda_1 = \begin{pmatrix} U(-\delta x_1 + \delta x_1 W + x_1^2 \widetilde{W}) + V(\delta^2 + x_1^2 W - \delta x_1 \widetilde{W}) \\ U(-\delta^2 - x_1^2 W + \delta x_1 \widetilde{W}) + V(-\delta x_1 + \delta x_1 W + x_1^2 \widetilde{W}) \end{pmatrix},$$

$$\lambda_2 = \begin{pmatrix} U(x_1^2 + \delta^2 W + \delta x_1 \widetilde{W}) + V(-\delta x_1 + \delta x_1 W - \delta^2 \widetilde{W}) \\ U(\delta x_1 - \delta x_1 W + \delta^2 \widetilde{W}) + V(x_1^2 + \delta^2 W + \delta x_1 \widetilde{W}) \end{pmatrix},$$

$$U = \sin\omega_3(t - t_1), \qquad V = \cos\omega_3(t - t_1),$$

$$W = e^{-2\eta(t-t_1)}, \qquad \widetilde{W} = \frac{b}{a}(1 - W),$$

and saltation matrices

$$S_1 = \begin{pmatrix} 1 & -\frac{\delta(\omega_1 + \omega_3)}{\omega_1 x_1} \\ 0 & \frac{\omega_3}{\omega_1} \end{pmatrix}, \qquad S_2 = \begin{pmatrix} 1 & -\frac{\delta(\omega_1 + \omega_3)}{\omega_3 x_1} \\ 0 & \frac{\omega_1}{\omega_3} \end{pmatrix}$$

defined by (I.1.16), (I.1.20), respectively.

Proof. Matrices $X_1(t)$ and $X_3(t)$ are derived easily because of the linearity of function

$f_+(x, y)$. Using

$$f_+(x_1, y_1) = \begin{pmatrix} -\omega_1\delta \\ -\omega_1\sqrt{\frac{\eta}{a} - \delta^2} \end{pmatrix}, \qquad f_-(x_1, y_1) = \begin{pmatrix} \omega_3\delta \\ -\omega_3\sqrt{\frac{\eta}{a} - \delta^2} \end{pmatrix},$$

$$f_+(x_2, y_2) = \begin{pmatrix} -\omega_1\delta \\ \omega_1\sqrt{\frac{\eta}{a} - \delta^2} \end{pmatrix}, \qquad f_-(x_2, y_2) = \begin{pmatrix} \omega_3\delta \\ \omega_3\sqrt{\frac{\eta}{a} - \delta^2} \end{pmatrix},$$

(I.1.35)

saltation matrices are obtained directly from their definitions. Since $(I.1.30)_0$ is 2-dimensional and one solution of the second part is already known – the limit cycle, we can apply Theorem I.1.10 to derive the fundamental solution of this part. So we get a matrix

$$\widetilde{X}_2(t) = \omega_3 \begin{pmatrix} -x_1 U + \delta V & U(\delta W + x_1\widetilde{W}) + V(x_1 W - \delta\widetilde{W}) \\ -\delta U - x_1 V & U(-x_1 W + \delta\widetilde{W}) + V(\delta W + x_1\widetilde{W}) \end{pmatrix}$$

such that

$$\widetilde{X}_2^{-1}(t_1) = \frac{a}{\eta\omega_3} \begin{pmatrix} \delta & -x_1 \\ x_1 & \delta \end{pmatrix}$$

and $\det\widetilde{X}_2(t_1) = \|f_-(x_1, y_1)\|^2 = \frac{\eta}{a}\omega_3^2$. If $X_2(t)$ has to satisfy (I.1.15), then clearly $X_2(t) = \widetilde{X}_2(t)\widetilde{X}_2^{-1}(t_1)$. □

Now, we can verify the basic assumptions.

Proposition I.1.12. *Assuming (I.1.31), unperturbed system $(I.1.30)_0$ has a T-periodic solution with initial point $(x_0, y_0) = \left(0, \delta + \sqrt{\frac{\eta}{a}}\right)$, defined by (I.1.2) with branches $\gamma_1(t)$, $\gamma_2(t)$ and $\gamma_3(t)$ given by (I.1.32), (I.1.33) and (I.1.34), respectively. Moreover, conditions H1), H2) and H3) are satisfied.*

Proof. Condition H1) was already verified. Since $\nabla h(x, y) = (0, 1)$ for all $(x, y) \in \mathbb{R}^2$ and (I.1.35) holds, also condition H2) is fulfilled.

Now suppose that $\dim \mathcal{N}(\mathbb{I} - \widetilde{P}_\xi(x_0, 0, \mu, \alpha)) > 1$. We recall that $f_+(x_0, y_0) \in \mathcal{N}(\mathbb{I} - \widetilde{P}_\xi(x_0, 0, \mu, \alpha))$. Since $\mathcal{N}(\mathbb{I} - \widetilde{P}_\xi(x_0, 0, \mu, \alpha))$ is linear, there is a vector

$$\bar{v} \in \mathcal{N}(\mathbb{I} - \widetilde{P}_\xi(x_0, 0, \mu, \alpha))$$

such that $\langle \bar{v}, f_+(x_0, y_0) \rangle = 0$. Then we can write $\bar{v} = (0, v)^*$. Using formula (I.1.18) for $\widetilde{P}_\xi$ we look for the image of $\bar{v}$ under the mapping $\widetilde{P}_\xi(x_0, 0, \mu, \alpha)$. We subsequently obtain

$$S_1 X_1(t_1)\bar{v} = \frac{v}{\omega_1}\sqrt{\frac{a}{\eta}}\begin{pmatrix} \omega_1 x_1 + \frac{\delta^2(\omega_1+\omega_3)}{x_1} \\ -\delta\omega_3 \end{pmatrix},$$

$$X_2(t_2)S_1 X_1(t_1)\bar{v} = \frac{v}{\omega_1}\sqrt{\frac{a}{\eta}}\begin{pmatrix} \frac{\delta^2}{x_1}(\omega_1+\omega_3) - x_1\omega_1 Z - \delta\omega_1\widetilde{Z} \\ \delta(\omega_1+\omega_3) + \delta\omega_1 Z - x_1\omega_1\widetilde{Z} \end{pmatrix}$$

where $Z = e^{-2\eta(t_2-t_1)}$ and $\widetilde{Z} = \frac{b}{a}(1 - Z)$ are values of W and $\widetilde{W}$ at $t = t_2$,

$$S_2 X_2(t_2)S_1 X_1(t_1)\bar{v} = \frac{v}{\omega_3}\sqrt{\frac{a}{\eta}}\begin{pmatrix} -\frac{\delta^2}{x_1}(\omega_1+\omega_3) - (\frac{\delta^2\omega_1}{x_1} + \frac{\eta\omega_3}{ax_1})Z + \delta\omega_1\widetilde{Z} \\ \delta(\omega_1+\omega_3) + \delta\omega_1 Z - x_1\omega_1\widetilde{Z} \end{pmatrix}$$

and finally

$$X_3(T)S_2 X_2(t_2)S_1 X_1(t_1)\bar{v} = v\begin{pmatrix} \frac{\delta}{x_1}\frac{\omega_1+\omega_3}{\omega_3} + \frac{\delta}{x_1}\frac{\omega_1+\omega_3}{\omega_3}Z - \frac{\omega_1}{\omega_3}\widetilde{Z} \\ Z \end{pmatrix}.$$

Since $Z \le \exp\{-\frac{2\eta}{\omega_3}\pi\} < 1$, it is obvious that $\bar{v} = \widetilde{P}_\xi(x_0, 0, \mu, \alpha)\bar{v}$ if and only if $\bar{v} = (0, 0)^*$. Hence the verification of condition H3) is finished.   □

Because, in general, the formula for $A(t)$ is rather awkward, we move to examples with concrete parameters.

Example I.1.13. Consider system $(I.1.30)_\varepsilon$ with

$$a = b = \delta = 1, \ \eta = 2, \ \omega_1 = 1, \ \omega_2 = 5,$$

$$g(x, y, t, \varepsilon, \mu) = \begin{cases} (\sin\omega t, 0)^* & \text{if } y > 0, \\ (0, 0)^* & \text{if } y < 0. \end{cases} \tag{I.1.36}$$

Then we have $\omega_3 = 3$, $T = 2\pi$, initial point $(x_0, y_0) = (0, 1 + \sqrt{2})$, saltation matrices

$$S_1 = \begin{pmatrix} 1 & -4 \\ 0 & 3 \end{pmatrix}, \qquad S_2 = \begin{pmatrix} 1 & -\frac{4}{3} \\ 0 & \frac{1}{3} \end{pmatrix}$$

and

$$\widetilde{P}_\xi(x_0, y_0, 0, \mu, \alpha) = \begin{pmatrix} 1 & 1 + \frac{5}{3}e^{-2\pi} \\ 0 & e^{-2\pi} \end{pmatrix}.$$

Therefore $R_1 = \left[\left(1 + \frac{5}{3}e^{-2\pi}, e^{-2\pi} - 1\right)^*\right]$ and $\psi = \left(1 - e^{-2\pi}, 1 + \frac{5}{3}e^{-2\pi}\right)^* \in R_2$.

After some algebra we obtain

$$M(\alpha) = \frac{1}{3} \frac{e^{-2\pi}}{\omega^2 - 1} \left[(\omega A + B) \sin \omega\alpha + (\omega C + D) \cos \omega\alpha \right]$$

for $M(\alpha) = M^\omega(\alpha)$ of (I.1.26), where

$$A = 4\sqrt{2} \sin\left(\frac{3}{4}\pi\omega\right) + \left(3e^{2\pi}\sqrt{2} + \sqrt{2}\right) \sin\left(\frac{5}{4}\pi\omega\right) + \left(3e^{2\pi} - 3\right) \sin(2\pi\omega),$$

$$B = -5 - 3e^{2\pi} - \left(\sqrt{2} + 3\sqrt{2}e^{2\pi}\right) \cos\left(\frac{3}{4}\pi\omega\right) + 4\sqrt{2} \cos\left(\frac{5}{4}\pi\omega\right) + \left(5 + 3e^{2\pi}\right) \cos(2\pi\omega),$$

$$C = 3e^{2\pi} - 3 - 4\sqrt{2} \cos\left(\frac{3}{4}\pi\omega\right) - \left(\sqrt{2} + 3\sqrt{2}e^{2\pi}\right) \cos\left(\frac{5}{4}\pi\omega\right) + \left(3 - 3e^{2\pi}\right) \cos(2\pi\omega),$$

$$D = -\left(\sqrt{2} + 3\sqrt{2}e^{2\pi}\right) \sin\left(\frac{3}{4}\pi\omega\right) + 4\sqrt{2} \sin\left(\frac{5}{4}\pi\omega\right) + \left(5 + 3e^{2\pi}\right) \sin(2\pi\omega).$$

$$\text{(I.1.37)}$$

For $\omega > 0$, $\omega \neq 1$, $M(\alpha)$ has a simple root if and only if $(\omega A + B)^2 + (\omega C + D)^2 > 0$. Since A and C are 8-periodic functions,

$$\sqrt{B^2 + D^2} \leq \left(\left(5 + 3e^{2\pi} + \sqrt{2} + 3\sqrt{2}e^{2\pi} + 4\sqrt{2} + 5 + 3e^{2\pi}\right)^2 \right.$$

$$\left. + \left(5 + 3e^{2\pi} + \sqrt{2} + 3\sqrt{2}e^{2\pi} + 4\sqrt{2}\right)^2 \right)^{\frac{1}{2}}$$

$$= \sqrt{3}\left(1 + \sqrt{2}\right) \sqrt{9e^{4\pi} + 30e^{2\pi} + 25} \leq 6739,$$

and according to Figure I.1.2, we have the estimate

$$\sqrt{(\omega A + B)^2 + (\omega C + D)^2} \geq \omega \sqrt{A^2 + C^2} - \sqrt{B^2 + D^2} \geq 400\omega - 6739,$$

and one can see that for $\omega \geq 17$, the T-periodic orbit in the perturbed system $(I.1.30)_\varepsilon$ persists for all $\varepsilon \neq 0$ small. It can be proved numerically (see Figure I.1.2) that

$$\frac{1}{|\omega^2 - 1|} \sqrt{(\omega A + B)^2 + (\omega C + D)^2} > 0 \qquad \text{(I.1.38)}$$

for $\omega \in (0, 17)$. We conclude:

Corollary I.1.14. *Consider* $(I.1.30)_\varepsilon$ *with parameters* (I.1.36). *Then* 2π-*periodic orbit persists for all* $\omega > 0$ *and* $\varepsilon \neq 0$ *small.*

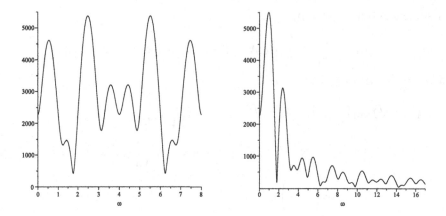

Figure I.1.2 Graphs of the functions $\sqrt{A^2 + C^2}$ and the left-hand side of (I.1.38)

Example I.1.15. Consider system $(I.1.30)_\varepsilon$ with

$$a = b = \delta = 1, \ \eta = 2, \ \omega_1 = 1, \ \omega_2 = 5,$$

$$g(x, y, t, \varepsilon, \mu) = \begin{cases} \mu_1(\sin \omega t, 0)^* & \text{if } y > 0, \\ \mu_2(x + y, 0)^* & \text{if } y < 0. \end{cases} \qquad (I.1.39)$$

Consequently, the Poincaré-Andronov-Melnikov function of (I.1.26) is

$$M(\alpha) = \mu_1 \frac{1}{3} \frac{e^{-2\pi}}{\omega^2 - 1} [(\omega A + B) \sin \omega\alpha + (\omega C + D) \cos \omega\alpha] + \mu_2 E$$

where A, B, C, D are given by (I.1.37) and

$$E = \frac{\sqrt{2}}{975} \left(739 - 223 e^{-2\pi} \right).$$

Function $M(\alpha)$ possesses a simple root if and only if

$$|\mu_2| < \frac{1}{3} \frac{e^{-2\pi}}{|\omega^2 - 1|} \frac{\sqrt{(\omega A + B)^2 + (\omega C + D)^2}}{E} |\mu_1|. \qquad (I.1.40)$$

Applying Theorem I.1.6 we obtain the next result.

Corollary I.1.16. *Consider $(I.1.30)_\varepsilon$ with parameters (I.1.39). If μ_1, μ_2 and ω satisfy (I.1.40), then the 2π-periodic orbit persists for $\varepsilon \neq 0$ small.*

Remark I.1.17. Inequality (I.1.40) means that if the periodic perturbation is sufficiently large (with respect to the non-periodic part of the perturbation), then the T-

periodic trajectory persists. Note that the right-hand side of (I.1.40) can be estimated from above by

$$\frac{\sqrt{c_1\omega^2 + c_2\omega + c_3}}{|\omega^2 - 1|}$$

for appropriate constants c_1, c_2, c_3, which tends to 0, if ω tends to $+\infty$. Hence the bigger frequency ω, the bigger $|\mu_1|$ is needed for fixed $\mu_2 \neq 0$ for persistence of the T-periodic orbit after Theorem I.1.6.

I.1.5. Piecewise-linear planar application

Now we consider the following piecewise-linear planar problem

$$\begin{aligned}
\dot{x} &= b_1 + \varepsilon\mu_1 \sin \omega t \\
\dot{y} &= -2a_1b_1x + \varepsilon\mu_2 \cos \omega t
\end{aligned} \quad \text{if } y > 0,$$

$$\begin{aligned}
\dot{x} &= -b_2 + \varepsilon\mu_1 \sin \omega t \\
\dot{y} &= -2a_2b_2x + \varepsilon\mu_2 \cos \omega t
\end{aligned} \quad \text{if } y < 0$$

$$(I.1.41)_\varepsilon$$

where all constants a_i, b_i for $i = 1, 2$ are assumed to be positive and $(\mu_1, \mu_2) \neq (0, 0)$, $\omega > 0$.

The starting point can be chosen in the form $(x_0, y_0) = (0, y_0)$ with $y_0 > 0$. Then with $h(x, y) = y$ we obtain results similar to those of the previous nonlinear case.

First, one can easily find the periodic trajectory starting at $(0, y_0) \in \Omega_+ = \{(x, y) \in \mathbb{R}^2 \mid y > 0\}$, intersecting transversally $\Omega_0 = \{(x, y) \in \mathbb{R}^2 \mid y = 0\}$ to $\Omega_- = \{(x, y) \in \mathbb{R}^2 \mid y < 0\}$, and returning back to Ω_+ transversally through Ω_0.

Lemma I.1.18. *For any $y_0 > 0$, the unperturbed system $(I.1.41)_0$ possesses a unique periodic solution starting at $(0, y_0)$ given by*

$$\gamma(t) = \begin{cases}
\gamma_1(t) = \left(b_1t, -a_1b_1^2t^2 + y_0\right) & \text{if } t \in [0, t_1], \\
\gamma_2(t) = \left(x_1 - b_2(t - t_1), a_2(x_1 - b_2(t - t_1))^2 - a_2x_1^2\right) & \text{if } t \in [t_1, t_2], \\
\gamma_3(t) = \left(x_2 + b_1(t - t_2), -a_1(x_2 + b_1(t - t_2))^2 + a_1x_2^2\right) & \text{if } t \in [t_2, T]
\end{cases}$$

where

$$t_1 = \frac{1}{b_1}\sqrt{\frac{y_0}{a_1}}, \qquad (x_1, y_1) = \left(\sqrt{\frac{y_0}{a_1}}, 0\right), \qquad t_2 = \frac{2}{b_2}\sqrt{\frac{y_0}{a_1}} + t_1,$$

$$(x_2, y_2) = (-x_1, 0), \qquad T = \frac{1}{b_1}\sqrt{\frac{y_0}{a_1}} + t_2.$$

Fundamental and saltation matrices are described in the next lemma.

Lemma I.1.19. *The unperturbed system* (I.1.41)$_0$ *has the corresponding fundamental matrices*

$$X_1(t) = \begin{pmatrix} 1 & 0 \\ -2a_1b_1t & 1 \end{pmatrix}, \qquad X_2(t) = \begin{pmatrix} 1 & 0 \\ -2a_2b_2(t - t_1) & 1 \end{pmatrix},$$

$$X_3(t) = \begin{pmatrix} 1 & 0 \\ -2a_1b_1(t - t_2) & 1 \end{pmatrix}$$

and saltation matrices

$$S_1 = \begin{pmatrix} 1 & \frac{\sqrt{a_1}(b_1+b_2)}{2a_1b_1\sqrt{y_0}} \\ 0 & \frac{a_2b_2}{a_1b_1} \end{pmatrix}, \qquad S_2 = \begin{pmatrix} 1 & \frac{\sqrt{a_1}(b_1+b_2)}{2a_2b_2\sqrt{y_0}} \\ 0 & \frac{a_1b_1}{a_2b_2} \end{pmatrix}.$$

Proof. Because of the linearity of this case, fundamental matrices are obtained from equalities

$$X_1(t) = e^{At}, \qquad X_2(t) = e^{B(t-t_1)}, \qquad X_3(t) = e^{A(t-t_2)}$$

where

$$A = \begin{pmatrix} 0 & 0 \\ -2a_1b_1 & 0 \end{pmatrix}, \qquad B = \begin{pmatrix} 0 & 0 \\ -2a_2b_2 & 0 \end{pmatrix}$$

are Jacobi matrices of the functions $f_+(x, y)$ and $f_-(x, y)$, respectively. Saltation matrices are given by their definitions in (I.1.16) and (I.1.20) where $\nabla h(x, y) = (0, 1)$ in $\overline{\Omega}$ and

$$f_+(x_1, y_1) = \begin{pmatrix} b_1 \\ -2a_1b_1\sqrt{\frac{y_0}{a_1}} \end{pmatrix}, \quad f_-(x_1, y_1) = \begin{pmatrix} -b_2 \\ -2a_2b_2\sqrt{\frac{y_0}{a_1}} \end{pmatrix},$$

$$f_+(x_2, y_2) = \begin{pmatrix} b_1 \\ 2a_1b_1\sqrt{\frac{y_0}{a_1}} \end{pmatrix}, \quad f_-(x_2, y_2) = \begin{pmatrix} -b_2 \\ 2a_2b_2\sqrt{\frac{y_0}{a_1}} \end{pmatrix}. \qquad \text{(I.1.42)}$$

□

In this case, the corresponding matrices can be easily multiplied to derive the following result.

Lemma I.1.20. *Function $A(t)$ of* (I.1.21) *for the system* (I.1.41)$_\varepsilon$ *possesses the form*

$$A(t) = \begin{cases} \begin{pmatrix} 1 - \dfrac{2\sqrt{a_1}b_1 t(b_1+b_2)}{b_2\sqrt{y_0}} & -\dfrac{b_1+b_2}{b_2\sqrt{a_1 y_0}} \\ 2a_1 b_1 t & 1 \end{pmatrix} & \text{if } t \in [0,t_1), \\[2em] \begin{pmatrix} -1 - \dfrac{2b_1}{b_2} + \dfrac{\sqrt{a_1}(t-t_1)(b_1+b_2)}{\sqrt{y_0}} & \dfrac{\sqrt{a_1}(b_1+b_2)}{2a_2 b_2\sqrt{y_0}} \\ 2(\sqrt{a_1 y_0} - a_1 b_2(t-t_1)) & -\dfrac{a_1}{a_2} \end{pmatrix} & \text{if } t \in [t_1,t_2), \\[2em] \begin{pmatrix} 1 & 0 \\ 2(a_1 b_1(t-t_2) - \sqrt{a_1 y_0}) & 1 \end{pmatrix} & \text{if } t \in [t_2, T]. \end{cases} \qquad (\text{I.1.43})$$

It remains to verify the basic assumptions.

Proposition I.1.21. *Conditions* H1), H2) *and* H3) *are satisfied.*

Proof. From Lemma I.1.18 and using (I.1.42), conditions H1) and H2) are immediately satisfied.

Now let $\dim \mathcal{N}(\mathbb{I} - \widetilde{P}_\xi(x_0, y_0, 0, \mu, \alpha)) > 1$. Then there exists

$$\bar{v} \in \mathcal{N}(\mathbb{I} - \widetilde{P}_\xi(x_0, y_0, 0, \mu, \alpha))$$

such that $\langle \bar{v}, f_+(x_0, y_0) \rangle = 0$, and we can write $\bar{v} = (0, v)^*$. Since

$$\widetilde{P}_\xi(x_0, y_0, 0, \mu, \alpha)\bar{v} = A(0)\bar{v} = \begin{pmatrix} -\dfrac{v(b_1+b_2)}{b_2\sqrt{a_1 y_0}} \\ v \end{pmatrix},$$

then $v = 0$, $\dim \mathcal{N}(\mathbb{I} - \widetilde{P}_\xi(x_0, y_0, 0, \mu, \alpha)) = 1$ and the condition H3) is verified as well. $\qquad \square$

Note that there are a lot of periodic trajectories in the neighborhood of $\gamma(t)$ but none of them has the same period, because the period $T = 2\sqrt{\dfrac{y_0}{a_1}}\left(\dfrac{1}{b_1} + \dfrac{1}{b_2}\right)$ depends on the initial point (x_0, y_0).

We have

$$\mathcal{R}(\mathbb{I} - \widetilde{P}_\xi(x_0, y_0, 0, \mu, \alpha)) = \mathcal{R}(\mathbb{I} - A(0)) = \mathbb{R} \times \{0\}.$$

Accordingly, we set $\psi = (0, 1)^*$ and $A^*(t)\psi = \begin{pmatrix} a_{21}(t) \\ a_{22}(t) \end{pmatrix}$, i.e. the second column of matrix $A(t)$. The assumptions of Theorem I.1.6 are equivalent to saying that

$$M(\alpha) = \sin \omega\alpha \left(\mu_1 \int_0^T a_{21}(t) \cos \omega t\, dt - \mu_2 \int_0^T a_{22}(t) \sin \omega t\, dt \right)$$

$$+ \cos \omega\alpha \left(\mu_1 \int_0^T a_{21}(t) \sin \omega t\, dt + \mu_2 \int_0^T a_{22}(t) \cos \omega t\, dt \right)$$

has a simple root. It is easy to see that this happens if and only if

$$\Phi(\omega) = \int_0^T e^{-\iota\omega t}(\mu_1 a_{21}(t) - \iota\mu_2 a_{22}(t))dt \neq 0 \qquad (I.1.44)$$

where $\iota = \sqrt{-1}$.

Similarly to [18], function $\Phi(\omega)$ is analytic for $\omega > 0$ and hence the following theorem holds (see [18, Theorem 4.2] and [19]).

Theorem I.1.22. *When $\Phi(\omega)$ is not identically equal to zero, then there is at most a countable set $\{\omega_j\} \subset (0, \infty)$ with possible accumulating point at $+\infty$ such that for any $\omega \in (0, \infty)\setminus\{\omega_j\}$, the T-periodic orbit $\gamma(t)$ persists for (I.1.41)$_\varepsilon$ under perturbations for $\varepsilon \neq 0$ small.*

Because for general parameters, the conditions on μ_1 and μ_2 that would allow us to decide when $\Phi(\omega)$ is identically zero or the set of roots is finite or countable are too complicated, we rather provide an example with concrete numerical values of parameters.

Example I.1.23. Consider system (I.1.41)$_\varepsilon$ with parameters

$$a_1 = a_2 = b_1 = b_2 = y_0 = 1. \qquad (I.1.45)$$

Now, from (I.1.44) we have

$$\Phi(\omega) = -4\iota\frac{e^{-2\iota\omega}}{\omega^2}(2\mu_1 + \omega\mu_2)\sin\omega(\cos\omega - 1).$$

Thence for $\omega \in (0, \infty)$ it holds: if $\omega = k\pi$ for some $k \in \mathbb{N}$ or $\omega = -\frac{2\mu_1}{\mu_2} > 0$ then $\Phi(\omega) = 0$. Applying Theorem I.1.6 we arrive at the following result.

Corollary I.1.24. *Consider* (I.1.41)$_\varepsilon$ *with parameters* (I.1.45). *If $\omega > 0$ is such that $\omega \neq k\pi$ for all $k \in \mathbb{N}$ and $\omega \neq -\frac{2\mu_1}{\mu_2}$ with $\mu_2 \neq 0$, then the T-periodic orbit $\gamma(t)$ persists under perturbations for $\varepsilon \neq 0$ small.*

Finally, if $\Phi(\omega)$ is identically zero then a higher order Melnikov function must be derived [20]. We omit those computations in our case, because they are very awkward.

I.1.6. Non-smooth electronic circuits

One of the interesting areas where non-smooth systems occur is that of electronic circuits. There are many interesting books dealing with such problems [2, 9, 21]. We list here several of this type of equation:

1. Valve generator of the form [9, p. 447]

$$\ddot{x} + 2h_1\dot{x} + x = 0, \quad x < -1,$$
$$\ddot{x} - 2h_2\dot{x} + x = 0, \quad x > -1,$$

where $h_1 > 0$ and $h_2 \in \mathbb{R}$ are constants.

2. Symmetric valve generator of the form [9, p. 461]

$$\ddot{x} + 2h_1\dot{x} + x = 0, \quad |x| > 1,$$
$$\ddot{x} - 2h_2\dot{x} + x = 0, \quad |x| < 1,$$

where $h_1 > 0$ and $h_2 \in \mathbb{R}$ are constants.

3. Valve generator with a so-called biased characteristic and a hard mode of excitation of the form [9, p. 469]

$$\ddot{x} + 2h\dot{x} + x = 1, \quad \dot{x} > b,$$
$$\ddot{x} + 2h\dot{x} + x = 0, \quad \dot{x} < b,$$

where $h > 0$ and $b > 0$ are constants.

4. DC-DC converters are used to change one DC voltage to another of the form [2, p. 36]

$$\dot{V} = -\frac{1}{RC}V + \frac{I}{C},$$
$$\dot{I} = -\frac{V}{L} + \begin{cases} 0, & V \geq V_r(t), \\ \frac{E}{L}, & V < V_r(t), \end{cases}$$

where V is the output voltage, I is the corresponding current and C, E, L, R are positive constants representing the capacitance, battery voltage, inductance and resistance, respectively. The reference voltage V_r is a piecewise-linear ramp signal given by

$$V_r(t) = \gamma + \eta(t \bmod T), \quad \gamma, \eta, T > 0.$$

5. Circuits with Zener diodes are studied in [21, pp. 7–8] when the simplest model is given by

$$\dot{x} = -\frac{R}{L}x + \frac{u}{L} + \frac{v}{L},$$
$$v = F(-x),$$

where

$$F(x) = \begin{cases} \alpha x + V_z, & x > 0, \\ \beta x - a, & x < 0 \end{cases}$$

for constants $\alpha \geq 0, \beta \geq 0, a > 0$ and $V_z > 0$.

We could study periodic perturbations of these systems or related ones but we do not go into detail, since computations are similar to those in previous sections, so they are rather complicated. We only mention that many similar models can be written as planar systems of the form

$$\dot{x} = Ax + \text{sgn}(w^*x)v \qquad (\text{I.1.46})$$

where A is a 2×2 real matrix and $w, v \in \mathbb{R}^2$ given vectors. The non-smooth system has been completely investigated in [22], so we refer the reader to that paper for more details, where transverse as well as sliding periodic solutions are established.

CHAPTER I.2

Bifurcation from family of periodic orbits in autonomous systems

I.2.1. Setting of the problem and main results

In the previous chapter, we studied discontinuous systems with time-periodic perturbation. Now, we move our attention to the case of autonomous perturbation. In this chapter, we investigate the persistence of a single periodic solution from a bunch of transverse periodic solutions of an unperturbed system. In comparison to Chapter I.1, here we seek the periodic solution with period close to the period of the original trajectory.

Let $\Omega \subset \mathbb{R}^n$ be an open set in $\mathbb{R}^n$ and $h(x)$ be a C^r-function on $\overline{\Omega}$, with $r \geq 3$. We set $\Omega_{\pm} := \{x \in \Omega \mid \pm h(x) > 0\}$, $\Omega_0 := \{x \in \Omega \mid h(x) = 0\}$. Let $f_{\pm} \in C^r_b(\overline{\Omega})$, $g \in C^r_b(\overline{\Omega} \times \mathbb{R} \times \mathbb{R}^p)$ and $h \in C^r_b(\overline{\Omega}, \mathbb{R})$. Let $\varepsilon \in \mathbb{R}$ and $\mu \in \mathbb{R}^p$, $p \geq 1$ be parameters. Furthermore, we suppose that 0 is a regular value of h.

We say that a function $x(t)$ is a solution of equation

$$\dot{x} = f_{\pm}(x) + \varepsilon g(x, \varepsilon, \mu), \quad x \in \overline{\Omega}_{\pm}, \tag{I.2.1}$$

if it is a solution of this equation in the sense analogous to Definition I.1.1.

Let us assume

H1) For $\varepsilon = 0$ equation (I.2.1) has a smooth family of T^{β}-periodic orbits $\{\gamma(\beta, t)\}$ parametrized by $\beta \in V \subset \mathbb{R}^k$, $0 < k < n$, V is an open set in $\mathbb{R}^k$. Each of the orbits is uniquely determined by its initial point $x_0(\beta) \in \Omega_+$, $x_0 \in C^r_b$, and consists of three branches

$$\gamma(\beta, t) = \begin{cases} \gamma_1(\beta, t) & \text{if } t \in [0, t_1^{\beta}], \\ \gamma_2(\beta, t) & \text{if } t \in [t_1^{\beta}, t_2^{\beta}], \\ \gamma_3(\beta, t) & \text{if } t \in [t_2^{\beta}, T^{\beta}], \end{cases} \tag{I.2.2}$$

where $0 < t_1^{\beta} < t_2^{\beta} < T^{\beta}$, $\gamma_1(\beta, t) \in \Omega_+$ for $t \in [0, t_1^{\beta})$, $\gamma_2(\beta, t) \in \Omega_-$ for $t \in (t_1^{\beta}, t_2^{\beta})$,

Poincaré-Andronov-Melnikov Analysis for Non-Smooth Systems.
http://dx.doi.org/10.1016/B978-0-12-804294-6.50004-7

$\gamma_3(\beta, t) \in \Omega_+$ for $t \in (t_2^\beta, T^\beta]$, and

$$
\begin{aligned}
x_1(\beta) &:= \gamma_1(\beta, t_1^\beta) = \gamma_2(\beta, t_1^\beta) \in \Omega_0, \\
x_2(\beta) &:= \gamma_2(\beta, t_2^\beta) = \gamma_3(\beta, t_2^\beta) \in \Omega_0, \\
x_0(\beta) &:= \gamma_3(\beta, T^\beta) = \gamma_1(\beta, 0) \in \Omega_+.
\end{aligned}
\tag{I.2.3}
$$

We suppose in addition that vectors

$$
\frac{\partial x_0(\beta)}{\partial \beta_1}, \ldots, \frac{\partial x_0(\beta)}{\partial \beta_k}, f_+(x_0(\beta))
$$

are linearly independent whenever $\beta \in V$.

H2) Moreover, we also assume that

$$
Dh(x_1(\beta))f_\pm(x_1(\beta)) < 0 \quad \text{and} \quad Dh(x_2(\beta))f_\pm(x_2(\beta)) > 0, \quad \beta \in V.
$$

Note by H1), H2) and the implicit function theorem (IFT) it can be shown that t_1^β, t_2^β and T^β are C_b^r-functions of β [23].

Now we study local bifurcations for $\gamma(\beta, t)$, so we fix $\beta_0 \in V$ and set $x_0^0 = x_0(\beta_0)$, $t_1^0 = t_1^{\beta_0}$, etc. Note by H1), $x_0(V)$ is an immersed C^r-submanifold of $\mathbb{R}^n$.

Let $x_+(\tau, \xi)(t, \varepsilon, \mu)$ and $x_-(\tau, \xi)(t, \varepsilon, \mu)$ denote the solution of the initial value problem

$$
\begin{aligned}
\dot{x} &= f_\pm(x) + \varepsilon g(x, \varepsilon, \mu) \\
x(\tau) &= \xi
\end{aligned}
\tag{I.2.4}
$$

with corresponding sign.

As in Lemma I.1.2, conditions H1) and H2) establish the existence of a Poincaré mapping.

Lemma I.2.1. *Assume* H1) *and* H2). *Then there exist* $\varepsilon_0, r_0 > 0$, *a neighborhood* $W \subset V$ *of* β_0 *in* $\mathbb{R}^k$ *and a Poincaré mapping (cf. Figure I.2.1)*

$$
P(\cdot, \beta, \varepsilon, \mu): B(x_0^0, r_0) \to \Sigma_\beta
$$

for all fixed $\beta \in W$, $\varepsilon \in (-\varepsilon_0, \varepsilon_0)$, $\mu \in \mathbb{R}^p$, *where*

$$
\Sigma_\beta = \{y \in \mathbb{R}^n \mid \langle y - x_0(\beta), f_+(x_0(\beta)) \rangle = 0\}.
$$

Moreover, $P: B(x_0^0, r_0) \times W \times (-\varepsilon_0, \varepsilon_0) \times \mathbb{R}^p \to \mathbb{R}^n$ *is* C^r-*smooth in all arguments and* $x_0(W) \subset B(x_0^0, r_0) \subset \Omega_+$.

Proof. IFT implies the existence of positive constants τ_1, r_1, δ_1, ε_1 and C^r-function

$$t_1(\cdot,\cdot,\cdot,\cdot)\colon (-\tau_1,\tau_1) \times B(x_0^0, r_1) \times (-\varepsilon_1,\varepsilon_1) \times \mathbb{R}^p \to (t_1^0 - \delta_1, t_1^0 + \delta_1)$$

such that $h(x_+(\tau,\xi)(t,\varepsilon,\mu)) = 0$ for $\tau \in (-\tau_1,\tau_1)$, $\xi \in B(x_0^0, r_1) \subset \Omega_+$, $\varepsilon \in (-\varepsilon_1,\varepsilon_1)$, $\mu \in \mathbb{R}^p$ and $t \in (t_1^0 - \delta_1, t_1^0 + \delta_1)$ if and only if $t = t_1(\tau,\xi,\varepsilon,\mu)$. Moreover, $t_1(0, x_0^0, 0, \mu) = t_1^0$ and

$$x_+(0, x_0(\beta))(t_1(0, x_0(\beta), 0, \mu), 0, \mu) \in \Omega_0 \cap \{\gamma(\beta, t) \mid t \in \mathbb{R}\},$$

thus $t_1(0, x_0(\beta), 0, \mu) = t_1^\beta$. Similarly, we derive functions t_2 and t_3 satisfying, respectively,

$$h(x_-(t_1(\tau,\xi,\varepsilon,\mu), x_+(\tau,\xi)(t_1(\tau,\xi,\varepsilon,\mu),\varepsilon,\mu),\varepsilon,\mu)(t_2(\tau,\xi,\varepsilon,\mu),\varepsilon,\mu)) = 0,$$

$$\langle x_+(t_2(\tau,\xi,\varepsilon,\mu), x_-(t_1(\tau,\xi,\varepsilon,\mu), x_+(\tau,\xi)(t_1(\tau,\xi,\varepsilon,\mu),\varepsilon,\mu))$$

$$(t_2(\tau,\xi,\varepsilon,\mu),\varepsilon,\mu))(t_3(\tau,\xi,\beta,\varepsilon,\mu),\varepsilon,\mu) - x_0(\beta), f_+(x_0(\beta))\rangle = 0.$$

Moreover, we have $t_2(0, x_0(\beta), 0, \mu) = t_2^\beta$ and $t_3(0, x_0(\beta), \beta, 0, \mu) = T^\beta$. Poincaré mapping is then defined as

$$P(\xi,\beta,\varepsilon,\mu) = x_+(t_2(0,\xi,\varepsilon,\mu), x_-(t_1(0,\xi,\varepsilon,\mu), x_+(0,\xi)(t_1(0,\xi,\varepsilon,\mu),\varepsilon,\mu))$$

$$(t_2(0,\xi,\varepsilon,\mu),\varepsilon,\mu))(t_3(0,\xi,\beta,\varepsilon,\mu),\varepsilon,\mu). \tag{I.2.5}$$

□

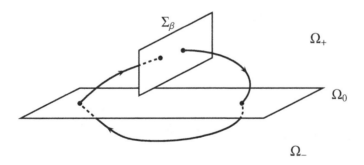

Figure I.2.1 Discontinuous Poincaré mapping

The next lemma describes some properties of derived Poincaré mapping.

Lemma I.2.2. *Let $P(\xi, \beta, \varepsilon, \mu)$ be defined by (I.2.5). Then*

$$P_\xi(x_0(\beta), \beta, 0, \mu) = (\mathbb{I} - S_\beta)A(\beta, 0), \tag{I.2.6}$$

$$P_\beta(x_0(\beta), \beta, 0, \mu) = S_\beta D x_0(\beta), \tag{I.2.7}$$

$$P_\varepsilon(x_0(\beta), \beta, 0, \mu) = (\mathbb{I} - S_\beta)\left(\int_0^{T^\beta} A(\beta, s)g(\gamma(\beta, s), 0, \mu)ds\right), \tag{I.2.8}$$

where P_ξ, P_β, P_ε are partial derivatives of P with respect to ξ, β, ε, respectively.

Here S_β is the orthogonal projection onto the 1-dimensional space $[f_+(x_0(\beta))]$ defined by

$$S_\beta u = \frac{\langle u, f_+(x_0(\beta)) \rangle f_+(x_0(\beta))}{\|f_+(x_0(\beta))\|^2}, \tag{I.2.9}$$

and $A(\beta, t)$ is given by

$$A(\beta, t) = \begin{cases} X_3(\beta, T^\beta)S_2(\beta)X_2(\beta, t_2^\beta)S_1(\beta)X_1(\beta, t_1^\beta)X_1^{-1}(\beta, t) & \text{if } t \in [0, t_1^\beta), \\ X_3(\beta, T^\beta)S_2(\beta)X_2(\beta, t_2^\beta)X_2^{-1}(\beta, t) & \text{if } t \in [t_1^\beta, t_2^\beta), \\ X_3(\beta, T^\beta)X_3^{-1}(\beta, t) & \text{if } t \in [t_2^\beta, T^\beta], \end{cases} \tag{I.2.10}$$

where

$$S_1(\beta) = \mathbb{I} + \frac{(f_-(x_1(\beta)) - f_+(x_1(\beta)))Dh(x_1(\beta))}{Dh(x_1(\beta))f_+(x_1(\beta))}, \tag{I.2.11}$$

$$S_2(\beta) = \mathbb{I} + \frac{(f_+(x_2(\beta)) - f_-(x_2(\beta)))Dh(x_2(\beta))}{Dh(x_2(\beta))f_-(x_2(\beta))}, \tag{I.2.12}$$

and finally, $X_1(\beta, t)$, $X_2(\beta, t)$ and $X_3(\beta, t)$ solve the following linear initial value problems

$$\begin{aligned} \dot{X}_1(\beta, t) &= Df_+(\gamma(\beta, t))X_1(\beta, t) \\ X_1(\beta, 0) &= \mathbb{I}, \end{aligned} \tag{I.2.13}$$

$$\begin{aligned} \dot{X}_2(\beta, t) &= Df_-(\gamma(\beta, t))X_2(\beta, t) \\ X_2(\beta, t_1^\beta) &= \mathbb{I}, \end{aligned} \tag{I.2.14}$$

$$\begin{aligned} \dot{X}_3(\beta, t) &= Df_+(\gamma(\beta, t))X_3(\beta, t) \\ X_3(\beta, t_2^\beta) &= \mathbb{I}, \end{aligned} \tag{I.2.15}$$

respectively.

Note saltation matrices $S_1(\beta)$, $S_2(\beta)$ are invertible (cf. [11]) with

$$S_1^{-1}(\beta) = \mathbb{I} + \frac{(f_+(x_1(\beta)) - f_-(x_1(\beta)))Dh(x_1(\beta))}{Dh(x_1(\beta))f_-(x_1(\beta))},$$

$$S_2^{-1}(\beta) = \mathbb{I} + \frac{(f_-(x_2(\beta)) - f_+(x_2(\beta)))Dh(x_2(\beta))}{Dh(x_2(\beta))f_+(x_2(\beta))}.$$

Considering the inner product $\langle a, b \rangle = b^* a$, it is possible to introduce matrix notation for operator $\mathcal{S}_\beta$ of (I.2.9):

$$S_\beta u = \frac{f_+(x_0(\beta))(f_+(x_0(\beta)))^*}{\|f_+(x_0(\beta))\|^2} u \tag{I.2.16}$$

which is symmetric, i.e. $S_\beta^* = S_\beta$. The derivative of mapping (I.2.5) has an important property:

Lemma I.2.3. *For any $\xi \in B(x_0^0, r_0)$, $\beta \in W$, $\varepsilon \in (-\varepsilon_0, \varepsilon_0)$ and $\mu \in \mathbb{R}^p$, $P_\xi(\xi, \beta, \varepsilon, \mu)$ has an eigenvalue 0 with corresponding eigenvector $f_+(\xi) + \varepsilon g(\xi, \varepsilon, \mu)$, i.e.*

$$P_\xi(\xi, \beta, \varepsilon, \mu)[f_+(\xi) + \varepsilon g(\xi, \varepsilon, \mu)] = 0.$$

Proof. Similarly to the unperturbed case in Lemma I.1.3 (see also [11]) we have

$$x_+(0, x_+(0, \xi)(t, \varepsilon, \mu))(t_1(0, x_+(0, \xi)(t, \varepsilon, \mu), \varepsilon, \mu), \varepsilon, \mu)$$
$$= x_+(0, \xi)(t_1(0, x_+(0, \xi)(t, \varepsilon, \mu), \varepsilon, \mu) + t, \varepsilon, \mu)$$

as the first intersection point of the trajectory of the perturbed system (I.2.1) and discontinuity boundary Ω_0. Hence

$$t_1(0, x_+(0, \xi)(t, \varepsilon, \mu), \varepsilon, \mu) + t = t_1(0, \xi, \varepsilon, \mu)$$

for any t sufficiently close to 0. Analogously we get

$$t_2(0, x_+(0, \xi)(t, \varepsilon, \mu), \varepsilon, \mu) + t = t_2(0, \xi, \varepsilon, \mu).$$

Next,

$$P(x_+(0, \xi)(t, \varepsilon, \mu), \beta, \varepsilon, \mu) = x_+(t_2(0, x_+(0, \xi)(t, \varepsilon, \mu), \varepsilon, \mu),$$
$$x_-(t_1(0, x_+(0, \xi)(t, \varepsilon, \mu), \varepsilon, \mu), x_+(0, x_+(0, \xi)(t, \varepsilon, \mu))$$
$$(t_1(0, x_+(0, \xi)(t, \varepsilon, \mu), \varepsilon, \mu), \varepsilon, \mu))(t_2(0, x_+(0, \xi)(t, \varepsilon, \mu), \varepsilon, \mu), \varepsilon, \mu))$$
$$(t_3(0, x_+(0, \xi)(t, \varepsilon, \mu), \beta, \varepsilon, \mu), \varepsilon, \mu)$$
$$= x_+(t_2(0, \xi, \varepsilon, \mu), x_-(t_1(0, \xi, \varepsilon, \mu), x_+(0, \xi)(t_1(0, \xi, \varepsilon, \mu), \varepsilon, \mu))$$
$$(t_2(0, \xi, \varepsilon, \mu), \varepsilon, \mu))(t_3(0, x_+(0, \xi)(t, \varepsilon, \mu), \beta, \varepsilon, \mu) + t, \varepsilon, \mu)$$

for any t close to 0. The most left-hand side of the latter equation is from Σ_β, and the most right-hand side is a point of a trajectory of (I.2.1) with given ε starting at

ξ. Therefore it is a fixed point $\zeta \in \Sigma_\beta$, and $t_3(0, x_+(0, \xi)(t, \varepsilon, \mu), \beta, \varepsilon, \mu) + t$ is constant. Consequently,

$$P_\xi(\xi, \beta, \varepsilon, \mu)[f_+(\xi) + \varepsilon g(\xi, \varepsilon, \mu)] = D_t P(x_+(0, \xi)(t, \varepsilon, \mu), \beta, \varepsilon, \mu)\big|_{t=0} = D_t \zeta\big|_{t=0} = 0.$$

$\square$

Note if we take $\xi = x_0(\beta)$ and $\varepsilon = 0$ in the above lemma then $\zeta = x_0(\beta)$ and

$$t_3(0, x_+(0, x_0(\beta))(t, 0, \mu), \beta, 0, \mu) + t = T^\beta.$$

For any $\xi \in \mathbb{R}^n$ we define orthogonal projection $\widetilde{S}_\beta$ onto Σ_β,

$$\widetilde{S}_\beta : \xi \mapsto \xi - S_\beta(\xi - x_0(\beta)).$$

Denoting $F(\xi, \beta, \varepsilon, \mu) := \widetilde{S}_\beta(\xi) - P(\xi, \beta, \varepsilon, \mu)$, ξ is an initial point from Σ_β of periodic orbit of perturbed system (I.2.1) if and only if it satisfies

$$F(\xi, \beta, \varepsilon, \mu) = 0, \qquad \xi \in \Sigma_\beta. \tag{I.2.17}$$

For $\xi \in \mathbb{R}^n$

$$F_\xi(x_0(\beta), \beta, 0, \mu)\xi = (\mathbb{I} - S_\beta)\xi - P_\xi(x_0(\beta), \beta, 0, \mu)\xi.$$

Thus from Lemma I.2.3,

$$F_\xi(x_0(\beta), \beta, 0, \mu)f_+(x_0(\beta)) = 0, \tag{I.2.18}$$

where F_ξ is the partial derivative of F with respect to ξ. On the other side,

$$F(x_0(\beta), \beta, 0, \mu) = x_0(\beta) - P(x_0(\beta), \beta, 0, \mu) = 0, \quad \forall \beta \in W.$$

Hence from (I.2.7)

$$P_\xi(x_0(\beta), \beta, 0, \mu)Dx_0(\beta) = (\mathbb{I} - S_\beta)Dx_0(\beta),$$

and therefore

$$F_\xi(x_0(\beta), \beta, 0, \mu)Dx_0(\beta) = 0, \qquad \forall \beta \in W.$$

Here we state the third condition:

H3) The set

$$\left\{ \frac{\partial x_0(\beta)}{\partial \beta_1}, \dots, \frac{\partial x_0(\beta)}{\partial \beta_k}, f_+(x_0(\beta)) \right\}$$

spans the null space of the operator $F_\xi(x_0(\beta), \beta, 0, \mu)$.
Note $\Sigma_\beta = [f_+(x_0(\beta))]^\perp + x_0(\beta)$. Let us denote

$$Z_\beta = \mathcal{N}F_\xi(x_0(\beta), \beta, 0, \mu) \cap [f_+(x_0(\beta))]^\perp, \qquad Y_\beta = \mathcal{R}F_\xi(x_0(\beta), \beta, 0, \mu) \tag{I.2.19}$$

the restricted null space and the range of the corresponding operator, respectively. Now from condition H3) we have

$$Z_\beta = \left[\frac{\partial x_0(\beta_0)}{\partial \beta_1}, \ldots, \frac{\partial x_0(\beta_0)}{\partial \beta_k}, f_+(x_0(\beta)) \right] \cap [f_+(x_0(\beta))]^\perp = (\mathbb{I} - S_\beta)Dx_0(\beta).$$

Using Gram-Schmidt orthogonalization we find an orthonormal basis $\{y_1, \ldots, y_{n-k-1}\}$ for vector space $Z_\beta^\perp$

such that $Z_\beta \perp Z_\beta^\perp$ and $Z_\beta \oplus Z_\beta^\perp = [f_+(x_0(\beta))]^\perp$.

We can define orthogonal projections

$$Q_\beta : \Sigma_\beta \to Y_\beta, \qquad \mathcal{P}_\beta : \Sigma_\beta \to Y_\beta^\perp, \qquad (I.2.20)$$

where $Y_\beta^\perp$ is an orthogonal complement to Y_β in $[f_+(x_0(\beta))]^\perp$, and the decomposition for any z sufficiently close to manifold $x_0(W)$,

$$z = x_0(\beta) + \xi \quad \text{for} \quad \beta \in W, \, \xi \in \left[\frac{\partial x_0(\beta)}{\partial \beta_1}, \ldots, \frac{\partial x_0(\beta)}{\partial \beta_k} \right]^\perp.$$

The second condition of (I.2.17) gives another restriction on ξ, i.e.

$$\xi \in \left[\frac{\partial x_0(\beta)}{\partial \beta_1}, \ldots, \frac{\partial x_0(\beta)}{\partial \beta_k}, f_+(x_0(\beta)) \right]^\perp.$$

Therefore

$$z = x_0(\beta) + \xi, \qquad \beta \in W, \, \xi \in Z_\beta^\perp. \qquad (I.2.21)$$

Note Z_β, Q_β and $\mathcal{P}_\beta$ are C^{r-1}-smooth with respect to β. Consequently, applying the Lyapunov-Schmidt reduction, equation (I.2.17) is equivalent to the next couple of equations

$$Q_\beta F(x_0(\beta) + \xi, \beta, \varepsilon, \mu) = 0, \qquad (I.2.22)$$
$$\mathcal{P}_\beta F(x_0(\beta) + \xi, \beta, \varepsilon, \mu) = 0. \qquad (I.2.23)$$

Considering the first one as the equation for $\xi \in Z_\beta^\perp$, it can be solved via IFT, since $Q_\beta F(x_0(\beta), \beta, 0, \mu) = 0$ and

$$D_\xi Q_\beta F(x_0(\beta) + \xi, \beta, 0, \mu)\big|_{\xi=0} = Q_\beta F_\xi(x_0(\beta), \beta, 0, \mu)\big|_{Z_\beta^\perp},$$

where $\big|_{Z_\beta^\perp}$ denotes the restriction on $Z_\beta^\perp$, is an isomorphism $Z_\beta^\perp$ onto Y_β for all $\mu \in \mathbb{R}^p$. So there exist positive constants ε_2, r_2 and C^{r-1}-function

$$\xi(\cdot, \cdot, \cdot) : B(\beta_0, r_2) \times (-\varepsilon_2, \varepsilon_2) \times \mathbb{R}^p \to Z_\beta^\perp$$

such that $Q_\beta F(x_0(\beta) + \xi, \beta, \varepsilon, \mu) = 0$ for $\beta \in B(\beta_0, r_2)$, $\varepsilon \in (-\varepsilon_2, \varepsilon_2)$ and $\mu \in \mathbb{R}^p$ if and only if $\xi = \xi(\beta, \varepsilon, \mu)$. Moreover $\xi(\beta, 0, \mu) = 0$, since $Q_\beta F(x_0(\beta), \beta, 0, \mu) = 0$.

Equation (I.2.23) now has the form

$$G(\beta, \varepsilon, \mu) := \mathcal{P}_\beta F(x_0(\beta) + \xi(\beta, \varepsilon, \mu), \beta, \varepsilon, \mu) = 0 \qquad (I.2.24)$$

for β close to β_0, ε to 0 and $\mu \in \mathbb{R}^p$. In the unperturbed case, this is easily solved, i.e. $G(\beta, 0, \mu) = 0$. In order to have a persisting periodic orbit, equation (I.2.24) has to be satisfied for all ε sufficiently close to 0. The next condition follows

$$
\begin{aligned}
0 = G_\varepsilon(\beta, 0, \mu) &= \mathcal{P}_\beta D_\varepsilon F(x_0(\beta) + \xi(\beta, \varepsilon, \mu), \beta, \varepsilon, \mu)\big|_{\varepsilon=0} \\
&= \mathcal{P}_\beta \left[F_\xi(x_0(\beta), \beta, 0, \mu)\xi_\varepsilon(\beta, 0, \mu) + F_\varepsilon(x_0(\beta), \beta, 0, \mu) \right] \\
&= \mathcal{P}_\beta F_\varepsilon(x_0(\beta), \beta, 0, \mu) = -\mathcal{P}_\beta P_\varepsilon(x_0(\beta), \beta, 0, \mu),
\end{aligned}
\qquad (I.2.25)
$$

where G_ε, F_ε are the partial derivatives of G, F with respect to ε, respectively. We note [24] that there exists an orthogonal basis $\{\psi_1(\beta), \ldots, \psi_k(\beta)\}$ of $Y_\beta^\perp$, i.e. $Y_\beta^\perp = [\psi_1(\beta), \ldots, \psi_k(\beta)]$ for each $\beta \in W$ and ψ_i are C^{r-1}-smooth. Then projection $\mathcal{P}_\beta$ of (I.2.20) can be written in the form

$$\mathcal{P}_\beta y = \sum_{i=1}^k \frac{\langle y, \psi_i(\beta) \rangle \psi_i(\beta)}{\|\psi_i(\beta)\|^2}.$$

Equation (I.2.25) can be rewritten as follows

$$\sum_{i=1}^k \frac{\langle P_\varepsilon(x_0(\beta), \beta, 0, \mu), \psi_i(\beta) \rangle \psi_i(\beta)}{\|\psi_i(\beta)\|^2} = 0. \qquad (I.2.26)$$

Using linear independence of $\psi_1(\beta), \ldots, \psi_k(\beta)$ and Lemma I.2.2 together with (I.2.26), we arrive at

$$M^\mu(\beta) = 0 \quad \text{if and only if} \quad G_\varepsilon(\beta, 0, \mu) = 0, \qquad (I.2.27)$$

where

$$
\begin{aligned}
M^\mu(\beta) &= (M_1^\mu(\beta), \ldots, M_k^\mu(\beta)), \\
M_i^\mu(\beta) &= \int_0^{T^\beta} \langle g(\gamma(\beta, t), 0, \mu), A^*(\beta, t)\psi_i(\beta) \rangle dt, \, i = 1, \ldots, k.
\end{aligned}
\qquad (I.2.28)
$$

Note by (I.2.10), we have

$$
A^*(\beta, t) = \begin{cases}
X_1^{-1*}(\beta, t)X_1^*(\beta, t_1^\beta)S_1^*(\beta)X_2^*(\beta, t_2^\beta)S_2^*(\beta)X_3^*(\beta, T^\beta) & \text{if } t \in [0, t_1^\beta), \\
X_2^{-1*}(\beta, t)X_2^*(\beta, t_2^\beta)S_2^*(\beta)X_3^*(\beta, T^\beta) & \text{if } t \in [t_1^\beta, t_2^\beta), \\
X_3^{-1*}(\beta, t)X_3^*(\beta, T^\beta) & \text{if } t \in [t_2^\beta, T^\beta].
\end{cases}
\qquad (I.2.29)
$$

We shall call the function M^μ defined by (I.2.28) a Poincaré-Andronov-Melnikov function for the discontinuous system (I.2.1) (see Remark I.2.6 below).

We know [11] that the linearization of (I.2.1) with $\varepsilon = 0$ along T^β-periodic solution

$\gamma(\beta, t)$ is given by

$$\dot{x} = Df_{\pm}(\gamma(\beta, t))x \tag{I.2.30}$$

which splits into two unperturbed equations

$$\dot{x} = Df_+(\gamma(\beta, t))x \qquad \text{if } t \in [0, t_1^\beta) \cup [t_2^\beta, T^\beta],$$
$$\dot{x} = Df_-(\gamma(\beta, t))x \qquad \text{if } t \in [t_1^\beta, t_2^\beta)$$

satisfying impulsive conditions

$$x(t_1^\beta+) = S_1(\beta)x(t_1^\beta-), \qquad x(t_2^\beta+) = S_2(\beta)x(t_2^\beta-) \tag{I.2.31}$$

and periodic condition

$$(\mathbb{I} - S_\beta)(x(T^\beta) - x(0)) = 0 \tag{I.2.32}$$

as well, where $x(t\pm) = \lim_{s \to t^{\pm}} x(s)$. Corresponding fundamental matrices are $X_1(\beta, t)$, $X_3(\beta, t)$ to the plus-equation and $X_2(\beta, t)$ to the minus-equation (cf. Lemma I.2.2). It follows that the fundamental matrix solution of the unperturbed variational equation (I.2.30) is given by

$$X(\beta, t) = \begin{cases} X_1(\beta, t) & \text{if } t \in [0, t_1^\beta), \\ X_2(\beta, t)S_1(\beta)X_1(\beta, t_1^\beta) & \text{if } t \in [t_1^\beta, t_2^\beta), \\ X_3(\beta, t)S_2(\beta)X_2(\beta, t_2^\beta)S_1(\beta)X_1(\beta, t_1^\beta) & \text{if } t \in [t_2^\beta, T^\beta]. \end{cases} \tag{I.2.33}$$

Especially, $X(\beta, T^\beta) = A(\beta, 0)$. To proceed, we need the following Fredholm like result for linear impulsive boundary value problems of the form (I.2.30), (I.2.31) and (I.2.32).

Lemma I.2.4. *Let* $A(t) \in C([0, T], L(\mathbb{R}^n))$, $B_1, B_2, B_3 \in L(\mathbb{R}^n)$, $0 < t_1 < t_2 < T$ *and* $h \in C := C([0, t_1], \mathbb{R}^n) \cap C([t_1, t_2], \mathbb{R}^n) \cap C([t_2, T], \mathbb{R}^n)$. *Then the nonhomogeneous problem*

$$\dot{x} = A(t)x + h(t),$$
$$x(t_i+) = B_i x(t_i-), \ i = 1, 2, \tag{I.2.34}$$
$$B_3(x(T) - x(0)) = 0$$

has a solution $x \in C^1 := C^1([0, t_1], \mathbb{R}^n) \cap C^1([t_1, t_2], \mathbb{R}^n) \cap C^1([t_2, T], \mathbb{R}^n)$ *if and only if* $\int_0^T \langle h(t), v(t) \rangle dt = 0$ *for any solution* $v \in C^1$ *of the adjoint system given by*

$$\dot{v} = -A^*(t)v,$$
$$v(t_i-) = B_i^* v(t_i+), \ i = 1, 2, \tag{I.2.35}$$
$$v(T) = v(0) \in \mathcal{N}B_3^\perp.$$

Proof. If $h \in C$ satisfies (I.2.34) for a $x \in C^1$, then for any $v \in C^1$ fulfilling (I.2.35), we derive

$$\int_0^T \langle h(t), v(t) \rangle dt = \int_0^T \langle \dot{x}(t) - A(t)x(t), v(t) \rangle dt$$

$$= \langle x(T), v(T) \rangle - \langle x(t_2+), v(t_2+) \rangle + \langle x(t_2-), v(t_2-) \rangle$$

$$-\langle x(t_1+), v(t_1+) \rangle + \langle x(t_1-), v(t_1-) \rangle - \langle x(0), v(0) \rangle - \int_0^T \langle x(t), \dot{v}(t) + A^*(t)v(t) \rangle dt$$

$$= \langle x(T) - x(0), v(T) \rangle + \langle x(0), v(T) - v(0) \rangle + \langle x(t_1-), v(t_1-) - B_1^* v(t_1+) \rangle$$

$$+\langle x(t_2-), v(t_2-) - B_2^* v(t_2+) \rangle - \int_0^T \langle x(t), \dot{v}(t) + A^*(t)v(t) \rangle dt = 0.$$

Reversely, we suppose that $h \in C$ satisfies $\int_0^T \langle h(t), v(t) \rangle dt = 0$ for any $v \in C^1$ fulfilling (I.2.35). Let $X(t)$ be the fundamental solution of $\dot{x} = A(t)x$ with $X(0) = \mathbb{I}$. Then

$$v(t) = \begin{cases} X^{-1*}(t)X^*(t_1)B_1^* X^{-1*}(t_1)X^*(t_2)B_2^* X^{-1*}(t_2)X^*(T)v(T) & \text{if } t \in [0, t_1), \\ X^{-1*}(t)X^*(t_2)B_2^* X^{-1*}(t_2)X^*(T)v(T) & \text{if } t \in [t_1, t_2), \\ X^{-1*}(t)X^*(T)v(T) & \text{if } t \in [t_2, T] \end{cases}$$

and

$$X^*(t_1)B_1^* X^{-1*}(t_1)X^*(t_2)B_2^* X^{-1*}(t_2)X^*(T)v(T) = v(T) \in NB_3^{\perp} = \mathcal{R}B_3^*. \qquad (\text{I}.2.36)$$

So

$$0 = \int_0^T \langle h(t), v(t) \rangle dt = \left\langle \int_{t_2}^T X(T)X^{-1}(s)h(s)ds \right.$$

$$+ \int_{t_1}^{t_2} X(T)X^{-1}(t_2)B_2 X(t_2)X^{-1}(s)h(s)ds \qquad (\text{I}.2.37)$$

$$+ \left. \int_0^{t_1} X(T)X^{-1}(t_2)B_2 X(t_2)X^{-1}(t_1)B_1 X(t_1)X^{-1}(s)h(s)ds, v(T) \right\rangle.$$

Next, by (I.2.36) and (I.2.37), we obtain $v(T) = B_3^* w$ for a $w \in \mathbb{R}^n$ satisfying

$$\left(X^*(t_1)B_1^* X^{-1*}(t_1)X^*(t_2)B_2^* X^{-1*}(t_2)X^*(T) - \mathbb{I} \right) B_3^* w = 0 \qquad (\text{I}.2.38)$$

along with

$$0 = \langle B_3 u, w \rangle,$$

$$u := \int_{t_2}^{T} X(T)X^{-1}(s)h(s)ds + \int_{t_1}^{t_2} X(T)X^{-1}(t_2)B_2X(t_2)X^{-1}(s)h(s)ds \qquad (\text{I}.2.39)$$

$$+ \int_0^{t_1} X(T)X^{-1}(t_2)B_2X(t_2)X^{-1}(t_1)B_1X(t_1)X^{-1}(s)h(s)ds.$$

But (I.2.38) and (I.2.39) imply the existence of $x_0 \in \mathbb{R}^n$ such that

$$B_3 \left(X(T)X^{-1}(t_2)B_2X(t_2)X^{-1}(t_1)B_1X(t_1)x_0 + u - x_0 \right) = 0. \qquad (\text{I}.2.40)$$

Then using (I.2.40), we see that the function

$$x(t) = \begin{cases} X(t)x_0 + \int_0^t X(t)X^{-1}(s)h(s)ds & \text{if } t \in [0, t_1), \\[2mm] X(t)X^{-1}(t_1)B_1\left(X(t_1)x_0 + \int_0^{t_1} X(t_1)X^{-1}(s)h(s)ds \right) & \\[1mm] \quad + \int_{t_1}^t X(t)X^{-1}(s)h(s)ds & \text{if } t \in [t_1, t_2), \\[2mm] X(t)X^{-1}(t_2)B_2\Big(X(t_2)X^{-1}(t_1)B_1\big(X(t_1)x_0 & \\[1mm] \quad + \int_0^{t_1} X(t_1)X^{-1}(s)h(s)ds \big) + \int_{t_1}^{t_2} X(t_2)X^{-1}(s)h(s)ds \Big) & \\[1mm] \quad + \int_{t_2}^t X(t)X^{-1}(s)h(s)ds & \text{if } t \in [t_2, T] \end{cases}$$

is a solution of (I.2.34). The proof is finished. $\qquad \square$

Applying the preceding lemma to (I.2.30), (I.2.31) and (I.2.32), we see that the adjoint variational system of (I.2.1) is given by the following linear impulsive boundary value problem

$$\begin{aligned} \dot X &= -Df_+^*(\gamma(\beta, t))X && \text{if } t \in [0, t_1^\beta], \\ \dot X &= -Df_-^*(\gamma(\beta, t))X && \text{if } t \in [t_1^\beta, t_2^\beta], \\ \dot X &= -Df_+^*(\gamma(\beta, t))X && \text{if } t \in [t_2^\beta, T^\beta], \\ X(t_i^\beta +) &= S_i^*(\beta)^{-1}X(t_i^\beta -), \ i = 1, 2, && \\ X(T^\beta) &= X(0) \in [f_+(x_0(\beta))]^\perp. && \end{aligned} \qquad (\text{I}.2.41)$$

Note for each $i = 1, \ldots, k$ and for any $\xi \in [f_+(x_0(\beta))]^\perp$, we have $S_\beta \psi_i(\beta) = 0$ and

$$0 = \langle F_\xi(x_0(\beta), \beta, 0, \mu)\xi, \psi_i(\beta) \rangle = \langle (\mathbb{I} - P_\xi(x_0(\beta), \beta, 0, \mu))\xi, \psi_i(\beta) \rangle$$
$$= \langle \xi, \psi_i(\beta) - A^*(\beta, 0)(\mathbb{I} - S_\beta)\psi_i(\beta) \rangle = \langle \xi, (\mathbb{I} - A^*(\beta, 0))\psi_i(\beta) \rangle,$$

and for $\xi \in [f_+(x_0(\beta))]$ (cf. (I.2.18)),

$$0 = \langle F_\xi(x_0(\beta), \beta, 0, \mu)\xi, \psi_i(\beta)\rangle = \langle(\mathbb{I} - S_\beta - P_\xi(x_0(\beta), \beta, 0, \mu))\xi, \psi_i(\beta)\rangle$$
$$= \langle\xi, (\mathbb{I} - A^*(\beta, 0))(\mathbb{I} - S_\beta)\psi_i(\beta)\rangle = \langle\xi, (\mathbb{I} - A^*(\beta, 0))\psi_i(\beta)\rangle.$$

Hence $A^*(\beta, 0)\psi_i(\beta) = \psi_i(\beta)$.

Next, from Lemma I.1.5 it is easy to see that for each $i = 1, \ldots, k$, $A^*(\beta, t)\psi_i(\beta)$ is the solution of (I.2.41).

Consequently, we can take in (I.2.28) any basis of T^β-periodic solutions of the adjoint variational equation.

Note, the condition $M^\mu(\beta) = 0$ is a necessary one for the persistence of periodic orbit.

The following theorem states the sufficient condition for the existence of a unique periodic solution of equation (I.2.1) with $\varepsilon \neq 0$.

Theorem I.2.5. *Let conditions* H1), H2), H3) *be satisfied and $M^\mu(\beta)$ be defined by* (I.2.28). *If $\beta_0 \in V$ is a simple root of M^{μ_0}, i.e.*

$$M^{\mu_0}(\beta_0) = 0, \qquad \det \mathrm{D}M^{\mu_0}(\beta_0) \neq 0,$$

then there exists a neighborhood U of the point $(0, \mu_0)$ in $\mathbb{R} \times \mathbb{R}^p$ and a C^{r-2}-function $\beta(\varepsilon, \mu)$, with $\beta(0, \mu_0) = \beta_0$, such that perturbed equation (I.2.1) possesses a unique persisting closed trajectory. Moreover, it contains a point

$$x^*(\varepsilon, \mu) := x_0(\beta(\varepsilon, \mu)) + \xi(\beta(\varepsilon, \mu), \varepsilon, \mu) \in \Sigma_{\beta(\varepsilon,\mu)}, \tag{I.2.42}$$

and it has a period $t_3(0, x^*(\varepsilon, \mu), \beta(\varepsilon, \mu), \varepsilon, \mu)$.

Proof. We introduce the function

$$H(\beta, \varepsilon, \mu) = \begin{cases} \frac{1}{\varepsilon}G(\beta, \varepsilon, \mu) & \text{if } \varepsilon \neq 0, \\ G_\varepsilon(\beta, \varepsilon, \mu) & \text{if } \varepsilon = 0. \end{cases}$$

We recall that G_ε is the partial derivative of G with respect to ε. Clearly, H is C^{r-2}-smooth and $H(\beta, \varepsilon, \mu) = 0$ gives the desired periodic solution. Note that $H(\beta, 0, \mu) = -\Psi(\beta)M^\mu(\beta)$, where $\Psi(\beta)$ is an $n \times k$ matrix with i-th column $\frac{\psi_i(\beta)}{\langle\psi_i(\beta),\psi_i(\beta)\rangle}$. Next, for the partial derivative H_β of H with respect to β, we derive

$$H_\beta(\beta_0, 0, \mu_0) = -\mathrm{D}\Psi(\beta_0)M^{\mu_0}(\beta_0) - \Psi(\beta_0)\mathrm{D}M^{\mu_0}(\beta_0) = -\Psi(\beta_0)\mathrm{D}M^{\mu_0}(\beta_0);$$

thus $H_\beta(\beta_0, 0, \mu_0)$ is an isomorphism, and from (I.2.27), IFT implies the existence of neighborhood U and function $\beta(\varepsilon, \mu)$ from the statement of the theorem. Results on x^* and the period of the persisting orbit follow immediately from the preceding arguments. $\qquad\square$

Let us denote the persisting periodic trajectory from the latter theorem by $\gamma^*(\varepsilon, \mu, t)$. Then clearly $\gamma^*(0, \mu, t) = \gamma(\beta_0, t)$.

Note that if function g is discontinuous in x, the preceding theorem remains true.

Remark I.2.6. If (I.2.1) is smooth, i.e. $f_\pm = f$, then $S_i(\beta) = \mathbb{I}$, $i = 1, 2$ and (I.2.41) has the form

$$\dot{X} = -Df^*(\gamma(\beta, t))X \qquad \text{if } t \in [0, T^\beta],$$
$$X(T^\beta) = X(0) \in [f_+(x_0(\beta))]^\perp. \tag{I.2.43}$$

So the Poincaré-Andronov-Melnikov function (I.2.28) possesses the form

$$M^\mu(\beta) = (M_1^\mu(\beta), \ldots, M_k^\mu(\beta)),$$
$$M_i^\mu(\beta) = \int_0^{T^\beta} \langle g(\gamma(\beta, t), 0, \mu), \psi_i(\beta, t) \rangle dt, i = 1, \ldots, k \tag{I.2.44}$$

for any smooth basis $\{\psi_i(\beta, t)\}_{i=1}^k$ of solutions of the adjoint periodic linear problem (I.2.43). Note (I.2.44) is the usual Poincaré-Andronov-Melnikov function [16, 25] for bifurcation of periodic orbits for NDS. Again, this is a reason why we call (I.2.28) the Poincaré-Andronov-Melnikov function for perturbed piecewise-smooth NDS (I.2.1).

I.2.2. Geometric interpretation of required assumptions

In this section, we look at the investigated problem from a geometric point of view. For any $\beta, \widetilde{\beta} \in W$, we can solve $t(\beta, \widetilde{\beta})$ from

$$\langle \gamma(\widetilde{\beta}, t(\beta, \widetilde{\beta})) - x_0(\beta), f_+(x_0(\beta)) \rangle = 0$$
$$t(\beta, \beta) = 0.$$

Note

$$\frac{\partial t(\beta, \beta)}{\partial \widetilde{\beta}} = -\frac{\langle Dx_0(\beta), f_+(x_0(\beta)) \rangle}{\|f_+(x_0(\beta))\|^2}. \tag{I.2.45}$$

Set $\widetilde{x}_\beta(\widetilde{\beta}) = \gamma(\widetilde{\beta}, t(\beta, \widetilde{\beta}))$, $\gamma_\beta(\widetilde{\beta}, t) = \gamma(\widetilde{\beta}, t(\beta, \widetilde{\beta}) + t)$. So $\gamma_\beta(\widetilde{\beta}, 0) = \widetilde{x}_\beta(\widetilde{\beta})$, $\widetilde{x}_\beta(\beta) = x_0(\beta)$, $\gamma_\beta(\beta, t) = \gamma(\beta, t)$ and

$$F(\widetilde{x}_\beta(\widetilde{\beta}), \beta, 0, \mu) = 0. \tag{I.2.46}$$

Indeed, analogously to the proof of Lemma I.2.3, $t_i(0, \widetilde{x}_\beta(\widetilde{\beta}), 0, \mu) + t(\beta, \widetilde{\beta}) = t_i^{\widetilde{\beta}}$ for $i = 1, 2$. Therefore

$$P(\widetilde{x}_\beta(\widetilde{\beta}), \beta, 0, \mu) = x_+(t_2^{\widetilde{\beta}}, x_2(\widetilde{\beta}))(t_3(0, \widetilde{x}_\beta(\widetilde{\beta}), \beta, 0, \mu) + t(\beta, \widetilde{\beta}), 0, \mu).$$

Since $\widetilde{x}_\beta(\widetilde{\beta}) \in \Sigma_\beta$, we obtain $t_3(0, \widetilde{x}_\beta(\widetilde{\beta}), \beta, 0, \mu) + t(\beta, \widetilde{\beta}) = T^{\widetilde{\beta}} + t(\beta, \widetilde{\beta})$ and consequently

$$F(\widetilde{x}_\beta(\widetilde{\beta}), \beta, 0, \mu) = \widetilde{x}_\beta(\widetilde{\beta}) - P(\widetilde{x}_\beta(\widetilde{\beta}), \beta, 0, \mu)$$
$$= \widetilde{x}_\beta(\widetilde{\beta}) - x_+(t_2^\beta, x_2(\widetilde{\beta}))(T^{\widetilde{\beta}} + t(\beta, \widetilde{\beta}), 0, \mu) = \widetilde{x}_\beta(\widetilde{\beta}) - \gamma(\widetilde{\beta}, T^{\widetilde{\beta}} + t(\beta, \widetilde{\beta})) = 0.$$

Equation (I.2.46) implies $F_\xi(x_0(\beta), \beta, 0, \mu) D\widetilde{x}_\beta(\beta) = 0$. Note by (I.2.45), $D\widetilde{x}_\beta(\beta) = (\mathbb{I} - S_\beta) Dx_0(\beta)$ and hence $F_\xi(x_0(\beta), \beta, 0, \mu)(\mathbb{I} - S_\beta) Dx_0(\beta) = 0$ which, by (I.2.6), is equivalent to

$$(\mathbb{I} - S_\beta) Dx_0(\beta) = (\mathbb{I} - S_\beta) A(\beta, 0)(\mathbb{I} - S_\beta) Dx_0(\beta).$$

Note $(\mathbb{I} - S_\beta) Dx_0(\beta)$ is k-dimensional.

Some solutions of (I.2.30) are described in the next lemma.

Lemma I.2.7. *Vectors* $\frac{\partial \gamma(\beta,t)}{\partial \beta_1}, \ldots, \frac{\partial \gamma(\beta,t)}{\partial \beta_k}, \frac{\partial \gamma(\beta,t)}{\partial t}$ *satisfy equation* (I.2.30) *as well as conditions* (I.2.31) *and* (I.2.32).

Proof. Using the notation from (I.2.2) we obtain

$$\gamma(\beta, t) = \begin{cases} x_+(0, x_0(\beta))(t, 0, \mu) & \text{if } t \in [0, t_1^\beta], \\ x_-(t_1^\beta, x_+(0, x_0(\beta))(t_1^\beta, 0, \mu))(t, 0, \mu) & \text{if } t \in [t_1^\beta, t_2^\beta], \quad (\text{I.2.47}) \\ x_+(t_2^\beta, x_-(t_1^\beta, x_+(0, x_0(\beta))(t_1^\beta, 0, \mu))(t_2^\beta, 0, \mu))(t, 0, \mu) & \text{if } t \in [t_2^\beta, T^\beta]. \end{cases}$$

Direct differentiation of (I.2.47) gives equation (I.2.30) with $x = \frac{\partial \gamma(\beta,t)}{\partial \beta_i}$ for $i = 1, \ldots, k$ or $x = \frac{\partial \gamma(\beta,t)}{\partial t}$. Relations (I.2.31) are also easily obtained from (I.2.47). Next, differentiating $\gamma(\beta, t + T^\beta) = \gamma(\beta, t)$ for any $t \in \mathbb{R}$ we derive $\dot{\gamma}(\beta, t + T^\beta) = \dot{\gamma}(\beta, t)$ and

$$\frac{\partial \gamma(\beta, t + T^\beta)}{\partial \beta_i} + \frac{\partial T^\beta}{\partial \beta_i} \dot{\gamma}(\beta, t + T^\beta) = \frac{\partial \gamma(\beta, t)}{\partial \beta_i},$$

which implies

$$(\mathbb{I} - S_\beta) \dot{\gamma}(\beta, T^\beta) = (\mathbb{I} - S_\beta) \dot{\gamma}(\beta, 0) = (\mathbb{I} - S_\beta) f_+(x_0(\beta)) = 0$$

and

$$(\mathbb{I} - S_\beta) \left[\frac{\partial \gamma(\beta, T^\beta)}{\partial \beta_i} - \frac{\partial \gamma(\beta, 0)}{\partial \beta_i} \right] = -\frac{\partial T^\beta}{\partial \beta_i} (\mathbb{I} - S_\beta) f_+(x_0(\beta)) = 0.$$

Consequently, (I.2.32) holds as well. The proof is finished.   □

We get the following equivalence to condition H3):

Proposition I.2.8. *Condition* H3) *is equivalent to saying that the set*

$$\left\{ \frac{\partial\gamma(\beta,t)}{\partial\beta_1}, \ldots, \frac{\partial\gamma(\beta,t)}{\partial\beta_k}, \frac{\partial\gamma(\beta,t)}{\partial t} \right\}$$

is a basis of linearly independent solutions of (I.2.30), (I.2.31) *and* (I.2.32).

I.2.3. On the hyperbolicity of persisting orbits

Here we state the sufficient condition for the hyperbolicity of trajectory $\gamma^*(\varepsilon, \mu, t)$ with $\varepsilon \neq 0$. First, we recall the result from [26] (see also [27]):

Lemma I.2.9. *Let* $E(\varepsilon) = \begin{pmatrix} A_\varepsilon & 0 \\ 0 & B_\varepsilon \end{pmatrix}$ *and* $D(\varepsilon)$ *be the continuous matrix functions* $\mathbb{R}^k \to \mathbb{R}^k$ *for* $\varepsilon \geq 0$ *such that* $\|A_\varepsilon\| \leq 1 - c\varepsilon$, $\|B_\varepsilon^{-1}\| \leq 1 - c\varepsilon$, *where* c *is a positive constant,* A_ε *and* B_ε *are* $k_1 \times k_1$ *and* $(k - k_1) \times (k - k_1)$ *blocks, respectively, i.e.* $E(\varepsilon)$ *is strongly* 1-*hyperbolic. Then* $E(\varepsilon) + \varepsilon^2 D(\varepsilon)$ *has no eigenvalues on* S^1 *for* $\varepsilon \neq 0$ *sufficiently small.*

We shall also need the following lemma.

Lemma I.2.10. *Let* U *be a neighborhood of* 0 *in* $\mathbb{R}$ *and* $M \in C^1(U, L(\mathbb{R}^n))$, *i.e.* $M(\varepsilon)$ *is a real matrix of the form* $n \times n$ *with an eigenvalue* $\lambda(\varepsilon) = \exp(\alpha(\varepsilon) + \imath\beta(\varepsilon))$ *and the corresponding eigenvector* $u(\varepsilon)$. *Suppose that* $\lambda(0) \neq 0$ *is simple. Then*

$$\alpha'(0) = \Re \frac{(M'(0)u(0), u(0))_M}{\lambda(0)\|u(0)\|_M^2}, \qquad \beta'(0) = \Im \frac{(M'(0)u(0), u(0))_M}{\lambda(0)\|u(0)\|_M^2},$$

where $(\cdot, \cdot)_M$ *is an inner product in* $\mathbb{C}^n$ *such that* $(M(0)q, u(0))_M = 0$ *whenever* $(q, u(0))_M = 0$, *and* $\|q\|_M^2 = (q, q)_M$.

Proof. If we denote

$$v(\varepsilon) = \frac{u(\varepsilon)\|u(0)\|_M}{(u(\varepsilon), u(0))_M}$$

for each ε sufficiently small, then $(v(\varepsilon), v(0))_M = 1$, and therefore $(v'(\varepsilon), v(0))_M = 0$. Differentiation of the identity $M(\varepsilon)v(\varepsilon) = \lambda(\varepsilon)v(\varepsilon)$ at $\varepsilon = 0$ gives

$$M'(0)v(0) + M(0)v'(0) = \lambda(0)[\alpha'(0) + \imath\beta'(0)]v(0) + \lambda(0)v'(0).$$

Applying the inner product in the form $(\cdot, v(0))_M$ on this equality gives

$$(M'(0)v(0), v(0))_M = \lambda(0)[\alpha'(0) + \imath\beta'(0)].$$

When one returns to $u(0)$, the proof is finished. □

Let (β_0, μ_0) determine the persisting trajectory $\gamma^*(\varepsilon, \mu, t)$ (see Theorem I.2.5). We shall study the hyperbolicity of the orbit on the fixed hyperplane Σ_{β_0} instead of changing Σ_β with $\beta = \beta(\varepsilon, \mu)$. Take $w_0(\varepsilon, \mu) = x_+(0, x^*(\varepsilon, \mu))(t_4(\varepsilon, \mu), \varepsilon, \mu)$ where $t_4(\varepsilon, \mu)$ is the nearest return time ($|t_4(\varepsilon, \mu)|$ is small) such that $w_0(\varepsilon, \mu) \in \Sigma_{\beta_0}$. Note $t_4(0, \mu) = 0$ and $w_0(0, \mu) = x_0(\beta_0)$. Moreover, $x_+(0, w_0(\varepsilon, \mu))(\cdot, \varepsilon, \mu)$ is a unique persisting solution of (I.2.1) with $\varepsilon \neq 0$ small and μ close to μ_0, which is in a neighborhood of $\gamma(\beta_0, \cdot)$. More precisely,

$$x_+(0, w_0(\varepsilon, \mu))(t, \varepsilon, \mu) = \gamma^*(\varepsilon, \mu, t_4(\varepsilon, \mu) + t).$$

Theorem I.2.11. *Let (β_0, μ_0) be as in Theorem I.2.5, $\xi \in \mathbb{R}^n$ and C be a regular $n \times n$ matrix such that*

$$C^{-1} P_\xi(x_0(\beta_0), \beta_0, 0, \mu_0) C = \begin{pmatrix} \mathbb{I}_k & 0 \\ 0 & B \end{pmatrix} =: A,$$

where $\mathbb{I}_k$ is the $k \times k$ identity matrix and B has simple eigenvalues $\lambda_1, \ldots, \lambda_l \in S^1 \backslash \{1\}$ with corresponding eigenvectors $v_1, \ldots, v_l$ and none of the other eigenvalues on S^1. Suppose that if we denote

$$C^{-1} D_\varepsilon \, P_\xi(w_0(\varepsilon, \mu_0), \beta_0, \varepsilon, \mu_0)\big|_{\varepsilon=0} C =: A_1 = \begin{pmatrix} A_{11} & A_{12} \\ A_{21} & A_{22} \end{pmatrix},$$

where A_{11} is a $k \times k$ block, then A_{11} has no eigenvalues on the imaginary axis. Moreover, let

$$\Re \frac{(A_{22} v_i, v_i)_B}{\lambda_i \|v_i\|_B^2} \neq 0, \qquad \forall i = 1, \ldots, l,$$

where $(\cdot, \cdot)_B$ is an inner product in $\mathbb{C}^n$ such that $(Bq, v_i)_B = 0$ whenever $(q, v_i)_B = 0$ for each $i = 1, \ldots, l$, and $\|q\|_B^2 = (q, q)_B$. Then the persisting orbit $\gamma^*(\varepsilon, \mu, t)$ is hyperbolic.

Proof. Let us denote $A^\mu(\varepsilon) = C^{-1} P_\xi(w_0(\varepsilon, \mu), \beta_0, \varepsilon, \mu), \varepsilon, \mu)C$ for ε small. Now from the Taylor expansion we have $A^\mu(\varepsilon) = A + \varepsilon A_1^\mu + O(\varepsilon^2)$, where $A_1^{\mu_0} = A_1$, and the same for its parts (note if $\varepsilon = 0$, the dependency on μ is lost). If we take $P^\mu(\varepsilon) = \mathbb{I} + \varepsilon B_1^\mu + \varepsilon^2 B_2^\mu(\varepsilon)$ with continuous matrix function B_2^μ,

$$B_1^\mu = \begin{pmatrix} 0 & B_{12}^\mu \\ B_{21}^\mu & 0 \end{pmatrix}, \qquad B_{12}^\mu = A_{12}^\mu(\mathbb{I} - B)^{-1}, \qquad B_{21}^\mu = (B - \mathbb{I})^{-1} A_{21}^\mu,$$

then

$$\widetilde{A}^\mu(\varepsilon) := P^\mu(\varepsilon) A^\mu(\varepsilon) P^{\mu-1}(\varepsilon) = \begin{pmatrix} \mathbb{I}_k & 0 \\ 0 & B \end{pmatrix} + \varepsilon \begin{pmatrix} A_{11}^\mu & 0 \\ 0 & A_{22}^\mu \end{pmatrix} + O(\varepsilon^2).$$

Similarly, without the change of eigenvalues, we could transform A^μ_{11} into the form

$$\begin{pmatrix} A^\mu_{111} & 0 \\ 0 & A^\mu_{112} \end{pmatrix},$$

where A^μ_{111} is $k_1 \times k_1$ with $0 \le k_1 \le k$, $\mathcal{R}\sigma(A^\mu_{111}) \subset (0, \infty)$ and $\mathcal{R}\sigma(A^\mu_{112}) \subset (-\infty, 0)$. Hence we shall suppose that A^μ_{11} is already in this form. Consequently,

$$\begin{aligned}
\widetilde{A}^\mu(\varepsilon) &= \begin{pmatrix} \mathbb{I}_{k_1} + \varepsilon A^\mu_{111} & 0 & 0 \\ 0 & \mathbb{I}_{k-k_1} + \varepsilon A^\mu_{112} & 0 \\ 0 & 0 & B + \varepsilon A^\mu_{22} \end{pmatrix} + O(\varepsilon^2) \\
&= \begin{pmatrix} E^\mu(\varepsilon) & 0 \\ 0 & B + \varepsilon A^\mu_{22} \end{pmatrix} + O(\varepsilon^2).
\end{aligned}$$

It can be shown [26, 27] that $E^\mu(\varepsilon)$ is strongly 1-hyperbolic for $\varepsilon > 0$ sufficiently small. Next, Lemma I.2.10 applied on the matrix function $\widetilde{A}^\mu_{22}(\varepsilon) = B + \varepsilon A^\mu_{22}$ with eigenvalues $\lambda^\mu_1(\varepsilon), \ldots, \lambda^\mu_l(\varepsilon)$ such that $\lambda^\mu_1(0) = \lambda_1, \ldots, \lambda^\mu_l(0) = \lambda_l \in S^1$, and eigenvectors $v_1, \ldots, v_l$ corresponding to $\widetilde{A}^\mu_{22}(0)$ and $\lambda_1, \ldots, \lambda_l$, respectively, implies that $\widetilde{A}^\mu_{22}(\varepsilon)$ is also strongly 1-hyperbolic for $\varepsilon > 0$ sufficiently small. Consequently, $\begin{pmatrix} E^\mu(\varepsilon) & 0 \\ 0 & B+\varepsilon A^\mu_{22} \end{pmatrix}$ is strongly 1-hyperbolic and Lemma I.2.9 applies to $\widetilde{A}^\mu(\varepsilon)$.

Note from Lemma I.2.3, $A^\mu(\varepsilon)$ has an eigenvalue 0 with the corresponding eigenvector $f_+(w_0(\varepsilon, \mu)) + \varepsilon g(w_0(\varepsilon, \mu), \varepsilon, \mu)$. In conclusion by Lemma I.2.9, $A^\mu(\varepsilon)$ has no eigenvalues on S^1 for $\varepsilon > 0$ sufficiently small. The analogous result can be proved for $\varepsilon < 0$. That means that for $|\varepsilon| > 0$ the sufficiently small perturbed trajectory $\gamma^*(\varepsilon, \mu, t)$ is hyperbolic. $\qquad\square$

By special assumptions, the sufficient condition for stability of persisting periodic orbit may be easily obtained from the previously stated theorem.

Corollary I.2.12. *Let the assumptions of Theorem I.2.11 be fulfilled. Furthermore, let B have no eigenvalues outside the unit circle, A_{11} have all eigenvalues with negative real part and*

$$\mathcal{R} \frac{(A_{22}v_i, v_i)_B}{\lambda_i \|v_i\|^2_B} < 0, \qquad \forall i = 1, \ldots, l.$$

Then $\gamma^*(\varepsilon, \mu, t)$ is stable (repeller) for $\varepsilon > 0$ ($\varepsilon < 0$) small.

Generally, the formula for $\mathrm{D}_\varepsilon\, P_\xi(w_0(\varepsilon, \mu), \beta_0, \varepsilon, \mu)\big|_{\varepsilon=0}$ is really complicated. However, in concrete examples it may be more easily found using computer software. Now we describe how to do that.

Differentiating the expansion

$$P(w, \beta_0, \varepsilon, \mu_0) = P(w, \beta_0, 0, \mu_0) + \varepsilon P_1(w, \beta_0, 0, \mu_0) + O(\varepsilon^2)$$

for any $w \in \Sigma_{\beta_0}$ gives

$$P_{\xi}(w, \beta_0, \varepsilon, \mu_0) = P_{\xi}(w, \beta_0, 0, \mu_0) + \varepsilon P_{1\xi}(w, \beta_0, 0, \mu_0) + O(\varepsilon^2).$$

Thence

$$D_{\varepsilon} \, P_{\xi}(w_0(\varepsilon, \mu_0), \varepsilon, \mu_0)\big|_{\varepsilon=0}$$

$$= P_{1\xi}(x_0(\beta_0), \beta_0, 0, \mu_0) + P_{\xi\xi}(x_0(\beta_0), \beta_0, 0, \mu_0) \frac{\partial w_0(0, \mu_0)}{\partial \varepsilon}, \qquad (\text{I}.2.48)$$

where $P_{1\xi}$ is the partial derivative of P_1 with respect to ξ, while $P_{\xi\xi}$ is the second partial derivative of P with respect to ξ.

Now we note that $P_1(\xi, \beta, 0, \mu)$ can be obtained by linearization of (I.2.4) as follows:

For the sense of simplicity, we shall omit some arguments. Let us denote $y_1(s)$, $y_2(s)$ and $y_3(s)$ the solutions of equations

$$\dot{y}_1 = f_+(y_1) + \varepsilon g(y_1, \varepsilon, \mu) \quad \text{on } [0, s_1] \qquad \dot{y}_2 = f_-(y_2) + \varepsilon g(y_2, \varepsilon, \mu) \quad \text{on } [s_1, s_2]$$
$$y_1(0) = \xi, \qquad\qquad\qquad\qquad y_2(s_1) = y_1(s_1),$$

$$\dot{y}_3 = f_+(y_3) + \varepsilon g(y_3, \varepsilon, \mu) \quad \text{on } [s_2, \infty)$$
$$y_3(s_2) = y_2(s_2),$$

respectively, where the first one has a general initial condition $\xi \in \Sigma_{\beta}$, and $s_1 < s_2$ satisfy $h(y_1(s_1)) = h(y_2(s_2)) = 0$. Moreover, let $s_3 > s_2$ be such that

$$\langle y_3(s_3) - x_0(\beta), f_+(x_0(\beta)) \rangle = 0.$$

Taylor expansions with respect to ε

$$y_i(t) = y_i^0(t) + \varepsilon y_i^1(t) + O(\varepsilon^2), \qquad i = 1, 2, 3,$$
$$s_i = s_i^0 + \varepsilon s_i^1 + O(\varepsilon^2), \qquad i = 1, 2, 3$$

imply

$$\dot{y}_1^0 = f_+(y_1^0)$$
$$y_1^0(0) = \xi \in \Sigma_\beta,$$

$$\dot{y}_1^1 = Df_+(y_1^0)y_1^1 + g(y_1^0, 0, \mu)$$
$$y_1^1(0) = 0,$$

$$h(y_1^0(s_1^0)) = 0,$$

$$s_1^1 = -\frac{Dh(y_1^0(s_1^0))y_1^1(s_1^0)}{Dh(y_1^0(s_1^0))f_+(y_1^0(s_1^0))},$$

$$\dot{y}_2^0 = f_-(y_2^0)$$
$$y_2^0(s_1^0) = y_1^0(s_1^0),$$

$$\dot{y}_2^1 = Df_-(y_2^0)y_2^1 + g(y_2^0, 0, \mu)$$
$$y_2^1(s_1^0) = \left[\mathbb{I} + \frac{(f_-(y_1^0(s_1^0)) - f_+(y_1^0(s_1^0)))Dh(y_1^0(s_1^0))}{Dh(y_1^0(s_1^0))f_+(y_1^0(s_1^0))}\right] y_1^1(s_1^0),$$

$$h(y_2^0(s_2^0)) = 0,$$

$$s_2^1 = -\frac{Dh(y_2^0(s_2^0))y_2^1(s_2^0)}{Dh(y_2^0(s_2^0))f_-(y_2^0(s_2^0))},$$

$$\dot{y}_3^0 = f_+(y_3^0)$$
$$y_3^0(s_2^0) = y_2^0(s_2^0),$$

$$\dot{y}_3^1 = Df_+(y_3^0)y_3^1 + g(y_3^0, 0, \mu)$$
$$y_3^1(s_2^0) = \left[\mathbb{I} + \frac{(f_+(y_2^0(s_2^0)) - f_-(y_2^0(s_2^0)))Dh(y_2^0(s_2^0))}{Dh(y_2^0(s_2^0))f_-(y_2^0(s_2^0))}\right] y_2^1(s_2^0),$$

$$y_3^0(s_3^0) \in \Sigma_\beta,$$

$$s_3^1 = -\frac{\langle y_3^1(s_3^0), f_+(x_0(\beta))\rangle}{\langle f_+(y_3^0(s_3^0)), f_+(x_0(\beta))\rangle}.$$

Note that all y_i^j and s_i^j depend on ξ. Since

$$P(\xi, \beta, \varepsilon, \mu) = y_3(s_3) = y_3^0(s_3^0) + \varepsilon\left[\dot{y}_3^0(s_3^0)s_3^1 + y_3^1(s_3^0)\right] + O(\varepsilon^2),$$

we get

$$P_1(\xi, \beta, 0, \mu) = f_+(y_3^0(s_3^0))s_3^1 + y_3^1(s_3^0). \tag{I.2.49}$$

Remark I.2.13. If $\xi = x_0(\beta)$ then $s_1^0 = t_1^\beta$, $s_2^0 = t_2^\beta$, $s_3^0 = T^\beta$, $y_1^0(s_1^0) = x_1(\beta)$, $y_2^0(s_2^0) = x_2(\beta)$, $y_3^0(s_3^0) = x_0(\beta)$ and $P_1(x_0(\beta), \beta, 0, \mu) = P_\varepsilon(x_0(\beta), \beta, 0, \mu)$ (see (I.2.8)).

Hence the algorithm consists of subsequent computation of y_1^0, y_1^1, s_1^0, s_1^1, y_2^0, y_2^1, s_2^0, s_2^1, y_3^0, y_3^1, s_3^0, s_3^1, applying formula (I.2.49), and computing leftover terms in (I.2.48).

I.2.4. The particular case of the initial manifold

Here we consider the special case of the manifold of initial points when $k = n - 1$, i.e. $x_0(V)$ is an immersed submanifold of codimension 1. Related problems have been studied in [28] for smooth dynamical systems. Then we can suppose that $x_0(\beta) = (\beta_1, \ldots, \beta_{n-1}, 0) = (\beta, 0)$. Let us denote $\bar{\xi} = (\xi, 0)$ for $\xi \in V$. We take $\Sigma =$

$\mathbb{R}^{n-1} \times \{0\} \cap \Omega_+$ and a new Poincaré mapping $\widetilde{P}: V \times (-\varepsilon_0, \varepsilon_0) \times \mathbb{R}^p \to \Sigma$ defined as

$$\widetilde{P}(\xi, \varepsilon, \mu) = x_+(t_2(0, \bar{\xi}, \varepsilon, \mu), x_-(t_1(0, \bar{\xi}, \varepsilon, \mu), x_+(0, \bar{\xi})(t_1(0, \bar{\xi}, \varepsilon, \mu), \varepsilon, \mu))$$
$$(t_2(0, \bar{\xi}, \varepsilon, \mu), \varepsilon, \mu))(t_3(0, \xi, \varepsilon, \mu), \varepsilon, \mu),$$

where $t_3(\cdot, \cdot, \cdot, \cdot)$ is a solution close to T^{ξ} of equation

$$\langle x_+(t_2(0, \bar{\xi}, \varepsilon, \mu), x_-(t_1(0, \bar{\xi}, \varepsilon, \mu), x_+(0, \bar{\xi})$$
$$(t_1(0, \bar{\xi}, \varepsilon, \mu), \varepsilon, \mu))(t_2(0, \bar{\xi}, \varepsilon, \mu), \varepsilon, \mu))(t, \varepsilon, \mu), e_n \rangle = 0$$

with $e_n = (0, \ldots, 0, 1) \in \mathbb{R}^n$. Moreover, $t_3(0, \xi, 0, \mu) = T^{\xi}$. Then one can easily derive the following formulae for the partial derivatives $\widetilde{P}_{\xi}$, $\widetilde{P}_{\varepsilon}$ of $\widetilde{P}$ with respect to ξ and ε, respectively,

$$\widetilde{P}_{\xi}(\xi, 0, \mu) = (\mathbb{I} - T_{\xi})A(\xi, 0),$$

$$\widetilde{P}_{\varepsilon}(\xi, 0, \mu) = (\mathbb{I} - T_{\xi})\left(\int_0^{T^{\xi}} A(\xi, s)g(\gamma(\xi, s), 0, \mu)ds\right)$$

with $T_{\xi}u = \frac{\langle u, e_n \rangle f_+(\bar{\xi})}{\langle f_+(\bar{\xi}), e_n \rangle}$ and A given by (I.2.10). Note $\widetilde{P}(\xi, 0, \mu) = \bar{\xi} \in \Sigma$ and

$$\widetilde{P}(\xi, \varepsilon, \mu) = \bar{\xi} + \varepsilon(\mathbb{I} - T_{\xi})\int_0^{T^{\xi}} A(\xi, t)g(\gamma(\xi, t), 0, \mu)dt + O(\varepsilon^2).$$

Consequently, we have the following theorem [26]:

Theorem I.2.14. *Let conditions H1), H2) be satisfied and $k = n - 1$. Let there be $(\xi_0, \mu_0) \in V \times \mathbb{R}^p$ such that $M^{\mu_0}(\xi_0) = 0$ and $\det DM^{\mu_0}(\xi_0) \neq 0$, where*

$$M^{\mu}(\xi) = \left[(\mathbb{I} - T_{\xi})\int_0^{T^{\xi}} A(\xi, t)g(\gamma(\xi, t), 0, \mu)dt\right]_{\mathbb{R}^{n-1}}$$

and the lower index $\mathbb{R}^{n-1}$ denotes the restriction on the first $n - 1$ coordinates. Then there is a unique periodic solution $x^*(\varepsilon, \mu, t)$ near $\gamma(\xi_0, t)$ of (I.2.1) with μ close to μ_0 and $\varepsilon \neq 0$ small. Moreover, for $\varepsilon > 0$ small

1. *if $\Re\sigma(DM^{\mu_0}(\xi_0)) \subset (-\infty, 0)$ then $x^*(\varepsilon, \mu, t)$ is stable,*
2. *if $\Re\sigma(DM^{\mu_0}(\xi_0)) \cap (0, \infty) \neq \emptyset$ then $x^*(\varepsilon, \mu, t)$ is unstable,*
3. *if $0 \notin \Re\sigma(DM^{\mu_0}(\xi_0))$ then $x^*(\varepsilon, \mu, t)$ is hyperbolic with the same hyperbolicity type as $DM^{\mu_0}(\xi_0)$.*

Proof. The existence part for $x^*(\varepsilon, \mu, t)$ follows as previously. The local asymptotic properties for $x^*(\varepsilon, \mu, t)$ are derived from standard arguments of [26, 27].   □

I.2.5. 3-dimensional piecewise-linear application

We shall consider the following piecewise-linear problem

$$
\begin{aligned}
\dot{x} &= \varepsilon(z - x^n) \\
\dot{y} &= b_1 \qquad\qquad\qquad\qquad\text{if } z > 0, \\
\dot{z} &= -2a_1 b_1 y + \varepsilon(\mu_1 - \mu_2 y^2)z
\end{aligned}
$$

$$(\text{I.2.50})_\varepsilon$$

$$
\begin{aligned}
\dot{x} &= 0 \\
\dot{y} &= -b_2 \qquad\qquad\qquad\qquad\text{if } z < 0 \\
\dot{z} &= -2a_2 b_2 y
\end{aligned}
$$

with positive constants a_1, a_2, b_1, b_2; $n \in \mathbb{N}$ and vector $\mu = (\mu_1, \mu_2)$ of real parameters.

Here we have $\Omega_\pm = \{(x, y, z) \in \mathbb{R}^3 \mid \pm z > 0\}$, $\Omega_0 = \{(x, y, 0) \in \mathbb{R}^3\}$ and $h(x, y, z) = z$. Let $x_0(\beta) = (\beta_1, 0, \beta_2)$, $\beta = (\beta_1, \beta_2)$, $\beta_2 > 0$ be an initial point. Then we have $\Sigma = \{(x, 0, z) \in \mathbb{R}^3 \mid z > 0\}$. Due to the linearity of problem $(\text{I.2.50})_0$ some results may be easily obtained. These are concluded in the following lemma.

Lemma I.2.15. *The unperturbed system* $(\text{I.2.50})_0$ *possesses a 2-parametrized system* $\{\gamma(\beta, t) \mid \beta_2 > 0\}$ *of periodic orbits starting at* $(\beta_1, 0, \beta_2)$ *and given by (cf.* $(\text{I.2.2}))$:

$$
\gamma_1(\beta, t) = (\beta_1, b_1 t, -a_1 b_1^2 t^2 + \beta_2),
$$

$$
\gamma_2(\beta, t) = (\beta_1, x_{12}(\beta) - b_2(t - t_1^\beta), a_2(x_{12}(\beta) - b_2(t - t_1^\beta))^2 - a_2 x_{12}^2(\beta)),
$$

$$
\gamma_3(\beta, t) = (\beta_1, x_{22}(\beta) + b_1(t - t_2^\beta), -a_1(x_{22}(\beta) + b_1(t - t_2^\beta))^2 + a_1 x_{22}^2(\beta)),
$$

where

$$
t_1^\beta = \frac{1}{b_1}\sqrt{\frac{\beta_2}{a_1}}, \qquad t_2^\beta = \frac{2}{b_2}\sqrt{\frac{\beta_2}{a_1}} + t_1^\beta, \qquad T^\beta = \frac{1}{b_1}\sqrt{\frac{\beta_2}{a_1}} + t_2^\beta,
$$

$$
x_i(\beta) = (x_{i1}(\beta), x_{i2}(\beta), x_{i3}(\beta)), \quad i = 1, 2,
$$

$$
x_1(\beta) = \left(\beta_1, \sqrt{\frac{\beta_2}{a_1}}, 0\right), \qquad x_2(\beta) = \left(\beta_1, -\sqrt{\frac{\beta_2}{a_1}}, 0\right).
$$

The corresponding fundamental matrices of (I.2.13), (I.2.14) *and* (I.2.15) *have now the forms*

$$X_1(\beta, t) = \begin{pmatrix} 1 & 0 & 0 \\ 0 & 1 & 0 \\ 0 & -2a_1b_1t & 1 \end{pmatrix}, \qquad X_2(\beta, t) = \begin{pmatrix} 1 & 0 & 0 \\ 0 & 1 & 0 \\ 0 & -2a_2b_2(t - t_1^\beta) & 1 \end{pmatrix},$$

$$X_3(\beta, t) = \begin{pmatrix} 1 & 0 & 0 \\ 0 & 1 & 0 \\ 0 & -2a_1b_1(t - t_2^\beta) & 1 \end{pmatrix}$$

and saltation matrices of (I.2.11) *and* (I.2.12) *are given by*

$$S_1(\beta) = \begin{pmatrix} 1 & 0 & 0 \\ 0 & 1 & \frac{\sqrt{a_1}(b_1+b_2)}{2a_1b_1\sqrt{\beta_2}} \\ 0 & 0 & \frac{a_2b_2}{a_1b_1} \end{pmatrix}, \qquad S_2(\beta) = \begin{pmatrix} 1 & 0 & 0 \\ 0 & 1 & \frac{\sqrt{a_1}(b_1+b_2)}{2a_2b_2\sqrt{\beta_2}} \\ 0 & 0 & \frac{a_1b_1}{a_2b_2} \end{pmatrix}.$$

Proof. The lemma can be proved exactly as Lemma I.1.19. $\qquad\qquad\square$

Another lemma shows that the derived theory can be applied to system (I.2.50)$_0$:

Lemma I.2.16. *System* (I.2.50)$_0$ *satisfies conditions* H1) *and* H2).

Proof. Since $\frac{\partial x_0(\beta)}{\partial \beta_1} = (1, 0, 0)^*$, $\frac{\partial x_0(\beta)}{\partial \beta_2} = (0, 0, 1)^*$, $f_+(x_0(\beta)) = (0, b_1, 0)^*$ and using Lemma I.2.15, H1) follows immediately. Since $Dh(x, y, z) = (0, 0, 1)$ and

$$f_+(x_1(\beta)) = \left(0, b_1, -2a_1b_1\sqrt{\frac{\beta_2}{a_1}}\right)^*, \qquad f_-(x_1(\beta)) = \left(0, -b_2, -2a_2b_2\sqrt{\frac{\beta_2}{a_1}}\right)^*,$$

$$f_+(x_2(\beta)) = \left(0, b_1, 2a_1b_1\sqrt{\frac{\beta_2}{a_1}}\right)^*, \qquad f_-(x_2(\beta)) = \left(0, -b_2, 2a_2b_2\sqrt{\frac{\beta_2}{a_1}}\right)^*,$$

H2) is also fulfilled. $\qquad\qquad\square$

Using Lemma I.2.15 we get

$$
A(\beta, t) = \begin{cases}
\begin{pmatrix}
1 & 0 & 0 \\
0 & 1 - \frac{2\sqrt{a_1}b_1 t(b_1+b_2)}{b_2\sqrt{\beta_2}} & -\frac{b_1+b_2}{b_2\sqrt{a_1\beta_2}} \\
0 & 2a_1 b_1 t & 1
\end{pmatrix} & \text{if } t \in [0, t_1^\beta), \\[2em]
\begin{pmatrix}
1 & 0 & 0 \\
0 & -1 - \frac{2b_1}{b_2} + \frac{\sqrt{a_1}(t-t_1^\beta)(b_1+b_2)}{\sqrt{\beta_2}} & \frac{\sqrt{a_1}(b_1+b_2)}{2a_2 b_2 \sqrt{\beta_2}} \\
0 & 2(\sqrt{a_1\beta_2} - a_1 b_2(t-t_1^\beta)) & -\frac{a_1}{a_2}
\end{pmatrix} & \text{if } t \in [t_1^\beta, t_2^\beta), \\[2em]
\begin{pmatrix}
1 & 0 & 0 \\
0 & 1 & 0 \\
0 & 2(a_1 b_1(t-t_2^\beta) - \sqrt{a_1\beta_2}) & 1
\end{pmatrix} & \text{if } t \in [t_2^\beta, T^\beta]
\end{cases}
\tag{I.2.51}
$$

from (I.2.10). Consequently, the mapping $M^\mu(\beta)$ from Section I.2.4 has the form

$$
M^\mu(\beta) = \left(-\frac{2}{3} \frac{\sqrt{\beta_2}(3\beta_1^n - 2\beta_2)}{\sqrt{a_1}b_1}, \; \frac{4}{15} \frac{\beta_2^{3/2}(5\mu_1 a_1 - \mu_2 \beta_2)}{a_1^{3/2} b_1} \right).
\tag{I.2.52}
$$

From Theorem I.2.14 we obtain the following result.

Proposition I.2.17. *For $\mu \in \mathbb{R}^2$ such that $\mu_1\mu_2 \leq 0$, $(\mu_1,\mu_2) \neq 0$ no periodic orbit persists. For $\mu_1\mu_2 > 0$, if $\varepsilon > 0$ and*
1. *n is odd, the only persisting periodic trajectory $\gamma(\beta_0, t)$ of system $(I.2.50)_0$ is determined by $\beta_0 = (\beta_{01}, \beta_{02})$ with $\beta_{01} = \left(\frac{2}{3}\beta_{02}\right)^{1/n}$, $\beta_{02} = \frac{5\mu_1 a_1}{\mu_2}$. Moreover, this trajectory is stable – it is a sink – for $\mu_1 > 0$ and unstable/hyperbolic for $\mu_1 < 0$,*
2. *n is even, there are exactly two persisting orbits γ_+, γ_- given by $\beta_{01} = \pm\left(\frac{2}{3}\beta_{02}\right)^{1/n}$, $\beta_{02} = \frac{5\mu_1 a_1}{\mu_2}$ with corresponding sign in β_{01}. Moreover, if*
   a. *$\mu_1 > 0$, then γ_+ is stable – it is a sink – and γ_- is unstable/hyperbolic,*
   b. *$\mu_1 < 0$, then γ_+ is unstable/hyperbolic and γ_- is unstable – it is a source.*
If $\varepsilon < 0$, the above statements remain true with sinks instead of sources and vice versa.

Proof. From (I.2.52) one can see that for $(\mu_1,\mu_2) \neq (0,0)$ the positive solution β_2 of

$$
\frac{4}{15} \frac{\beta_2^{3/2}(5\mu_1 a_1 - \mu_2 \beta_2)}{a_1^{3/2} b_1} = 0
$$

exists if and only if $\beta_2 = \frac{5\mu_1 a_1}{\mu_2}$ and $\mu_1\mu_2 > 0$. Then with respect to n we get one or two solutions β_1 of

$$-\frac{2}{3}\frac{\sqrt{\beta_2}(3\beta_1^n - 2\beta_2)}{\sqrt{a_1}b_1} = 0.$$

For arbitrary $n \in \mathbb{N}$ we have $\beta_{02} = \frac{5\mu_1 a_1}{\mu_2}$ and

$$DM^\mu\left(\left(\frac{2}{3}\beta_{02}\right)^{1/n}, \beta_{02}\right) = \begin{pmatrix} -\frac{2\sqrt{5}(\frac{10}{3})^{\frac{n-1}{n}}a_1^{\frac{3n-2}{2n}}n\left(\frac{\mu_1}{\mu_2}\right)^{\frac{3n-2}{2n}}}{\sqrt{a_1}b_1} & \frac{4\sqrt{5}}{3b_1}\sqrt{\frac{\mu_1}{\mu_2}} \\ 0 & -\frac{4\sqrt{5}\mu_1}{3b_1}\sqrt{\frac{\mu_1}{\mu_2}} \end{pmatrix}.$$

Therefore, this orbit persists. If n is even then

$$DM^\mu\left(-\left(\frac{2}{3}\beta_{02}\right)^{1/n}, \beta_{02}\right) = \begin{pmatrix} \frac{2\sqrt{5}(\frac{10}{3})^{\frac{n-1}{n}}a_1^{\frac{3n-2}{2n}}n\left(\frac{\mu_1}{\mu_2}\right)^{\frac{3n-2}{2n}}}{\sqrt{a_1}b_1} & \frac{4\sqrt{5}}{3b_1}\sqrt{\frac{\mu_1}{\mu_2}} \\ 0 & -\frac{4\sqrt{5}\mu_1}{3b_1}\sqrt{\frac{\mu_1}{\mu_2}} \end{pmatrix}.$$

Hence it also persists. The statements on the stability of persisting trajectories follow directly from $DM^\mu(\beta)$.   □

Numerical results illustrating the latter proposition are presented in Figure I.2.2.

Remark I.2.18. If $(\mu_1, \mu_2) = (0,0)$ it is not possible to determine the persisting orbit via Theorem I.2.14 since

$$M^{(0,0)}(\beta) = \left(-\frac{2}{3}\frac{\sqrt{\beta_2}(3\beta_1^n - 2\beta_2)}{\sqrt{a_1}b_1}, 0\right).$$

However, we know that if there is a persisting trajectory, then there exists $\beta_2 > 0$ such that the trajectory contains $\left(\left(\frac{2}{3}\beta_2\right)^n, 0, \beta_2\right) \in \mathbb{R}^3$ if n is odd, and $\left(\left(\frac{2}{3}\beta_2\right)^n, 0, \beta_2\right)$ or $\left(-\left(\frac{2}{3}\beta_2\right)^n, 0, \beta_2\right)$ if n is even. To find the persisting orbit, a higher order Melnikov function has to be computed (cf. [20]).

I.2.6. Coupled Van der Pol and harmonic oscillators at 1-1 resonance

In this section we shall consider two weakly coupled oscillators at resonance, one of which is a Van der Pol oscillator and the other a harmonic oscillator, given by equations

$$\begin{aligned} \ddot{x} + \varepsilon(1 - x^2)\dot{x} + a_\pm^2 x + \varepsilon\mu(x - y) &= 0 \\ \ddot{y} + \varepsilon\dot{y} + \omega^2 y - \varepsilon\mu(x - y) &= 0 \end{aligned} \qquad \text{for } \pm x > 0 \qquad (I.2.53)_\varepsilon$$

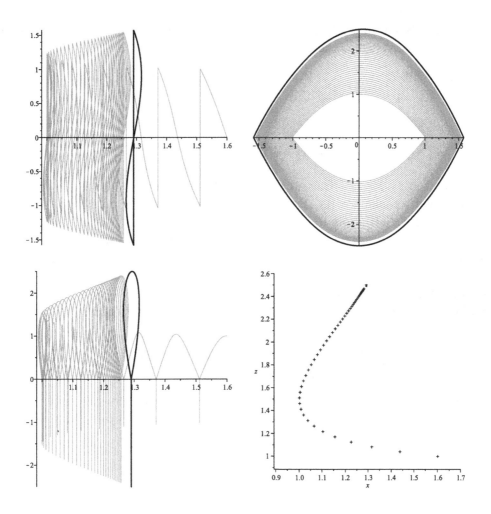

Figure I.2.2 Numerically computed trajectory projected onto xy-, yz- and xz-plane and the Poincaré mapping of the orbit of $(I.2.50)_\varepsilon$ with $a_1 = a_2 = b_1 = b_2 = 1$, $\mu_1 = 1$, $\mu_2 = 2$, $n = 2$, $\varepsilon = 0.05$, $\beta = (1.6, 1)$. Asterisk corresponds to persisting periodic orbit/stable limit cycle, denoted dark in previous figures

with positive constants a_+, a_-, ω such that

$$\frac{2}{\omega} = \frac{1}{a_+} + \frac{1}{a_-}, \qquad\qquad (I.2.54)$$

where $\mu > 0$ is a fixed parameter and $\varepsilon \neq 0$ is small. After transforming $(I.2.53)_\varepsilon$ into a 4-dimensional system we get

$$\dot{x}_1 = a_\pm x_2$$
$$\dot{x}_2 = -a_\pm x_1 - \varepsilon(1 - x_1^2)x_2 - \varepsilon\frac{\mu}{a_\pm}(x_1 - y_1)$$
$$\dot{y}_1 = \omega y_2 \qquad\qquad\qquad \text{for } \pm x_1 > 0 \qquad (I.2.55)_\varepsilon$$
$$\dot{y}_2 = -\omega y_1 - \varepsilon y_2 + \varepsilon\frac{\mu}{\omega}(x_1 - y_1)$$

and $h(x_1, x_2, y_1, y_2) = x_1$. Thus we have $\Omega_\pm = \{(x_1, x_2, y_1, y_2) \in \mathbb{R}^4 \mid \pm x_1 > 0\}$, $\Omega_0 = \{0\} \times \mathbb{R}^3$ and we take $\Sigma = \{(x_1, 0, y_1, y_2) \in \mathbb{R}^4 \mid x_1 > 0\}$ with the initial point $x_0(\beta) = (\beta_1, 0, \beta_2, \beta_3)$, $\beta_1 > 0$. From the linearity of the unperturbed system $(I.2.55)_0$ the next result follows immediately and can be proved using a matrix exponential (see e.g. Section I.1.4).

Lemma I.2.19. *System $(I.2.55)_0$ has a 3-parametrized family $\{\gamma(\beta, t) \mid \beta_1 > 0\}$ of periodic solutions such that $\gamma(\beta, 0) = (\beta_1, 0, \beta_2, \beta_3)$, given by (cf. (I.2.2)):*

$$\gamma_1(\beta, t) = (\beta_1 \cos a_+ t, -\beta_1 \sin a_+ t, \beta_2 \cos \omega t + \beta_3 \sin \omega t, -\beta_2 \sin \omega t + \beta_3 \cos \omega t),$$
$$\gamma_2(\beta, t) = \left(-\beta_1 \sin a_-(t - t_1^\beta), -\beta_1 \cos a_-(t - t_1^\beta), \beta_2 \cos \omega t + \beta_3 \sin \omega t,\right.$$
$$\left. -\beta_2 \sin \omega t + \beta_3 \cos \omega t\right),$$
$$\gamma_3(\beta, t) = \left(\beta_1 \sin a_+(t - t_2^\beta), \beta_1 \cos a_+(t - t_2^\beta), \beta_2 \cos \omega t + \beta_3 \sin \omega t,\right.$$
$$\left. -\beta_2 \sin \omega t + \beta_3 \cos \omega t\right),$$

where

$$t_1^\beta = \frac{\pi}{2a_+}, \qquad t_2^\beta = \frac{\pi}{a_-} + t_1^\beta, \qquad T^\beta = \frac{\pi}{2a_+} + t_2^\beta$$

and (cf. (I.2.3))

$$\bar{x}_i(\beta) = (\bar{x}_{i1}(\beta), \ldots, \bar{x}_{i4}(\beta)), \quad i = 1, 2,$$
$$\bar{x}_1(\beta) = \left(0, -\beta_1, \beta_2 \cos \omega t_1^\beta + \beta_3 \sin \omega t_1^\beta, -\beta_2 \sin \omega t_1^\beta + \beta_3 \cos \omega t_1^\beta\right),$$
$$\bar{x}_2(\beta) = \left(0, \beta_1, \beta_2 \cos \omega t_2^\beta + \beta_3 \sin \omega t_2^\beta, -\beta_2 \sin \omega t_2^\beta + \beta_3 \cos \omega t_2^\beta\right).$$

Using notation

$$X_\pm(t) = \begin{pmatrix} \cos a_\pm t & \sin a_\pm t & 0 & 0 \\ -\sin a_\pm t & \cos a_\pm t & 0 & 0 \\ 0 & 0 & \cos \omega t & \sin \omega t \\ 0 & 0 & -\sin \omega t & \cos \omega t \end{pmatrix}$$

the corresponding fundamental matrices of (I.2.13), (I.2.14) *and* (I.2.15) *are* $X_1(t) = X_+(t)$, $X_2(t) = X_-(t - t_1^\beta)$ *and* $X_3(t) = X_+(t - t_2^\beta)$, *respectively. Saltation matrices of* (I.2.11) *and* (I.2.12) *are diagonal:*

$$S_1(\beta) = \mathrm{diag}\{a_-/a_+, 1, 1, 1\}, \qquad S_2(\beta) = \mathrm{diag}\{a_+/a_-, 1, 1, 1\}.$$

Note that the assumed relation (I.2.54) between a_+, a_- and ω means $\omega T^\beta = 2\pi$. That explains the name 1-1 resonance in (I.2.53)$_\varepsilon$.

Lemma I.2.20. *System* (I.2.55)$_0$ *satisfies conditions* H1) *and* H2).

Proof. Since $x_0(\beta) = (\beta_1, 0, \beta_2, \beta_3)$, $\mathrm{D}h(x, y, z) = (1, 0, 0, 0)$ and

$$f_+(x_0(\beta)) = (0, -a_+\beta_1, \omega\beta_3, -\omega\beta_2)^*,$$
$$f_\pm(\bar{x}_1(\beta)) = (-a_\pm\beta_1, 0, \omega\bar{x}_{14}(\beta), -\omega\bar{x}_{13}(\beta))^*,$$
$$f_\pm(\bar{x}_2(\beta)) = (a_\pm\beta_1, 0, \omega\bar{x}_{24}(\beta), -\omega\bar{x}_{23}(\beta))^*,$$

both conditions are easy to verify.   □

Of course, it is possible to continue with general values of $a_\pm$ and ω, but the resulting formulae are rather awkward. Therefore we set $a_+ = 2$, $a_- = 6$, $\omega = 3$. Then following the procedure of Section I.2.4, we derive the discontinuous Poincaré-Andronov-Melnikov function $M^\mu(\beta) = (M_1^\mu(\beta), M_2^\mu(\beta), M_3^\mu(\beta))$ with

$$M_1^\mu(\beta) = \frac{\pi}{12}(\beta_1^2 - 4)\beta_1 - \frac{43\sqrt{2}}{135}\mu\beta_3,$$

$$M_2^\mu(\beta) = -\frac{\pi}{3}\beta_2 + \frac{43\sqrt{2}}{135}\mu\frac{\beta_2\beta_3}{\beta_1} - \frac{\pi}{12}\mu\beta_3, \qquad (I.2.56)$$

$$M_3^\mu(\beta) = \frac{\pi}{12}\mu\beta_2 - \frac{\pi}{3}\beta_3 - \frac{43\sqrt{2}}{135}\mu\frac{\beta_2^2}{\beta_1} + \frac{28\sqrt{2}}{135}\mu\beta_1.$$

Equation $M_1^\mu(\beta) = 0$ has a simple root

$$\beta_3(\mu, \beta_1) = \frac{45\sqrt{2}}{344}\frac{\pi(\beta_1^2 - 4)\beta_1}{\mu}. \qquad (I.2.57)$$

Similarly

$$\beta_2(\mu, \beta_1) = \frac{45\sqrt{2}}{344}\frac{\pi(\beta_1^2 - 4)\beta_1}{\beta_1^2 - 8} \qquad (I.2.58)$$

is a simple root of equation $M_2^\mu(\beta_1, \beta_2, \beta_3(\mu, \beta_1)) = 0$. The third equation $M_3^\mu(\beta) = 0$, now has the form

$$\frac{\sqrt{2}\beta_1}{(\beta_1^2 - 8)^2 \mu} \left(-\frac{15}{344}\pi^2\beta_1^6 + \left(\frac{75}{86}\pi^2 + \frac{28}{135}\mu^2\right)\beta_1^4 \right.$$

$$\left. -\left(\frac{240}{43}\pi^2 + \frac{15}{344}\pi^2\mu^2 + \frac{448}{135}\mu^2\right)\beta_1^2 + \frac{480}{43}\pi^2 + \frac{15}{86}\pi^2\mu^2 + \frac{1792}{135}\mu^2 \right) = 0. \quad (I.2.59)$$

It is not possible to find solutions β_1 of the last equation analytically. Nevertheless, we can derive μ^2 as a rational function of β_1 given by

$$\mu^2(\beta_1) = F(\beta_1) = \frac{p_1(\beta_1)}{p_2(\beta_2)},$$

$$p_1(\beta_1) = 2025\pi^2(\beta_1^2 - 4)(\beta_1^2 - 8)^2, \quad (I.2.60)$$

$$p_2(\beta_2) = 9632\beta_1^4 - (2025\pi^2 + 154112)\beta_1^2 + 8100\pi^2 + 616448,$$

which is plotted in Figure I.2.3.

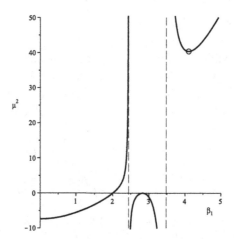

Figure I.2.3 The graph of the function $\mu^2(\beta_1)$ with respect to β_1. Asymptotes intersect β_1-axis at $\beta_{1-} \doteq 2.445$ and $\beta_{1+} \doteq 3.478$. The bifurcation point $(\bar{\beta}_1, F(\bar{\beta}_1)) \doteq (4.097, 40.423)$ is denoted by a circle

Note $p_2(\beta_1) = (\beta_1 - \beta_{1-})(\beta_1 + \beta_{1-})(\beta_1 - \beta_{1+})(\beta_1 + \beta_{1+})$ for

$$\beta_{1\pm} = \frac{1}{8}\sqrt{\frac{1}{301}\left(154112 + 2025\pi^2 \pm 45\pi\sqrt{154112 + 2025\pi^2}\right)}$$

with $\beta_{1-} \doteq 2.445$ and $\beta_{1+} \doteq 3.478$. Clearly, only for positive values of μ^2 a periodic orbit can persist, which is determined by the equation $\mu^2 = F(\beta_1), \beta_1 > 0$. Thus from now on we assume $\beta_1 > 2$. Moreover, from Figure I.2.3 one can see that for all $\mu > 0$ a periodic orbit can persist, and for $\mu^2 > F(\bar\beta_1) \doteq 40.423$ there exist three possible persisting periodic solutions. Here, $\bar\beta_1 \doteq 4.097$ is a unique solution of $F'(\beta_1) = 0$ greater than $2\sqrt{2} \doteq 2.828$. Note $F(2\sqrt{2}) = F'(2\sqrt{2}) = 0$. Since

$$\mu(\beta_1) = \sqrt{F(\beta_1)} > 0$$

for all

$$\begin{aligned}
\beta_1 \in I &:= I_1 \cup I_2 \cup I_3, \\
I_1 &= (2, \beta_{1-}) \doteq (2, 2.445), \\
I_2 &= (\beta_{1+}, \bar\beta_1) \doteq (3.478, 4.097), \\
I_3 &= (\bar\beta_1, \infty) \doteq (4.097, \infty)
\end{aligned} \tag{I.2.61}$$

is a simple root of equation (I.2.59), we know (see [29, Lemma 3.5.5]) even without calculating $DM^\mu(\beta)$ that $(\beta_1, \beta_2(\mu(\beta_1), \beta_1), \beta_3(\mu(\beta_1), \beta_1))$ is a simple root of $M^{\mu(\beta_1)}(\beta)$ for all $\beta_1 \in I$. Hence using the first part of Theorem I.2.14 we have just proved the following statement.

Proposition I.2.21. *Let* $\beta_2(\mu, \beta_1)$, $\beta_3(\mu, \beta_1)$ *be defined by* (I.2.58), (I.2.57) *and I by* (I.2.61), *respectively. For* $\mu \in \left(0, \sqrt{F(\bar\beta_1)}\right) \doteq (0, 6.358)$ *there is a unique persisting periodic solution* $x^*(\varepsilon, \mu, t)$ *of system* (I.2.55)$_\varepsilon$ *for any* $\varepsilon \neq 0$ *small. Moreover*

$$x^*(0, \mu, 0) = (\beta^*, 0, \beta_2(\mu, \beta^*), \beta_3(\mu, \beta^*)), \quad \beta^* \in I_1$$

for $\beta^* = \beta^*(\mu)$ *being a unique positive solution of equation* (I.2.59).

For $\mu \in \left(\sqrt{F(\bar\beta_1)}, \infty\right) \doteq (6.358, \infty)$ *system* (I.2.55)$_\varepsilon$ *possesses three persisting periodic solutions* $x_i^*(\varepsilon, \mu, t)$, $i = 1, 2, 3$ *for any* $\varepsilon \neq 0$ *small. In addition,*

$$x_i^*(0, \mu, 0) = (\beta_i^*, 0, \beta_2(\mu, \beta_i^*), \beta_3(\mu, \beta_i^*)), \quad \beta_i^* \in I_i, \quad i = 1, 2, 3$$

for positive solutions $\beta_1^* = \beta_1^*(\mu)$, $\beta_2^* = \beta_2^*(\mu)$ *and* $\beta_3^* = \beta_3^*(\mu)$ *of equation* (I.2.59).

To study asymptotic properties of these periodic solutions, we need the following result.

Lemma I.2.22. *The next statements hold for a cubic equation* $\lambda^3 + a_2\lambda^2 + a_1\lambda + a_0 = 0$:
(i) *All its roots have negative real parts if and only if* $a_2 > 0$ *and* $a_1 a_2 > a_0 > 0$.
(ii) *All its roots have positive real parts if and only if* $a_2 < 0$ *and* $a_1 a_2 < a_0 < 0$.
(iii) *It has a zero root if and only if* $a_0 = 0$, *and this root is simple if* $a_1 \neq 0$.
(iv) *It has a nonzero pure imaginary root if and only if* $a_1 > 0$ *and* $a_1 a_2 = a_0$, *and then it is given by* $\pm\sqrt{a_1}\iota$.

Proof. (i) follows from the Routh-Hurwitz criterion [30]. (ii) follows from (i) by exchanging $\lambda \leftrightarrow -\lambda$. (iii) and (iv) are elementary to prove.   □

The Jacobian matrix of function $M^\mu(\beta)$ of (I.2.56) at $\beta_2 = \beta_2(\mu, \beta_1)$, $\beta_3 = \beta_3(\mu, \beta_1)$ is given by

$$DM^\mu(\beta) = \begin{pmatrix} \frac{\pi}{4}\beta_1^2 - \frac{\pi}{3} & 0 & -\frac{43\sqrt{2}}{135}\mu \\ -\frac{15\sqrt{2}\pi^2}{1376}\frac{(\beta_1^2-4)^2}{\beta_1^2-8} & \frac{\pi}{12}(\beta_1^2 - 8) & \frac{\pi}{3}\frac{\mu}{\beta_1^2-8} \\ A & -\frac{\pi}{12}\frac{\mu\beta_1^2}{\beta_1^2-8} & -\frac{\pi}{3} \end{pmatrix},$$

$$A = \frac{\sqrt{2}}{185760}\frac{\mu((2025\pi^2 + 38528)\beta_1^4 - (16200\pi^2 + 616448)\beta_1^2 + 32400\pi^2 + 2465792)}{(\beta_1^2 - 8)^2}.$$

$$\tag{I.2.62}$$

Consequently, the characteristic polynomial of matrix (I.2.62) has the form

$$P(\lambda) = \lambda^3 + a_2\lambda^2 + a_1\lambda + a_0,$$

$$a_0(\beta_1) = \frac{\pi^3}{108B}\left(4816\beta_1^8 - (2025\pi^2 + 115584)\beta_1^6 + (16200\pi^2 + 924672)\beta_1^4\right.$$
$$\left. -(32400\pi^2 + 2465792)\beta_1^2\right),$$

$$a_1(\beta_1) = \frac{\pi^2}{72B}\left(14448\beta_1^8 - (2025\pi^2 + 423808)\beta_1^6 + (48600\pi^2 + 4315136)\beta^4\right. \tag{I.2.63}$$
$$\left. -(226800\pi^2 + 17260544)\beta_1^2 + 259200\pi^2 + 19726336\right),$$

$$a_2(\beta_1) = \frac{4\pi}{3} - \frac{\pi}{3}\beta_1^2,$$

$$B = 9632\beta_1^4 - (2025\pi^2 + 154112)\beta_1^2 + 8100\pi^2 + 616448.$$

Dependence of $a_0(\beta_1)$, $a_1(\beta_1)$ and $a_1(\beta_1)a_2(\beta_1) - a_0(\beta_1)$ on β_1 is illustrated in Figure I.2.4. Now, a result on stability of persisting trajectories follows.

Proposition I.2.23. *Let* $x^*(\varepsilon, \mu, t)$, $x_i^*(\varepsilon, \mu, t)$, $i = 1, 2, 3$ *be as in Proposition I.2.21. Then for any* $\varepsilon \neq 0$ *small all of these solutions of system* (I.2.55)$_\varepsilon$ *are hyperbolic, and for* $\varepsilon > 0$ *small none of them is stable and the only repeller is* $x_2^*(\varepsilon, \mu, t)$.

Proof. Using Theorem I.2.14 and Lemma I.2.22 applied on the characteristic polynomial (I.2.63) we directly obtain the statement. The sign of $a_2(\beta_1)$ is obvious, and those of $a_0(\beta_1)$, $a_1(\beta_1)$ and $a_1(\beta_1)a_2(\beta_1) - a_0(\beta_1)$ can be seen from Figure I.2.4 or computed from definitions in (I.2.63).   □

Note that from Lemma I.2.22 in bifurcation point (cf. Figure I.2.4) $\bar{\beta}_1 \doteq 4.097$, $\bar{\mu} =$

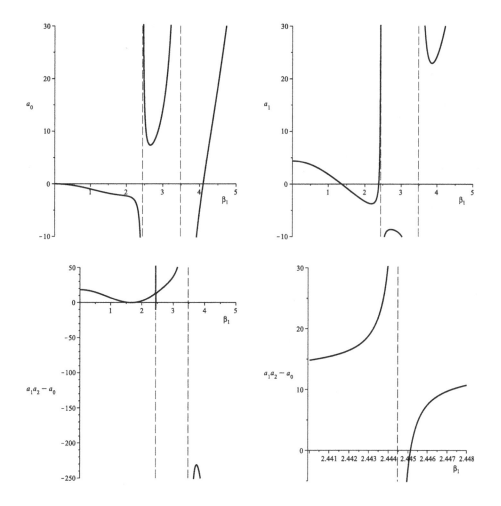

Figure I.2.4 Illustration of coefficients $a_0(\beta_1)$, $a_1(\beta_1)$ with respect to β_1, and the combination $a_1(\beta_1)a_2(\beta_1) - a_0(\beta_1)$ with a detailed view on the behavior in the neighborhood of the asymptote

$\sqrt{F(\bar{\beta}_1)} \doteq 6.358$ the matrix $DM^\mu(\beta)$ of (I.2.62) has an eigenvalue 0 of multiplicity 1. In this case we cannot apply Theorem I.2.14, and a higher order Melnikov function (cf. [20]) has to be used to determine if the solution $\gamma(\bar{\beta}_1, \beta_2(\bar{\mu}, \bar{\beta}_1), \beta_3(\bar{\mu}, \bar{\beta}_1), t)$ even persists.

CHAPTER I.3

Bifurcation from single periodic orbit in autonomous systems

I.3.1. Setting of the problem and main results

In this chapter, we consider the degenerated case of the manifold of initial points from the previous chapter, i.e. we shall assume that the unperturbed equation possesses a transverse periodic solution of period T, and we look for a sufficient condition on the perturbation such that the perturbed system has a periodic solution close to the original one with period close to T. So we do not suppose that a periodic solution is embedded into a family of periodic solutions. Related problems for smooth systems are studied in [14, 16, 25, 31–33].

Let $\Omega \subset \mathbb{R}^n$ be an open set in $\mathbb{R}^n$ and $h(x)$ be a C^r-function on $\overline{\Omega}$, with $r \geq 3$. We set $\Omega_\pm := \{x \in \Omega \mid \pm h(x) > 0\}$, $\Omega_0 := \{x \in \Omega \mid h(x) = 0\}$. Let $f_\pm \in C_b^r(\overline{\Omega})$, $g \in C_b^r(\overline{\Omega} \times \mathbb{R} \times \mathbb{R}^p)$ and $h \in C_b^r(\overline{\Omega}, \mathbb{R})$. Let $\varepsilon \in \mathbb{R}$ and $\mu \in \mathbb{R}^p$, $p \geq 1$ be parameters. Furthermore, we suppose that 0 is a regular value of h.

We say that a function $x(t)$ is a solution of the equation

$$\dot{x} = f_\pm(x) + \varepsilon g(x, \varepsilon, \mu), \quad x \in \overline{\Omega}_\pm, \tag{I.3.1}$$

if it is a solution of this equation in the sense analogous to Definition I.1.1.

Let us assume

H1) For $\varepsilon = 0$ equation (I.3.1) has a periodic orbit $\gamma(t)$ of period T. The orbit is given by its initial point $x_0 \in \Omega_+$ and consists of three branches

$$\gamma(t) = \begin{cases} \gamma_1(t) & \text{if } t \in [0, t_1], \\ \gamma_2(t) & \text{if } t \in [t_1, t_2], \\ \gamma_3(t) & \text{if } t \in [t_2, T], \end{cases} \tag{I.3.2}$$

where $0 < t_1 < t_2 < T$, $\gamma_1(t) \in \Omega_+$ for $t \in [0, t_1)$, $\gamma_2(t) \in \Omega_-$ for $t \in (t_1, t_2)$, $\gamma_3(t) \in \Omega_+$ for $t \in (t_2, T]$, and

$$\begin{aligned} x_1 &:= \gamma_1(t_1) = \gamma_2(t_1) \in \Omega_0, \\ x_2 &:= \gamma_2(t_2) = \gamma_3(t_2) \in \Omega_0, \\ x_0 &:= \gamma_3(T) = \gamma_1(0) \in \Omega_+. \end{aligned} \tag{I.3.3}$$

H2) Moreover, we also assume that

$$Dh(x_1)f_{\pm}(x_1) < 0 \qquad \text{and} \qquad Dh(x_2)f_{\pm}(x_2) > 0.$$

Let $x_+(\tau, \xi)(t, \varepsilon, \mu)$ and $x_-(\tau, \xi)(t, \varepsilon, \mu)$ denote the solution of the initial value problem

$$\begin{aligned}\dot{x} &= f_{\pm}(x) + \varepsilon g(x, \varepsilon, \mu) \\ x(\tau) &= \xi \end{aligned} \tag{I.3.4}$$

with corresponding sign. Note

$$x_{\pm}(\tau, \xi)(t, \varepsilon, \mu) = x_{\pm}(0, \xi)(t - \tau, \varepsilon, \mu). \tag{I.3.5}$$

First, we slightly modify Lemma I.1.2 for the autonomous case.

Lemma I.3.1. *Assume* H1) *and* H2). *Then there exist* $\varepsilon_0, r_0 > 0$ *and a Poincaré mapping (cf. Figure I.3.1)*

$$P(\cdot, \varepsilon, \mu)\colon B(x_0, r_0) \to \Sigma$$

for all fixed $\varepsilon \in (-\varepsilon_0, \varepsilon_0)$, $\mu \in \mathbb{R}^p$, *where*

$$\Sigma = \{y \in \mathbb{R}^n \mid \langle y - x_0, f_+(x_0) \rangle = 0\}.$$

Moreover, $P\colon B(x_0, r_0) \times (-\varepsilon_0, \varepsilon_0) \times \mathbb{R}^p \to \mathbb{R}^n$ *is* C^r*-smooth in all arguments and* $B(x_0, r_0) \subset \Omega_+$.

Proof. The lemma can be easily proved as Lemma I.1.2 using the implicit function theorem (IFT). We obtain the existence of C^r-functions t_1, t_2 and t_3 satisfying, respectively,

$$h(x_+(\tau, \xi)(t_1(\tau, \xi, \varepsilon, \mu), \varepsilon, \mu)) = 0,$$

$$h(x_-(t_1(\tau, \xi, \varepsilon, \mu), x_+(\tau, \xi)(t_1(\tau, \xi, \varepsilon, \mu), \varepsilon, \mu))(t_2(\tau, \xi, \varepsilon, \mu), \varepsilon, \mu)) = 0$$

and

$$\langle x_+(t_2(\tau, \xi, \varepsilon, \mu), x_-(t_1(\tau, \xi, \varepsilon, \mu), x_+(\tau, \xi)(t_1(\tau, \xi, \varepsilon, \mu), \varepsilon, \mu))$$
$$(t_2(\tau, \xi, \varepsilon, \mu), \varepsilon, \mu))(t_3(\tau, \xi, , \varepsilon, \mu), \varepsilon, \mu) - x_0, f_+(x_0) \rangle = 0$$

for (τ, ξ, ε) close to $(0, x_0, 0)$ and $\mu \in \mathbb{R}^p$. Moreover, we have

$$t_1(0, x_0, 0, \mu) = t_1, \quad t_2(0, x_0, 0, \mu) = t_2, \quad t_3(0, x_0, 0, \mu) = T.$$

Poincaré mapping is then defined as

$$P(\xi, \varepsilon, \mu) = x_+(t_2(0, \xi, \varepsilon, \mu), x_-(t_1(0, \xi, \varepsilon, \mu), x_+(0, \xi)(t_1(0, \xi, \varepsilon, \mu), \varepsilon, \mu))$$
$$(t_2(0, \xi, \varepsilon, \mu), \varepsilon, \mu))(t_3(0, \xi, \varepsilon, \mu), \varepsilon, \mu). \tag{I.3.6}$$

□

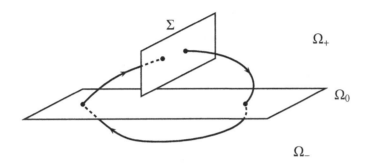

Figure I.3.1 Discontinuous Poincaré mapping

In contrast to Chapters I.1 and I.2, here we shall need to calculate the second order derivative of Poincaré mapping with respect to ξ. In order to do this, we construct linearization of P at a general point $(\xi, 0, \mu)$ with ξ sufficiently close to x_0. Note that in each of the next steps, dependence on μ is lost, since we set $\varepsilon = 0$ and μ occurs only as a parameter of g. Therefore, we omit the dependence on ε and μ in the following linearizations, and denote $x_\pm(\tau, \xi, t)$ the solution of the unperturbed system

$$\dot{x} = f_\pm(x)$$
$$x(\tau) = \xi. \tag{I.3.7}$$

For this time $t_i(\xi) = t_i(0, \xi, 0, \mu)$ for $i = 1, 2, 3$, and

$$x_1(\xi) = x_+(0, \xi, t_1(\xi)),$$
$$x_2(\xi) = x_-(t_1(\xi), x_1(\xi), t_2(\xi)), \tag{I.3.8}$$
$$x_3(\xi) = x_+(t_2(\xi), x_2(\xi), t_3(\xi)).$$

Differentiating of (I.3.7) with respect to ξ we get

$$\dot{x}_{\pm\xi}(\tau, \xi, t) = Df_\pm(x_\pm(\tau, \xi, t))x_{\pm\xi}(\tau, \xi, t)$$
$$x_{\pm\xi}(\tau, \xi, \tau) = \mathbb{I} \tag{I.3.9}$$

where the lower index ξ denotes the partial derivative with respect to ξ and $\mathbb{I}$ the $n \times n$

identity matrix. Let us denote

$$X_1^\xi(t) = x_{+\xi}(0, \xi, t), \tag{I.3.10}$$

$$X_2^\xi(t) = x_{-\xi}(t_1(\xi), x_1(\xi), t), \tag{I.3.11}$$

$$X_3^\xi(t) = x_{+\xi}(t_2(\xi), x_2(\xi), t). \tag{I.3.12}$$

From identities

$$h(x_1(\xi)) = 0, \qquad h(x_2(\xi)) = 0, \qquad \langle x_3(\xi) - x_0, f_+(x_0) \rangle = 0$$

using relations (I.3.8) we derive, respectively,

$$t_{1\xi}(\xi) = -\frac{Dh(x_1(\xi))X_1^\xi(t_1(\xi))}{Dh(x_1(\xi))f_+(x_1(\xi))}, \qquad t_{2\xi}(\xi) = -\frac{Dh(x_2(\xi))X_2^\xi(t_2(\xi))S_1^\xi X_1^\xi(t_1(\xi))}{Dh(x_2(\xi))f_-(x_2(\xi))},$$

$$t_{3\xi}(\xi) = -\frac{\langle X_3^\xi(t_3(\xi))S_2^\xi X_2^\xi(t_2(\xi))S_1^\xi X_1^\xi(t_1(\xi))\cdot, f_+(x_0) \rangle}{\langle f_+(x_3(\xi)), f_+(x_0) \rangle},$$

where

$$S_1^\xi = \mathbb{I} + \frac{(f_-(x_1(\xi)) - f_+(x_1(\xi)))Dh(x_1(\xi))}{Dh(x_1(\xi))f_+(x_1(\xi))}, \tag{I.3.13}$$

$$S_2^\xi = \mathbb{I} + \frac{(f_+(x_2(\xi)) - f_-(x_2(\xi)))Dh(x_2(\xi))}{Dh(x_2(\xi))f_-(x_2(\xi))} \tag{I.3.14}$$

are saltation matrices taken at a general initial point ξ. Considering the inner product $\langle a, b \rangle = b^* a$, we can write

$$t_{3\xi}(\xi) = -\frac{f_+(x_0)^* X_3^\xi(t_3(\xi))S_2^\xi X_2^\xi(t_2(\xi))S_1^\xi X_1^\xi(t_1(\xi))}{\langle f_+(x_3(\xi)), f_+(x_0) \rangle}.$$

In view of these facts, we can state the lemma concluding some properties of the Poincaré mapping.

Lemma I.3.2. *Let $P(\xi, \varepsilon, \mu)$ be defined by (I.3.6). Then for ξ sufficiently close to x_0,*

$$P_\xi(\xi, 0, \mu) = (\mathbb{I} - S^\xi)A(\xi, 0), \tag{I.3.15}$$

$$P_\varepsilon(x_0, 0, \mu) = (\mathbb{I} - S^{x_0})\left(\int_0^T A(x_0, s)g(\gamma(s), 0, \mu)ds \right), \tag{I.3.16}$$

where P_ξ, P_ε are partial derivatives of P with respect to ξ, ε, respectively. Here

$$S^\xi = \frac{f_+(x_3(\xi))f_+(x_0)^*}{\langle f_+(x_3(\xi)), f_+(x_0) \rangle} \tag{I.3.17}$$

is the projection onto $[f_+(x_3(\xi))]$ in the direction orthogonal to $f_+(x_0)$ (i.e. $S^\xi y = 0$ if *and only if $\langle y, f_+(x_0) \rangle = 0$) and $A(\xi, t)$ is given by*

$$A(\xi, t) = \begin{cases} X_3^\xi(t_3(\xi))S_2^\xi X_2^\xi(t_2(\xi))S_1^\xi X_1^\xi(t_1(\xi))X_1^\xi(t)^{-1} & \text{if } t \in [0, t_1(\xi)), \\ X_3^\xi(t_3(\xi))S_2^\xi X_2^\xi(t_2(\xi))X_2^\xi(t)^{-1} & \text{if } t \in [t_1(\xi), t_2(\xi)), \quad \text{(I.3.18)} \\ X_3^\xi(t_3(\xi))X_3^\xi(t)^{-1} & \text{if } t \in [t_2(\xi), t_3(\xi)] \end{cases}$$

with saltation matrices S_1^ξ of (I.3.13) and S_2^ξ of (I.3.14), and X_1^ξ, X_2^ξ, X_3^ξ being defined by (I.3.10), (I.3.11), (I.3.12), respectively. In addition, $P_\xi(x_0, 0, \mu)$ has an eigenvalue 0 with the corresponding eigenvector $f_+(x_0)$, i.e.

$$P_\xi(x_0, 0, \mu)f_+(x_0) = 0.$$

Proof. Since $P(\xi, 0, \mu) = x_3(\xi)$ where $x_3(\xi)$ is given by (I.3.8), the result on $P_\xi(\xi, 0, \mu)$ follows from the preceding discussion. Next, $P_\varepsilon(x_0, 0, \mu)$ is obtained by differentiating (I.3.4) with respect to ε (cf. Lemma I.1.4). The statement on the eigenvalue is proved in a more general form in Lemma I.2.3.   □

For simplicity, we shall drop the upper index ξ when $\xi = x_0$, i.e. $X_i(t) = X_i^{x_0}(t)$ for $i = 1, 2, 3$, $S_i = S_i^{x_0}$ for $i = 1, 2$, $S = S^{x_0}$. Clearly $t_1(x_0) = t_1$, $t_2(x_0) = t_2$, $t_3(x_0) = T$, $x_1(x_0) = x_1$, $x_2(x_0) = x_2$, $x_3(x_0) = x_0$ and S is the orthogonal projection onto $[f_+(x_0)]$.

Now we want to find a persisting periodic solution of (I.3.1) for $\varepsilon \neq 0$, i.e. we are looking for a periodic solution of the perturbed equation in the neighborhood of $\gamma(\cdot)$ such that if ε tends to 0 then the new solution tends to $\gamma(\cdot)$. This problem is equivalent to solving the next equation

$$F(x, \varepsilon, \mu) := x - P(x, \varepsilon, \mu) = 0 \qquad \text{(I.3.19)}$$

for $(x, \varepsilon) \in \Sigma \times \mathbb{R}$ close to $(x_0, 0)$. Thus

$$F : (B(x_0, r_0) \cap \Sigma) \times (-\varepsilon_0, \varepsilon_0) \times \mathbb{R}^p \to [f_+(x_0)]^\perp \subset \mathbb{R}^n$$

and $F(x_0, 0, \mu) = 0$. If we denote

$$Z = \mathcal{N}F_\xi(x_0, 0, \mu), \qquad\qquad Y = \mathcal{R}F_\xi(x_0, 0, \mu) \qquad \text{(I.3.20)}$$

the null space and the range of the operator $F_\xi(x_0, 0, \mu)$, then we take the orthogonal decomposition

$$[f_+(x_0)]^\perp = Z \oplus Z^\perp, \qquad\qquad [f_+(x_0)]^\perp = Y \oplus Y^\perp$$

with the orthogonal projections $Q : [f_+(x_0)]^\perp \to Y$, $\mathcal{P} : [f_+(x_0)]^\perp \to Y^\perp$. Note $\Sigma = x_0 + [f_+(x_0)]^\perp$ and $\dim Z = \dim Y^\perp$. For $\xi_1 \in Z$, $\xi_2 \in Z^\perp$ we consider the Taylor expansion

$$F(x_0 + \xi_1 + \xi_2, \varepsilon, \mu) = D_\xi F(x_0, 0, \mu)\xi_2 + \varepsilon D_\varepsilon F(x_0, 0, \mu)$$

$$+\frac{1}{2}D_{\xi\xi}F(x_0, 0, \mu)[\xi_1 + \xi_2]^2 + \varepsilon D_{\varepsilon\xi}F(x_0, 0, \mu)[\xi_1 + \xi_2] + \frac{1}{2}\varepsilon^2 D_{\varepsilon\varepsilon}F(x_0, 0, \mu)$$

$$+O(|\xi_1|^3 + |\xi_2|^3 + |\varepsilon|^3).$$

Equation (I.3.19) has the form $F(x_0 + \xi_1 + \xi_2, \varepsilon, \mu) = 0$ for $\xi_1 \in Z$ and $\xi_2 \in Z^\perp$ small. To solve it, we apply the Lyapunov-Schmidt decomposition

$$QF(x_0 + \xi_1 + \xi_2, \varepsilon, \mu) = 0, \tag{I.3.21}$$

$$\mathcal{P}F(x_0 + \xi_1 + \xi_2, \varepsilon, \mu) = 0. \tag{I.3.22}$$

The first of these equations solved via IFT gives the existence of a unique C^r-function $\xi_2(\xi_1, \varepsilon, \mu)$ for ξ_1, ε small such that equation (I.3.21) is satisfied for ξ_1, ξ_2, ε small if and only if $\xi_2 = \xi_2(\xi_1, \varepsilon, \mu)$. Moreover $\xi_2(0, 0, \mu) = 0$. Differentiating (I.3.21) for $\xi_2 = \xi_2(\xi_1, \varepsilon, \mu)$ with respect to ξ_1 at $(\xi_1, \varepsilon) = (0, 0)$ we derive $D_{\xi_1}\xi_2(0, 0, \mu) = 0$. Therefore $\xi_2 = O(\xi_1^2) + O(\varepsilon)$. So we scale

$$\xi_1 \longleftrightarrow \varepsilon\xi_1, \qquad \varepsilon \longleftrightarrow \pm\varepsilon^2.$$

Then we get

$$0 = \frac{1}{\varepsilon^2}\mathcal{P}F(x_0 + \varepsilon\xi_1 + \xi_2(\varepsilon\xi_1, \pm\varepsilon^2, \mu), \pm\varepsilon^2, \mu)$$

$$= \pm\mathcal{P}D_\varepsilon F(x_0, 0, \mu) + \frac{1}{2}\mathcal{P}D_{\xi\xi}F(x_0, 0, \mu)\xi_1^2 + O(\varepsilon) \tag{I.3.23}$$

as an equivalent problem to equation (I.3.22). Let $\dim Z = k > 0$ and $\{\psi_1, \ldots, \psi_k\}$ be an orthogonal basis of $Y^\perp$. Then applying Lemma I.3.2, equation (I.3.23) becomes

$$0 = \sum_{i=1}^{k} \frac{\psi_i}{\|\psi_i\|^2}\left[\pm \int_0^T \langle g(\gamma(s)), 0, \mu), A^*(x_0, s)\psi_i\rangle ds\right.$$

$$\left. +\frac{1}{2}\langle A^{-1}(x_0, 0)D_\xi((\mathbb{I} - S^\xi)A(\xi, 0)\xi_1)_{\xi=x_0}\xi_1, A^*(x_0, 0)\psi_i\rangle\right] + O(\varepsilon).$$

Linearization of equation (I.3.1) with $\varepsilon = 0$ along T-periodic solution $\gamma(t)$ gives the variational equation

$$\dot{x}(t) = Df_\pm(\gamma(t))x(t) \tag{I.3.24}$$

which splits into the couple of equations

$$\dot{x} = Df_+(\gamma(t))x \qquad \text{if } t \in [0, t_1) \cup [t_2, T],$$

$$\dot{x} = Df_-(\gamma(t))x \qquad \text{if } t \in [t_1, t_2)$$

satisfying the impulsive conditions

$$x(t_1+) = S_1 x(t_1-), \qquad x(t_2+) = S_2 x(t_2-), \qquad (I.3.25)$$

as well as the periodic condition

$$(\mathbb{I} - S)(x(T) - x(0)) = 0, \qquad (I.3.26)$$

where $x(t\pm) = \lim_{s \to t^\pm} x(s)$. From the definition of $X_i(t)$,

$$X(t) = \begin{cases} X_1(t) & \text{if } t \in [0, t_1), \\ X_2(t)S_1 X_1(t_1) & \text{if } t \in [t_1, t_2), \\ X_3(t)S_2 X_2(t_2)S_1 X_1(t_1) & \text{if } t \in [t_2, T] \end{cases}$$

solves the variational equation (I.3.24) and the conditions (I.3.25). So does $X(t)c$ for any $c \in \mathbb{R}^n$. Moreover, $X(t)v$ is a solution of periodic condition (I.3.26) if and only if $v \in [f_+(x_0), Z]$. Indeed, from Lemma I.3.2 and since $(\mathbb{I} - S)f_+(x_0) = 0$, we get

$$(\mathbb{I} - S)(\mathbb{I} - A(x_0, 0))f_+(x_0) = 0.$$

For $w \in Z \subset [f_+(x_0)]^\perp$,

$$0 = D_\xi F(x_0, 0, \mu)w = (\mathbb{I} - (\mathbb{I} - S)A(x_0, 0))w = (\mathbb{I} - S)(\mathbb{I} - A(x_0, 0))w,$$

and for $w \in Z^\perp \subset [f_+(x_0)]^\perp$,s

$$0 \neq D_\xi F(x_0, 0, \mu)w = (\mathbb{I} - S)(\mathbb{I} - A(x_0, 0))w.$$

From our result – Lemma I.2.4 – we know that the adjoint variational system of (I.3.1) with $\varepsilon = 0$ is given by the following linear impulsive boundary value problem

$$\begin{aligned} \dot{X} &= -Df_+^*(\gamma(t))X \quad \text{if } t \in [0, t_1], \\ \dot{X} &= -Df_-^*(\gamma(t))X \quad \text{if } t \in [t_1, t_2], \\ \dot{X} &= -Df_+^*(\gamma(t))X \quad \text{if } t \in [t_2, T], \\ X(t_i-) &= S_i^* X(t_i+), \quad i = 1, 2, \\ X(T) &= X(0) \in [f_+(x_0)]^\perp. \end{aligned} \qquad (I.3.27)$$

Since

$$A^*(x_0, t) = X^{-1*}(t)X_1^*(t_1)S_1^* X_2^*(t_2)S_2^* X_3^*(T),$$

Lemma I.1.5 implies that $A^*(x_0, t)\psi$ solves the adjoint variational equation with impulsive conditions. If moreover $\psi \in Y^\perp$ then also the boundary condition is satisfied, hence $A^*(x_0, t)\psi$ is a solution of the adjoint variational system (I.3.27) whenever

$\psi \in Y^\perp$. To see that $A^*(x_0, t)\psi$ satisfies the boundary condition, we consider

$$0 = \langle D_\xi F(x_0, 0, \mu)\xi, \psi \rangle = \langle (\mathbb{I} - (\mathbb{I} - S)A(x_0, 0))\xi, \psi \rangle$$
$$= \langle \xi, (\mathbb{I} - A^*(x_0, 0)(\mathbb{I} - S^*))\psi \rangle = \langle \xi, (\mathbb{I} - A^*(x_0, 0))\psi \rangle$$

for all $\xi \in [f_+(x_0)]^\perp$, and if $\xi \in [f_+(x_0)]$, Lemma I.3.2 yields

$$0 = \langle (\mathbb{I} - S)\xi - P_\xi(x_0, 0, \mu)\xi, \psi \rangle = \langle (\mathbb{I} - S)(\mathbb{I} - A(x_0, 0))\xi, \psi \rangle$$
$$= \langle \xi, (\mathbb{I} - A^*(x_0, 0))\psi \rangle$$

taking P_ξ as a partial derivative of P with respect to ξ. In conclusion, we get the following theorem.

Theorem I.3.3. *Let $\{\psi_1, \ldots, \psi_k\}$ be an orthogonal basis of $Y^\perp$ with Y given by (I.3.20) and $A(\xi, t)$ be defined by (I.3.18). If ξ_1^0 is a simple root of function $M_\pm^{\mu_0}(\xi_1)$ where $M_\pm^\mu(\xi_1) = (M_{1\pm}^\mu(\xi_1), \ldots, M_{k\pm}^\mu(\xi_1))$ and*

$$M_{i\pm}^\mu(\xi_1) = \pm \int_0^T \langle g(\gamma(s), 0, \mu), A^*(x_0, s)\psi_i \rangle ds$$

$$+ \frac{1}{2} \left\langle A^{-1}(x_0, 0) D_\xi((\mathbb{I} - S^\xi)A(\xi, 0)\xi_1)_{\xi=x_0}\xi_1, A^*(x_0, 0)\psi_i \right\rangle$$

for $i = 1, \ldots, k$ with "+" or "−" sign, i.e. $M_+^{\mu_0}(\xi_1^0) = 0$ and $\det D_{\xi_1} M_+^{\mu_0}(\xi_1^0) \neq 0$, or $M_-^{\mu_0}(\xi_1^0) = 0$ and $\det D_{\xi_1} M_-^{\mu_0}(\xi_1^0) \neq 0$, then there exists a unique (for each sign) C^r-function $\xi_1(\epsilon, \mu)$ with $\epsilon \sim 0$ small and $\mu \sim \mu_0$ such that there is a periodic solution of equation (I.3.1) with $\varepsilon = \pm\epsilon^2 \neq 0$ sufficiently small and μ close to μ_0. This solution has an initial point

$$x^* = x_0 + \epsilon\xi_1(\epsilon, \mu) + \xi_2(\epsilon\xi_1(\epsilon, \mu), \pm\epsilon^2, \mu)$$

and period $t_3(0, x^*, \pm\epsilon^2, \mu)$. Note $\xi_1(0, \mu_0) = \xi_1^0$.

I.3.2. The special case for linear switching manifold

If the function h has the form $h(x) = \langle a, x \rangle + c$ for given $a \in \mathbb{R}^n$, $c \in \mathbb{R}$, some of our results can be simplified. In this case we can take $\Sigma \subset \Omega_0$ and derive another Poincaré mapping $P(\cdot, \varepsilon, \mu): B(x_0, r_0) \subset \Sigma \to \Sigma$ given by (cf. (I.3.6))

$$P(\xi, \varepsilon, \mu) = x_-(x_+(\xi)(t_1(\xi, \varepsilon, \mu), \varepsilon, \mu))(t_2(\xi, \varepsilon, \mu) - t_1(\xi, \varepsilon, \mu), \varepsilon, \mu) \qquad (\text{I.3.28})$$

where we omitted the dependence on τ, since we always assume $\tau = 0$ and equation (I.3.1) is autonomous. This time

$$t_2(\xi, \varepsilon, \mu) = t_3(\xi, \varepsilon, \mu), \qquad x_2(\xi) = x_-(t_1(\xi), x_1(\xi), t_2(\xi)) = x_3(\xi)$$

(see (I.3.8)) and we have only one saltation matrix S_1^ξ. Hence the assumption H1) this time has the form:

H1') For $\varepsilon = 0$ equation (I.3.1) has a unique periodic orbit $\gamma(t)$ of period T. The orbit is given by its initial point $x_0 \in \Sigma \subset \Omega_0$ and consists of two branches

$$\gamma(t) = \begin{cases} \gamma_1(t) & \text{if } t \in [0, t_1], \\ \gamma_2(t) & \text{if } t \in [t_1, T], \end{cases}$$

where $0 < t_1 < T$, $\gamma_1(t) \in \Omega_+$ for $t \in (0, t_1)$, $\gamma_2(t) \in \Omega_-$ for $t \in (t_1, T)$, and

$$x_1 := \gamma_1(t_1) = \gamma_2(t_1) \in \Omega_0,$$
$$x_0 := \gamma_2(T) = \gamma_1(0) \in \Omega_0.$$

The new Poincaré mapping has slightly different properties.

Lemma I.3.4. *Let $h(x) = \langle a, x \rangle + c$ for given $a \in \mathbb{R}^n$, $c \in \mathbb{R}$ and $P(\xi, \varepsilon, \mu)$ be defined by (I.3.28). Then for $\xi \in \Sigma$ sufficiently close to x_0,*

$$P_\xi(\xi, 0, \mu) = (\mathbb{I} - S^\xi)A(\xi, 0),$$

$$P_\varepsilon(x_0, 0, \mu) = (\mathbb{I} - S^{x_0})\left(\int_0^T A(x_0, s)g(\gamma(s), 0, \mu)ds \right),$$

where P_ξ, P_ε are partial derivatives of P with respect to ξ, ε, respectively,

$$S^\xi w = \frac{\langle a, w \rangle}{\langle a, f_-(x_2(\xi)) \rangle} f_-(x_2(\xi)) \qquad w \in \mathbb{R}^n, \tag{I.3.29}$$

$$A(\xi, t) = \begin{cases} X_2^\xi(t_2(\xi))S_1^\xi X_1^\xi(t_1(\xi))X_1^\xi(t)^{-1} & \text{if } t \in [0, t_1(\xi)), \\ X_2^\xi(t_2(\xi))X_2^\xi(t)^{-1} & \text{if } t \in [t_1(\xi), t_2(\xi)] \end{cases} \tag{I.3.30}$$

with saltation matrix S_1^ξ of (I.3.13) now given by

$$S_1^\xi w = w + \frac{\langle a, w \rangle}{\langle a, f_+(x_1(\xi)) \rangle}(f_-(x_1(\xi)) - f_+(x_1(\xi))) \qquad w \in \mathbb{R}^n \tag{I.3.31}$$

and X_1^ξ, X_2^ξ being defined by (I.3.10), (I.3.11), respectively.

We consider equation (I.3.19) with P given by (I.3.28). By the procedure described in the preceding section we derive the Poincaré-Andronov-Melnikov function and the following theorem analogous to Theorem I.3.3.

Theorem I.3.5. *Let $h(x) = \langle a, x \rangle + c$ for given $a \in \mathbb{R}^n$, $c \in \mathbb{R}$, $\{\psi_1, \ldots, \psi_k\}$ be an orthogonal basis of $Y^\perp$ with Y given by (I.3.20), $A(\xi, t)$ be defined by (I.3.30) and S^ξ*

by (I.3.29). If ξ_1^0 is a simple root of function $M_\pm^{\mu_0}(\xi_1)$ where

$$M_\pm^\mu(\xi_1) = (M_{1\pm}^\mu(\xi_1), \ldots, M_{k\pm}^\mu(\xi_1))$$

and

$$M_{i\pm}^\mu(\xi_1) = \pm \int_0^T \langle g(\gamma(s), 0, \mu), A^*(x_0, s)\psi_i \rangle ds$$

$$+ \frac{1}{2} \left\langle A^{-1}(x_0, 0) D_\xi((\mathbb{I} - S^\xi)A(\xi, 0)\xi_1)_{\xi=x_0}\xi_1, A^*(x_0, 0)\psi_i \right\rangle$$

for $i = 1, \ldots, k$ with "+" or "−" sign, i.e. $M_+^{\mu_0}(\xi_1^0) = 0$ and $\det D_{\xi_1} M_+^{\mu_0}(\xi_1^0) \neq 0$, or $M_-^{\mu_0}(\xi_1^0) = 0$ and $\det D_{\xi_1} M_-^{\mu_0}(\xi_1^0) \neq 0$, then there exists a unique (for each sign) C^r-function $\xi_1(\epsilon, \mu)$ with $\epsilon \sim 0$ and $\mu \sim \mu_0$ such that there is a periodic solution of equation (I.3.1) with $\varepsilon = \pm\epsilon^2 \neq 0$ sufficiently small and μ close to μ_0. This solution has an initial point

$$x^* = x_0 + \epsilon\xi_1(\epsilon, \mu) + \xi_2(\epsilon\xi_1(\epsilon, \mu), \pm\epsilon^2, \mu)$$

and period $t_2(x^*, \pm\epsilon^2, \mu)$. Note $\xi_1(0, \mu_0) = \xi_1^0$.

The method of Poincaré mapping can be used to determine the hyperbolicity and stability of the persisting orbit, similarly to Section I.2.3. For the degenerate case when $\dim Z = n - 1$ for Z defined in (I.3.20), i.e. $Y = \{0\} = Z^\perp$ and $Y^\perp = [\psi_1, \ldots, \psi_k] = Z$, we have the next result.

Theorem I.3.6. *Let μ_0, ξ_1^0 be as in Theorem I.3.5, $h(x) = \langle a, x \rangle + c$ for given $a \in \mathbb{R}^n$, $c \in \mathbb{R}$ and P be defined by (I.3.28). Assume that $D_{\xi\xi} P(x_0, 0, \mu_0)\xi_1^0 \big|_{[a]^\perp}$ has no eigenvalues on the imaginary axis. Then the persisting trajectory is hyperbolic.*

Proof. From the Taylor expansion with respect to ε at $\varepsilon = 0$ for the derivative of the Poincaré mapping at a general point x,

$$D_\xi P(x, \pm\varepsilon^2, \mu)v = D_\xi P(x, 0, \mu)v + O(\varepsilon^2),$$

we have at $x_0 + \varepsilon\xi_1(\varepsilon, \mu)$,

$$A^\mu(\varepsilon)v := D_\xi P(x_0 + \varepsilon\xi_1(\varepsilon, \mu), \pm\varepsilon^2, \mu)v = D_\xi P(x_0 + \varepsilon\xi_1(\varepsilon, \mu), 0, \mu)v + O(\varepsilon^2)$$

$$= D_\xi P(x_0, 0, \mu)v + \varepsilon D_{\xi\xi} P(x_0, 0, \mu)\xi_1^0 v + O(\varepsilon^2) = v + \varepsilon D_{\xi\xi} P(x_0, 0, \mu)\xi_1^0 v + O(\varepsilon^2).$$

Hence $A^\mu(\varepsilon) = \mathbb{I} + \varepsilon A_1^\mu + O(\varepsilon^2)$ where $A_1^\mu = D_{\xi\xi} P(x_0, 0, \mu)\xi_1^0 \big|_{[a]^\perp}$. Let k_1 be the number of all eigenvalues of A_1^μ with negative real parts and $k_2 := n - k_1 - 1$ with positive.

Then there exists a regular matrix $P^\mu(\varepsilon)$ such that

$$\widetilde{A}^\mu(\varepsilon) := P^\mu(\varepsilon)A^\mu(\varepsilon)P^{\mu-1}(\varepsilon) = \mathbb{I} + \varepsilon \begin{pmatrix} A^\mu_{11} & 0 \\ 0 & A^\mu_{22} \end{pmatrix} + O(\varepsilon^2) = E^\mu(\varepsilon) + O(\varepsilon^2)$$

where A^μ_{11}, A^μ_{22} are $k_1 \times k_1$, $k_2 \times k_2$ blocks, respectively, and $\mathcal{R}\sigma(A^\mu_{11}) \subset (-\infty, 0)$, $\mathcal{R}\sigma(A^\mu_{22}) \subset (0, \infty)$. It can be shown [26, 27] that $E^\mu(\varepsilon)$ is strongly 1-hyperbolic. Consequently, the statement follows from Lemma I.2.9. $\square$

We can slightly modify the assumptions of the last theorem to obtain a stability criterion.

Corollary I.3.7. *Let the assumptions of Theorem I.3.6 be fulfilled and, moreover, all eigenvalues of* $D_{\xi\xi} P(x_0, 0, \mu^0)\xi^0_1\big|_{[a]^\perp}$ *have negative real parts. Then the persisting periodic orbit is stable (repeller) for* $\varepsilon > 0$ ($\varepsilon < 0$) *sufficiently small.*

I.3.3. Planar application

In this section we consider the following system

$$\begin{aligned} \dot{x} &= y + \delta + \varepsilon x(2 - \mu_1 x^2 - \mu_2 y^2) \\ \dot{y} &= -x + \varepsilon(x + y(x - y^2)) \end{aligned} \qquad \text{if } y > 0,$$

$$\begin{aligned} \dot{x} &= x + y - \delta + (x^2 + (y-\delta)^2)(-x - (y-\delta)) \\ &\quad + (x^2 + (y-\delta)^2)^2(x/4 + (y-\delta)/2) \\ \dot{y} &= -x + y - \delta + (x^2 + (y-\delta)^2)(x - (y-\delta)) \\ &\quad + (x^2 + (y-\delta)^2)^2(-x/2 + (y-\delta)/4) \end{aligned} \qquad \text{if } y < 0 \qquad (I.3.32)_\varepsilon$$

with parameters $\mu_1, \mu_2 \in \mathbb{R}$ and constant $0 < \delta < \sqrt{2}$. We investigate the persistence of a periodic orbit under perturbation using the method described in Section I.3.2. In this case we have $h(x, y) = y$, $\Omega_\pm = \{(x, y) \in \mathbb{R}^2 \mid \pm y > 0\}$, $\Omega_0 = \mathbb{R} \times \{0\}$, and take $\Sigma = \{(x, 0) \in \mathbb{R}^2 \mid x < 0\}$. The phase portrait of the first part of the unperturbed problem $(I.3.32)_0$ considered in the whole plane consists of concentric circles with the common center at $(0, -\delta)$. In affine polar coordinates $x = \rho \cos\phi$, $y = \rho \sin\phi + \delta$, the second part of $(I.3.32)_0$ has the form

$$\dot{\rho} = \frac{1}{4}\rho(\rho^2 - 2)^2, \quad \dot{\phi} = -1 + \rho^2 - \frac{\rho^4}{2}.$$

So it possesses only one periodic solution which is described in the next statement.

Lemma I.3.8. *The unperturbed system* $(I.3.32)_0$ *possesses a unique periodic solution* *given by*

$$\gamma(t) = \begin{cases} \left(-\sqrt{2-\delta^2}\cos t + \delta\sin t, -\delta + \sqrt{2-\delta^2}\sin t + \delta\cos t\right) & \text{if } t \in [0, t_1], \\ \left(\sqrt{2-\delta^2}\cos(t-t_1) - \delta\sin(t-t_1),\right. \\ \qquad \left. \delta - \sqrt{2-\delta^2}\sin(t-t_1) - \delta\cos(t-t_1)\right) & \text{if } t \in [t_1, T] \end{cases}$$

$$\text{(I.3.33)}$$

with

$$t_1 = \arccos(\delta^2 - 1), \qquad T = 2t_1$$

and

$$\bar{x}_0 = \left(-\sqrt{2-\delta^2}, 0\right), \qquad \bar{x}_1 = \left(\sqrt{2-\delta^2}, 0\right).$$

Now we should calculate fundamental matrices $X_1^\xi(t)$, $X_2^\xi(t)$, saltation matrix $S_1(\xi)$ and projection S^ξ for general $\xi \in \Sigma$ close to $\bar{x}_0$. However, it is sufficient to derive the formula $D_\xi((\mathbb{I} - S^\xi)A(\xi, 0)\xi_1)\xi_1$ at $\xi = \bar{x}_0$ and for $\xi_1 \in \Sigma$. This formula is rather awkward and can be found at the end of this chapter. We underline that it is enough to derive $X_1(t)$, $X_2(t)$, S_1, and S, i.e. the matrices evaluated at $\xi = \bar{x}_0$. Nevertheless, Diliberto's Theorem I.1.10 has to be used to obtain the fundamental matrix $X_2(t)$. Applying this theorem, we get the mentioned matrices.

Lemma I.3.9. *System* (I.3.32)$_0$ *has fundamental matrices* $X_1(t)$, $X_2(t)$ *along* $\gamma(t)$ *of* (I.3.33), *satisfying* (I.3.10), (I.3.11) *for* $\xi = \bar{x}_0$, *respectively, given by*

$$X_1(t) = \begin{pmatrix} \cos t & \sin t \\ -\sin t & \cos t \end{pmatrix}, \qquad X_2(t) = \widetilde{X}_2(t)\widetilde{X}_2^{-1}(t_1)$$

where

$$\widetilde{X}_2(t) = \begin{pmatrix} -\sqrt{2-\delta^2}\sin(t-t_1) - \delta\cos(t-t_1) & A\sin(t-t_1) + B\cos(t-t_1) \\ -\sqrt{2-\delta^2}\cos(t-t_1) + \delta\sin(t-t_1) & A\cos(t-t_1) - B\sin(t-t_1) \end{pmatrix},$$

$$A = -4\sqrt{2-\delta^2}(t-t_1) - \delta, \qquad B = \sqrt{2-\delta^2} - 4\delta(t-t_1).$$

Saltation matrix S_1 *of* (I.3.31) *at* $\xi = \bar{x}_0$ *has the form*

$$S_1 = \begin{pmatrix} 1 & \frac{2\delta}{\sqrt{2-\delta^2}} \\ 0 & 1 \end{pmatrix}.$$

Projection S *defined by* (I.3.29) *at* $\xi = \bar{x}_0$ *is*

$$S = \begin{pmatrix} 0 & -\frac{\delta}{\sqrt{2-\delta^2}} \\ 0 & 1 \end{pmatrix}.$$

Proof. Matrices S_1 and S are obtained directly from their definitions. Since the first part of $(I.3.32)_0$ is linear, $X_1(t)$ is the matrix solutions of this system. Using Theorem I.1.10 we derive matrix $\widetilde{X}_2(t)$ and, consequently, $X_2(t)$ which has to fulfill $X_2(t_1) = \mathbb{I}$. □

Using the previous Lemma I.3.9, we derive

$$(\mathbb{I} - S^{\bar{x}_0})A(\bar{x}_0, 0) = \begin{pmatrix} 1 & -\frac{\delta}{\sqrt{2-\delta^2}} \\ 0 & 0 \end{pmatrix}, \quad \psi_1 = (1, 0)^*.$$

So (cf. Lemma I.3.4)

$$P_\xi(\bar{x}_0, 0, \mu) = (\mathbb{I} - S^{\bar{x}_0})A(\bar{x}_0, 0)\big|_{T_{\bar{x}_0}\Sigma} = 1$$

for the derivative of the Poincaré mapping and $|_{T_{\bar{x}_0}\Sigma}$ denoting the restriction onto the tangent space to Σ at $\bar{x}_0$.

For arbitrary $0 < \delta < \sqrt{2}$ the Poincaré-Andronov-Melnikov function is rather awkward. Therefore, we fix parameter $\delta = 1$, so after some algebra we get

$$M^\mu_\pm(u) = \pm G(\mu) - \pi u^2, \quad \mu = (\mu_1, \mu_2),$$
$$G(\mu) = \frac{68 - 135\pi + 30\pi^2}{24} + \frac{8 - 19\pi + 6\pi^2}{8}\mu_1 + \frac{28 - 65\pi + 18\pi^2}{24}\mu_2. \tag{I.3.34}$$

For simplicity, we shortened

$$M^\mu_\pm : T_{\bar{x}_0}\Sigma = \mathbb{R} \times \{0\} \to T_{\bar{x}_0}\Sigma, \qquad \xi_1 = (u, 0) \mapsto M^\mu_\pm(\xi_1)$$

to

$$M^\mu_\pm : \mathbb{R} \to \mathbb{R}, \qquad u \mapsto M^\mu_\pm(u).$$

Theorem I.3.5 implies the existence of periodic solutions after perturbation.

Proposition I.3.10. *Let $\mu^0 = (\mu_1^0, \mu_2^0)$ be such that $G(\mu^0) \neq 0$ for G given by (I.3.34). Then equation $(I.3.32)_\varepsilon$ with $\delta = 1$ has exactly two (zero) periodic solutions orbitally close to γ for $\varepsilon \neq 0$ sufficiently small with $G(\mu^0)\varepsilon > 0$ $(G(\mu^0)\varepsilon < 0)$ and μ close to μ^0.*

Proof. Let $G(\mu^0) > 0$. Then

$$u_\pm = \pm\sqrt{\frac{G(\mu^0)}{\pi}}$$

are simple roots of $M^{\mu^0}_+(u) = 0$ for M given by (I.3.34). Similarly, if $G(\mu^0) < 0$ then

$$v_\pm = \pm\sqrt{-\frac{G(\mu^0)}{\pi}}$$

are simple roots of $M_-^{\mu^0}(u) = 0$. Consequently, the statement follows from Theorem I.3.5.   □

One can compute that in this case $D_{\xi\xi}P(x_0, 0, \mu)uv = -2\pi uv$. So, from Corollary I.3.7 we get the result on stability of the persisting orbits (cf. Figure I.3.2). For simplicity, we denote $\gamma_{u_\pm}^\varepsilon(t)$ and $\gamma_{v_\pm}^\varepsilon(t)$ the persisting solutions containing points $x_0 + \varepsilon(u_\pm, 0) + O(\varepsilon^2)$ and $x_0 + \varepsilon(v_\pm, 0) + O(\varepsilon^2)$, respectively.

Proposition I.3.11. *Let μ^0 be such that $G(\mu^0) > 0$ $(G(\mu^0) < 0)$. Then $\gamma_{u_+}^\varepsilon(t)$ is stable and $\gamma_{u_-}^\varepsilon(t)$ is repelling $(\gamma_{v_+}^\varepsilon(t)$ is repelling and $\gamma_{v_-}^\varepsilon(t)$ is stable).*

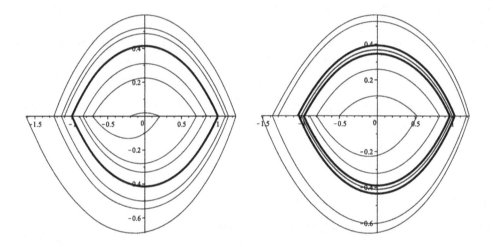

Figure I.3.2 Numerically computed trajectory of $(I.3.32)_\varepsilon$ with $\delta = 1$, $\mu_1 = 3$, $\mu_2 = 4$ and $\varepsilon = 0$ on the left, $\varepsilon = 0.1$ on the right. Periodic orbits are denoted with dark. In the perturbed case $G(\mu^0) \doteq 0.563$, $u_+ \doteq -0.958$, $u_- \doteq -1.042$

Related applications are presented in [4, 7, 22].

I.3.4. Formulae for the second derivatives

We calculate

$$D_\xi((\mathbb{I} - S^\xi)A(\xi, 0)u)_{\xi=x_0}u$$
$$= D_\xi\left((\mathbb{I} - S^\xi)X_3^\xi(t_3(\xi))S_2^\xi X_2^\xi(t_2(\xi))S_1^\xi X_1^\xi(t_1(\xi))u\right)_{\xi=x_0} u$$

as the derivative of a product, where

$$D_\xi \left[X_1^\xi(t_1(\xi)) \right]_{x_0} u = Y_1(t_1)u + Df_+(x_1)X_1(t_1)Dt_1(x_0)u,$$

$$D_\xi \left[X_2^\xi(t_2(\xi)) \right]_{x_0} u = Y_2(t_2)S_1X_1(t_1)u - X_2(t_2)Df_-(x_1)Dt_1(x_0)u$$
$$+ Df_-(x_2)X_2(t_2)Dt_2(x_0)u,$$

$$D_\xi \left[X_3^\xi(t_3(\xi)) \right]_{x_0} u = Y_3(T)S_2X_2(t_2)S_1X_1(t_1)u - X_3(T)Df_+(x_2)Dt_2(x_0)u$$
$$+ Df_+(x_0)X_3(T)Dt_3(x_0)u,$$

$$D S_1^\xi \Big|_{x_0} u = \frac{1}{(Dh(x_1)f_+(x_1))^2}[[(Df_-(x_1) - Df_+(x_1))Dx_1(x_0)uDh(x_1)$$
$$+ (f_-(x_1) - f_+(x_1))D^2h(x_1)Dx_1(x_0)u]Dh(x_1)f_+(x_1)$$
$$- (f_-(x_1) - f_+(x_1))Dh(x_1)[D^2h(x_1)Dx_1(x_0)uf_+(x_1)$$
$$+ Dh(x_1)Df_+(x_1)Dx_1(x_0)u]],$$

$$D S_2^\xi \Big|_{x_0} u = \frac{1}{(Dh(x_2)f_-(x_2))^2}[[(Df_+(x_2) - Df_-(x_2))Dx_2(x_0)uDh(x_2)$$
$$+ (f_+(x_2) - f_-(x_2))D^2h(x_2)Dx_2(x_0)u]Dh(x_2)f_-(x_2)$$
$$- (f_+(x_2) - f_-(x_2))Dh(x_2)[D^2h(x_2)Dx_2(x_0)uf_-(x_2)$$
$$+ Dh(x_2)Df_-(x_2)Dx_2(x_0)u]],$$

$$D S^\xi \Big|_{x_0} u = \frac{(\|f_+(x_0)\|^2 \mathbb{I} - f_+(x_0)f_+(x_0)^*)Df_+(x_0)Dx_3(x_0)uf_+(x_0)^*}{\|f_+(x_0)\|^4},$$

$$Dx_1(x_0)u = X_1(t_1)u + f_+(x_1)Dt_1(x_0)u,$$
$$Dx_2(x_0)u = X_2(t_2)S_1X_1(t_1)u + f_-(x_2)Dt_2(x_0)u,$$
$$Dx_3(x_0)u = (\mathbb{I} - S)A(x_0, 0)u,$$

$$Dt_1(x_0)u = -\frac{Dh(x_1)X_1(t_1)u}{Dh(x_1)f_+(x_1)}, \qquad Dt_2(x_0)u = -\frac{Dh(x_2)X_2(t_2)S_1X_1(t_1)u}{Dh(x_2)f_-(x_2)},$$

$$Dt_3(x_0)u = -\frac{f_+(x_0)^*X_3(T)S_2X_2(t_2)S_1X_1(t_1)u}{\|f_+(x_0)\|^2},$$

$$Y_1(t)uv = X_1(t) \int_0^t X_1^{-1}(s) \mathrm{D}^2 f_+(\gamma(s)) X_1(s) u X_1(s) v \, ds, \quad t \in [0, t_1],$$

$$Y_2(t)uv = X_2(t) \int_{t_1}^t X_2^{-1}(s) \mathrm{D}^2 f_-(\gamma(s)) X_2(s) u X_2(s) v \, ds, \quad t \in [t_1, t_2],$$

$$Y_3(t)uv = X_3(t) \int_{t_2}^t X_3^{-1}(s) \mathrm{D}^2 f_+(\gamma(s)) X_3(s) u X_3(s) v \, ds, \quad t \in [t_2, T].$$

CHAPTER I.4

Sliding solution of periodically perturbed systems

I.4.1. Setting of the problem and main results

Until now, we always assumed that the periodic trajectory of an unperturbed system transversally crosses the discontinuity boundary and later returns back, again transversally through the boundary. This time, we shall investigate the persistence of periodic trajectories that after a transverse impact remain on the boundary for some time and then return into the original region. We consider a discontinuous differential equation with a time periodic nonautonomous perturbation as in Chapter I.1. Related problems are studied in many works such as [2, 4, 5, 7, 8, 34–38].

Let $\Omega \subset \mathbb{R}^n$ be an open set in $\mathbb{R}^n$ and $h(x)$ be a C^r-function on $\overline{\Omega}$, with $r \geq 3$. We set $\Omega_{\pm} := \{x \in \Omega \mid \pm h(x) > 0\}$, $\Omega_0 := \{x \in \Omega \mid h(x) = 0\}$ for a regular value 0 of h. Let $f_{\pm} \in C_b^r(\overline{\Omega})$, $g_{\pm} \in C_b^r(\overline{\Omega} \times \mathbb{R} \times \mathbb{R} \times \mathbb{R}^p)$ and $h \in C_b^r(\overline{\Omega}, \mathbb{R})$. Moreover, let $g_{\pm}$ be T-periodic in t. Let $\varepsilon \in \mathbb{R}$ and $\mu \in \mathbb{R}^p$, $p \geq 1$ be parameters.

Definition I.4.1. We say that a function $x(t)$ is a sliding solution of the equation

$$\dot{x} = f_{\pm}(x) + \varepsilon g_{\pm}(x, t + \alpha, \varepsilon, \mu), \quad x \in \overline{\Omega}_{\pm}, \tag{I.4.1}$$

if it is continuous, piecewise C^1, satisfies equation (I.4.1) on $\Omega_{\pm}$, sliding mode equation

$$\dot{x} = F_0(x, t + \alpha, \varepsilon, \mu) \tag{I.4.2}$$

on Ω_0, where

$$F_0(x, t, \varepsilon, \mu) = (1 - \beta(x, t, \varepsilon, \mu))F_-(x, t, \varepsilon, \mu) + \beta(x, t, \varepsilon, \mu)F_+(x, t, \varepsilon, \mu),$$

$$F_{\pm}(x, t, \varepsilon, \mu) = f_{\pm}(x) + \varepsilon g_{\pm}(x, t, \varepsilon, \mu),$$

$$\beta(x, t, \varepsilon, \mu) = \frac{Dh(x)F_-(x, t, \varepsilon, \mu)}{Dh(x)(F_-(x, t, \varepsilon, \mu) - F_+(x, t, \varepsilon, \mu))}$$

(see [5]) and, moreover, the following holds: if $x(t_0) \in \Omega_0$ and there exists $\rho_1 > 0$ such that for any $0 < \rho < \rho_1$ we have $x(t_0 - \rho) \in \Omega_{\pm}$, then there exists $\rho_2 > 0$ such that $x(t_0 + \rho) \in \Omega_0$ for any $0 < \rho < \rho_2$ (the solution remains on Ω_0 for some nonzero time).

Note that the "sliding" of a sliding solution is assured since $Dh(x)F_0(x, t + \alpha, \varepsilon, \mu) =$

Poincaré-Andronov-Melnikov Analysis for Non-Smooth Systems.
http://dx.doi.org/10.1016/B978-0-12-804294-6.50006-0

0 for $x \in \Omega_0$.

Let us assume

H1) For $\varepsilon = 0$ equation (I.4.1) has a T-periodic orbit $\gamma(t)$. The orbit is given by its initial point $x_0 \in \Omega_+$ and consists of three branches

$$\gamma(t) = \begin{cases} \gamma_1(t) & \text{if } t \in [0, t_1], \\ \gamma_2(t) & \text{if } t \in [t_1, t_2], \\ \gamma_3(t) & \text{if } t \in [t_2, T], \end{cases} \tag{I.4.3}$$

where $0 < t_1 < t_2 < T$, $\gamma_1(t) \in \Omega_+$ for $t \in [0, t_1)$, $\gamma_2(t) \in \Omega_0$ for $t \in [t_1, t_2]$, $\gamma_3(t) \in \Omega_+$ for $t \in (t_2, T]$, and

$$\begin{aligned} x_1 &:= \gamma_1(t_1) = \gamma_2(t_1) \in \Omega_0, \\ x_2 &:= \gamma_2(t_2) = \gamma_3(t_2) \in \Omega_0, \\ x_0 &:= \gamma_3(T) = \gamma_1(0) \in \Omega_+. \end{aligned} \tag{I.4.4}$$

H2) Moreover, we also assume that

$$\begin{aligned} Dh(x)(f_-(x) - f_+(x)) &> 0 & \text{if } x \in \Omega_0, \\ Dh(\gamma(t))f_+(\gamma(t)) &< 0 & \text{if } t \in [t_1, t_2), \\ Dh(\gamma(t))f_-(\gamma(t)) &> 0 & \text{if } t \in [t_1, t_2], \end{aligned}$$

$$Dh(x_2)f_+(x_2) = 0, \qquad D_s^2[h(\gamma(t_2 + s))]_{s=0^+} \neq 0.$$

Later, it will be seen that it is sufficient to assume the first inequality in H2) to be satisfied only in the neighborhood of $\{\gamma(t) \mid t \in [t_1, t_2]\}$. Next, from H2) it follows that $Dh(x)(F_-(x, t, \varepsilon, \mu) - F_+(x, t, \varepsilon, \mu)) > 0$ for any $\varepsilon \neq 0$ sufficiently close to 0. Hence $F_0(x, t, \varepsilon, \mu)$ is well-defined and we get

$$F_0(x, t, \varepsilon, \mu) = f_0(x) + \varepsilon g_0(x, t, \varepsilon, \mu)$$

where

$$f_0(x) = \frac{f_+(x)Dh(x)f_-(x) - f_-(x)Dh(x)f_+(x)}{Dh(x)(f_-(x) - f_+(x))}$$

and

$$g_0(x, t, \varepsilon, \mu) = \frac{1}{[Dh(x)(f_-(x) - f_+(x))]^2}[(f_+(x) - f_-(x))$$

$$\times (Dh(x)g_+(x, t, 0, \mu)Dh(x)f_-(x) - Dh(x)g_-(x, t, 0, \mu)Dh(x)f_+(x))$$

$$+ (g_+(x, t, 0, \mu)Dh(x)f_-(x) - g_-(x, t, 0, \mu)Dh(x)f_+(x))Dh(x)(f_-(x) - f_+(x))] + O(\varepsilon).$$

Denote

$$K_{\varepsilon, \mu, \alpha} = \{x \in \Omega_0 \mid x \sim x_2, \ Dh(x)F_+(x, \varepsilon, \mu, \alpha) = 0\}.$$

Remark I.4.2.

1. We obtain that $K = K_{0,\mu,\alpha}$ is a C^{r-1}-submanifold of Ω_0 of codimension 1 in a neighborhood of x_2. So is $K_{\varepsilon,\mu,\alpha}$ for ε sufficiently small.
2. Since $\dot{\gamma}(t_2^-) = f_0(x_2) = f_+(x_2) = \dot{\gamma}(t_2^+)$ where $\dot{\gamma}(t^{\pm}) = \lim_{s \to t^{\pm}} \dot{\gamma}(s)$, then $\gamma(t)$ is C^1-smooth at $t = t_2$.
3. From identity

$$D_s^2[h(\gamma(t_2 + s))]_{s=0^+} = D_s[Dh(\gamma(t_2 + s))f_+(\gamma(t_2 + s))]_{s=0^+}$$

$$= D^2h(x_2)f_+(x_2)f_+(x_2) + Dh(x_2)Df_+(x_2)f_+(x_2) = D_x[Dh(x)f_+(x)]_{x=x_2}f_0(x_2)$$

we get that $\gamma(t)$ crosses K transversally. Then clearly the solution of the perturbed equation close to $\gamma(t)$ crosses $K_{\varepsilon,\mu,\alpha}$ transversally. Moreover, assumptions H1), H2) imply that

$$D_s^2[h(\gamma(t_2 + s))]_{s=0^+} > 0.$$

Let $x_i(\tau, \xi)(t, \varepsilon, \mu, \alpha)$ denote the solution of the initial value problem

$$\dot{x} = f_i(x) + \varepsilon g_i(x, t + \alpha, \varepsilon, \mu)$$
$$x(\tau) = \xi \qquad\qquad\qquad\qquad (I.4.5)$$

with i being an element of a set of lower indices $\{+, -, 0\}$. First, we modify Lemma I.1.2 for the case of a sliding trajectory and show the existence of a sliding Poincaré mapping.

Lemma I.4.3. *Assume* H1) *and* H2). *Then there exist* $\varepsilon_0, r_0 > 0$ *and a Poincaré mapping (cf. Figure I.4.1)*

$$P(\cdot, \varepsilon, \mu, \alpha): B(x_0, r_0) \to \Sigma$$

for all fixed $\varepsilon \in (-\varepsilon_0, \varepsilon_0)$, $\mu \in \mathbb{R}^p$, $\alpha \in \mathbb{R}$, *where*

$$\Sigma = \{y \in \mathbb{R}^n \mid \langle y - x_0, f_+(x_0)\rangle = 0\}.$$

Moreover, $P: B(x_0, r_0) \times (-\varepsilon_0, \varepsilon_0) \times \mathbb{R}^p \times \mathbb{R} \to \mathbb{R}^n$ *is* C^{r-1}-*smooth in all arguments and* $B(x_0, r_0) \subset \Omega_+$.

Proof. Implicit function theorem (IFT) yields the existence of positive constants τ_1, $r_1, \delta_1, \varepsilon_1$ and C^r-function

$$t_1(\cdot, \cdot, \cdot, \cdot, \cdot): (-\tau_1, \tau_1) \times B(x_0, r_1) \times (-\varepsilon_1, \varepsilon_1) \times \mathbb{R}^p \times \mathbb{R} \to (t_1 - \delta_1, t_1 + \delta_1)$$

such that $h(x_+(\tau, \xi)(t, \varepsilon, \mu, \alpha)) = 0$ for $\tau \in (-\tau_1, \tau_1)$, $\xi \in B(x_0, r_1) \subset \Omega_+$, $\varepsilon \in (-\varepsilon_1, \varepsilon_1)$, $\mu \in \mathbb{R}^p$, $\alpha \in \mathbb{R}$ and $t \in (t_1 - \delta_1, t_1 + \delta_1)$ if and only if $t = t_1(\tau, \xi, \varepsilon, \mu, \alpha)$. Moreover,

$t_1(0, x_0, 0, \mu, \alpha) = t_1$. Now, since

$$D_s h(x_0(t_1(0, x_0, 0, \mu, \alpha), x_+(0, x_0)(t_1(0, x_0, 0, \mu, \alpha), 0, \mu, \alpha)), 0, \mu, \alpha))(t_2 + s, 0, \mu, \alpha))_{s=0^+} = 0,$$
$$D_s^2 h(x_0(t_1(0, x_0, 0, \mu, \alpha), x_+(0, x_0)(t_1(0, x_0, 0, \mu, \alpha), 0, \mu, \alpha)), 0, \mu, \alpha))(t_2 + s, 0, \mu, \alpha))_{s=0^+} > 0,$$

IFT gives the existence of positive constants $\tau_2, r_2, \delta_2, \varepsilon_2$ and C^{r-1}-function

$$t_2(\cdot, \cdot, \cdot, \cdot, \cdot): (-\tau_2, \tau_2) \times B(x_0, r_2) \times (-\varepsilon_2, \varepsilon_2) \times \mathbb{R}^p \times \mathbb{R} \to (t_2 - \delta_2, t_2 + \delta_2)$$

such that

$$x_0(t_1(\tau, \xi, \varepsilon, \mu, \alpha), x_+(\tau, \xi)(t_1(\tau, \xi, \varepsilon, \mu, \alpha), \varepsilon, \mu, \alpha))(t, \varepsilon, \mu, \alpha) \in K_{\varepsilon, \mu, \alpha},$$

i.e.

$$D_s h(x_0(t_1(\tau, \xi, \varepsilon, \mu, \alpha), x_+(\tau, \xi)(t_1(\tau, \xi, \varepsilon, \mu, \alpha), \varepsilon, \mu, \alpha)), \varepsilon, \mu, \alpha))(t + s, \varepsilon, \mu, \alpha))_{s=0^+} = 0,$$
$$D_s^2 h(x_0(t_1(\tau, \xi, \varepsilon, \mu, \alpha), x_+(\tau, \xi)(t_1(\tau, \xi, \varepsilon, \mu, \alpha), \varepsilon, \mu, \alpha)), \varepsilon, \mu, \alpha))(t + s, \varepsilon, \mu, \alpha))_{s=0^+} > 0$$

for $\tau \in (-\tau_2, \tau_2)$, $\xi \in B(x_0, r_2) \subset B(x_0, r_1)$, $\varepsilon \in (-\varepsilon_2, \varepsilon_2)$, $\mu \in \mathbb{R}^p$, $\alpha \in \mathbb{R}$ and $t \in (t_2 - \delta_2, t_2 + \delta_2)$ if and only if $t = t_2(\tau, \xi, \varepsilon, \mu, \alpha)$. Moreover, $t_2(0, x_0, 0, \mu, \alpha) = t_2$. Similarly to $t_1(\tau, \xi, \varepsilon, \mu, \alpha)$, we obtain positive constants $\tau_0, r_0, \delta_0, \varepsilon_0$ and C^{r-1}-function

$$t_3(\cdot, \cdot, \cdot, \cdot, \cdot): (-\tau_0, \tau_0) \times B(x_0, r_0) \times (-\varepsilon_0, \varepsilon_0) \times \mathbb{R}^p \times \mathbb{R} \to (T - \delta_0, T + \delta_0)$$

satisfying

$$\langle x_+(t_2(\tau, \xi, \varepsilon, \mu, \alpha), x_0(t_1(\tau, \xi, \varepsilon, \mu, \alpha), x_+(\tau, \xi)(t_1(\tau, \xi, \varepsilon, \mu, \alpha), \varepsilon, \mu, \alpha))$$
$$(t_2(\tau, \xi, \varepsilon, \mu, \alpha), \varepsilon, \mu, \alpha))(t_3(\tau, \xi, \varepsilon, \mu, \alpha), \varepsilon, \mu, \alpha) - x_0, f_+(x_0) \rangle = 0$$

for any $\tau \in (-\tau_0, \tau_0)$, $\xi \in B(x_0, r_0) \subset B(x_0, r_2)$, $\varepsilon \in (-\varepsilon_0, \varepsilon_0)$, $\mu \in \mathbb{R}^p$, $\alpha \in \mathbb{R}$. In addition, $t_3(0, x_0, 0, \mu, \alpha) = T$. Consequently, the Poincaré mapping is defined as

$$P(\xi, \varepsilon, \mu, \alpha) = x_+(t_2(0, \xi, \varepsilon, \mu, \alpha), x_0(t_1(0, \xi, \varepsilon, \mu, \alpha), x_+(0, \xi)(t_1(0, \xi, \varepsilon, \mu, \alpha), \varepsilon, \mu, \alpha))$$
$$(t_2(0, \xi, \varepsilon, \mu, \alpha), \varepsilon, \mu, \alpha))(t_3(0, \xi, \varepsilon, \mu, \alpha), \varepsilon, \mu, \alpha).$$

□

To find T-periodic solutions of the perturbed equation (I.4.1) close to $\gamma(\cdot)$, we solve the couple of equations

$$P(\xi, \varepsilon, \mu, \alpha) = \xi,$$
$$t_3(0, \xi, \varepsilon, \mu, \alpha) = T.$$

We reduce this problem to the single equation

$$F(x, \varepsilon, \mu, \alpha) := x - \widetilde{P}(x, \varepsilon, \mu, \alpha) = 0 \tag{I.4.6}$$

for $(x, \alpha) \in \Sigma \times \mathbb{R}$, $x \sim x_0$ and parameters $\varepsilon \sim 0$, $\mu \in \mathbb{R}^p$ by introducing the strobo-

Figure I.4.1 Sliding Poincaré mapping

scopic Poincaré mapping (cf. (I.1.12))

$$\widetilde{P}(\xi, \varepsilon, \mu, \alpha) = x_+(t_2(0, \xi, \varepsilon, \mu, \alpha), x_0(t_1(0, \xi, \varepsilon, \mu, \alpha), x_+(0, \xi)(t_1(0, \xi, \varepsilon, \mu, \alpha), \varepsilon, \mu, \alpha))$$
$$(t_2(0, \xi, \varepsilon, \mu, \alpha), \varepsilon, \mu, \alpha))(T, \varepsilon, \mu, \alpha).$$

(I.4.7)

Properties of the mapping are concluded in the following lemma.

Lemma I.4.4. *Let* $\widetilde{P}(\xi, \varepsilon, \mu, \alpha)$ *be defined by* (I.4.7). *Then its derivatives fulfill*

$$D_\xi \widetilde{P}(x_0, 0, \mu, \alpha) = A(0),$$

(I.4.8)

$$D_\varepsilon \widetilde{P}(x_0, 0, \mu, \alpha) = \int_0^T A(s)g(\gamma(s), s + \alpha, \mu)ds,$$

(I.4.9)

where

$$g(x, t, \mu) = \begin{cases} g_+(x, t, 0, \mu) & \text{if } x \in \Omega_+, \\ g_0(x, t, 0, \mu) & \text{if } x \in \Omega_0, \end{cases}$$

(I.4.10)

A(t) is given by

$$A(t) = \begin{cases} X_3(T)X_2(t_2)S X_1(t_1)X_1(t)^{-1} & \text{if } t \in [0, t_1), \\ X_3(T)X_2(t_2)X_2(t)^{-1} & \text{if } t \in [t_1, t_2), \\ X_3(T)X_3(t)^{-1} & \text{if } t \in [t_2, T] \end{cases}$$

(I.4.11)

with saltation matrix

$$S = \mathbb{I} + \frac{(f_0(x_1) - f_+(x_1))Dh(x_1)}{Dh(x_1)f_+(x_1)}$$

(I.4.12)

and fundamental matrix solutions $X_1(t)$, $X_2(t)$, $X_3(t)$ satisfying, respectively,

$$\dot{X}_1(t) = Df_+(\gamma(t))X_1(t) \qquad \dot{X}_2(t) = Df_0(\gamma(t))X_2(t) \qquad \dot{X}_3(t) = Df_+(\gamma(t))X_3(t)$$
$$X_1(0) = \mathbb{I}, \qquad\qquad\qquad X_2(t_1) = \mathbb{I}, \qquad\qquad\qquad X_3(t_2) = \mathbb{I}.$$

(I.4.13)

In addition, $D_\xi \widetilde{P}(x_0, 0, \mu, \alpha)$ has an eigenvalue 1 with corresponding eigenvector $f_+(x_0)$, i.e.

$$D_\xi \widetilde{P}(x_0, 0, \mu, \alpha) f_+(x_0) = f_+(x_0).$$

Proof. Analogously to Chapter I.1 we derive the following identities

$$D_\xi x_+(0, x_0)(t, 0, \mu, \alpha) = X_1(t),$$

$$D_\varepsilon x_+(0, x_0)(t, 0, \mu, \alpha) = \int_0^t X_1(t)X_1^{-1}(s)g_+(\gamma(s), s + \alpha, 0, \mu)ds$$

for $t \in [0, t_1]$,

$$D_\xi x_0(t_1, x_1)(t, 0, \mu, \alpha) = X_2(t), \qquad D_\tau x_0(t_1, x_1)(t, 0, \mu, \alpha) = -X_2(t)f_0(x_1),$$

$$D_\varepsilon x_0(t_1, x_1)(t, 0, \mu, \alpha) = \int_{t_1}^t X_2(t)X_2^{-1}(s)g_0(\gamma(s), s + \alpha, 0, \mu)ds$$

for $t \in [t_1, t_2]$,

$$D_\xi x_+(t_2, x_2)(t, 0, \mu, \alpha) = X_3(t), \qquad D_\tau x_+(t_2, x_2)(t, 0, \mu, \alpha) = -X_3(t)f_+(x_2),$$

$$D_\varepsilon x_+(t_2, x_2)(t, 0, \mu, \alpha) = \int_{t_2}^t X_3(t)X_3^{-1}(s)g_+(\gamma(s), s + \alpha, 0, \mu)ds$$

for $t \in [t_2, T]$, and for times

$$D_\xi t_1(0, x_0, 0, \mu, \alpha) = -\frac{Dh(x_1)X_1(t_1)}{Dh(x_1)f_+(x_1)},$$

$$D_\varepsilon t_1(0, x_0, 0, \mu, \alpha) = -\frac{Dh(x_1) \int_0^{t_1} X_1(t_1)X_1^{-1}(s)g_+(\gamma(s), s + \alpha, 0, \mu)ds}{Dh(x_1)f_+(x_1)},$$

$$D_\xi t_2(0, x_0, 0, \mu, \alpha) = -\frac{(D^2 h(x_2) f_0(x_2) + Dh(x_2) D f_0(x_2)) X_2(t_2) S X_1(t_1)}{D^2 h(x_2) f_0(x_2) f_0(x_2) + Dh(x_2) D f_0(x_2) f_0(x_2)},$$

$$D_\varepsilon t_2(0, x_0, 0, \mu, \alpha) = -\frac{1}{D^2 h(x_2) f_0(x_2) f_0(x_2) + Dh(x_2) D f_0(x_2) f_0(x_2)}$$

$$\times \left[(D^2 h(x_2) f_0(x_2) + Dh(x_2) D f_0(x_2)) \left(\int_{t_1}^{t_2} X_2(t_2) X_2^{-1}(s) g_0(\gamma(s), s + \alpha, 0, \mu) ds \right. \right.$$

$$\left. \left. + \int_0^{t_1} X_2(t_2) S X_1(t_1) X_1^{-1}(s) g_+(\gamma(s), s + \alpha, 0, \mu) ds \right) + Dh(x_2) g_0(x_2, t_2 + \alpha, 0, \mu) \right].$$

Differentiating (I.4.7) with respect to ξ, or ε, we get the statements on the derivatives. Now let t be sufficiently small. Since

$$x_+(0, x_+(0, x_0)(t, 0, \mu, \alpha))(t_1(0, x_+(0, x_0)(t, 0, \mu, \alpha), 0, \mu, \alpha), 0, \mu, \alpha)$$

$$= x_+(0, x_0)(t_1(0, x_+(0, x_0)(t, 0, \mu, \alpha), 0, \mu, \alpha) + t, 0, \mu, \alpha)$$

is an element of Ω_0 and as well of $\{\gamma(t) \mid t \sim t_1\}$, we have

$$t_1(0, x(0, x_0)(t, 0, \mu, \alpha), 0, \mu, \alpha) + t = t_1$$

for all t close to 0. Consequently,

$$x_+(0, x_0)(t_1(0, x_+(0, x_0)(t, 0, \mu, \alpha), 0, \mu, \alpha) + t, 0, \mu, \alpha) = x_1.$$

Similarly, the left-hand side of the following identity is an element of K and the right-hand side is from $\{\gamma(t) \mid t \sim t_2\}$:

$$x_0(t_1(0, x_+(0, x_0)(t, 0, \mu, \alpha), 0, \mu, \alpha), x_+(0, x_+(0, x_0)(t, 0, \mu, \alpha))$$

$$(t_1(0, x_+(0, x_0)(t, 0, \mu, \alpha), 0, \mu, \alpha), 0, \mu, \alpha))(t_2(0, x_+(0, x_0)(t, 0, \mu, \alpha), 0, \mu, \alpha), 0, \mu, \alpha)$$

$$= x_0(t_1(0, x_+(0, x_0)(t, 0, \mu, \alpha), 0, \mu, \alpha), x_1)(t_2(0, x_+(0, x_0)(t, 0, \mu, \alpha), 0, \mu, \alpha), 0, \mu, \alpha)$$

$$= x_0(t_1 - t, x_1)(t_2(0, x_+(0, x_0)(t, 0, \mu, \alpha), 0, \mu, \alpha), 0, \mu, \alpha)$$

$$= x_0(t_1, x_1)(t_2(0, x_+(0, x_0)(t, 0, \mu, \alpha), 0, \mu, \alpha) + t, 0, \mu, \alpha).$$

Therefore,

$$t_2(0, x_+(0, x_0)(t, 0, \mu, \alpha), 0, \mu, \alpha) + t = t_2$$

for all t close to 0, and

$$x_0(t_1, x_1)(t_2(0, x_+(0, x_0)(t, 0, \mu, \alpha), 0, \mu, \alpha) + t, 0, \mu, \alpha) = x_2.$$

Finally,

$$x_+(t_2(0, x_+(0, x_0)(t, 0, \mu, \alpha), 0, \mu, \alpha), x_0(t_1(0, x_+(0, x_0)(t, 0, \mu, \alpha), 0, \mu, \alpha),$$
$$x_+(0, x_+(0, x_0)(t, 0, \mu, \alpha))(t_1(0, x_+(0, x_0)(t, 0, \mu, \alpha), 0, \mu, \alpha), 0, \mu, \alpha))$$
$$(t_2(0, x_+(0, x_0)(t, 0, \mu, \alpha), 0, \mu, \alpha), 0, \mu, \alpha))(T, 0, \mu, \alpha)$$
$$= x_+(t_2(0, x_+(0, x_0)(t, 0, \mu, \alpha), 0, \mu, \alpha), x_2)(T, 0, \mu, \alpha)$$
$$= x_+(t_2 - t, x_2)(T, 0, \mu, \alpha) = x_+(t_2, x_2)(T + t, 0, \mu, \alpha).$$

Hence,

$$D_\xi \widetilde{P}(x_0, 0, \mu, \alpha) f_+(x_0) = D_t[\widetilde{P}(x_+(0, x_0)(t, 0, \mu, \alpha), 0, \mu, \alpha)]_{t=0}$$
$$= D_t[x_+(t_2, x_2)(T + t, 0, \mu, \alpha)]_{t=0} = f_+(x_0)$$

and the proof is finished.   □

Note that in the case of sliding the second saltation matrix (see (I.1.20), (I.2.12), (I.3.14)) S_2 has the form

$$S_2 = \mathbb{I} + \frac{(f_+(x_2) - f_0(x_2))(D^2 h(x_2) f_0(x_2) + Dh(x_2) Df_0(x_2))}{D^2 h(x_2) f_0(x_2) f_0(x_2) + Dh(x_2) Df_0(x_2) f_0(x_2)} = \mathbb{I}$$

since $f_+(x_2) = f_0(x_2)$. This corresponds to the regularity of $\gamma(t)$ at t_2 (see Remark I.4.2).

Now, we solve equation (I.4.6) using Lyapunov-Schmidt reduction method. We denote

$$Z = \mathcal{N} D_\xi F(x_0, 0, \mu, \alpha), \qquad Y = \mathcal{R} D_\xi F(x_0, 0, \mu, \alpha) \qquad (I.4.14)$$

the null space and the range of the operator $D_\xi F(x_0, 0, \mu, \alpha)$, and

$$Q: \mathbb{R}^n \to Y, \qquad \mathcal{P}: \mathbb{R}^n \to Y^\perp \qquad (I.4.15)$$

orthogonal projections onto Y and $Y^\perp$, respectively, where $Y^\perp$ is the orthogonal complement to Y in $\mathbb{R}^n$. From Lemma I.4.4 we know that $f_+(x_0) \in Z$. For simplicity, we take the third assumption, the so-called non-degeneracy condition
H3) $\mathcal{N} D_\xi F(x_0, 0, \mu, \alpha) = [f_+(x_0)]$.
Using the orthogonal projections, we split equation (I.4.6) into the couple of equations

$$QF(\xi, \varepsilon, \mu, \alpha) = 0,$$
$$\mathcal{P}F(\xi, \varepsilon, \mu, \alpha) = 0. \qquad (I.4.16)$$

The first one of these can be solved using IFT, since

$$QF(x_0, 0, \mu, \alpha) = 0$$

and $QD_\xi F(x_0, 0, \mu, \alpha)$ is an isomorphism $[f_+(x_0)]^\perp$ onto Y for all $(\mu, \alpha) \in \mathbb{R}^p \times \mathbb{R}$. Thus we get the existence of a unique C^{r-1}-function $\xi = \xi(\varepsilon, \mu, \alpha)$ for ε close to

0 and $(\mu, \alpha) \in \mathbb{R}^p \times \mathbb{R}$ satisfying $\mathcal{Q}F(\xi(\varepsilon, \mu, \alpha), \varepsilon, \mu, \alpha) = 0$ for all such $(\varepsilon, \mu, \alpha)$ and $\xi(0, \mu, \alpha) = x_0$. The second equation is the so-called bifurcation equation for $\alpha \in \mathbb{R}$,

$$\mathcal{P}F(\xi(\varepsilon, \mu, \alpha), \varepsilon, \mu, \alpha) = 0. \tag{I.4.17}$$

Let $Y^{\perp} = [\psi]$ for arbitrary and fixed ψ. Then we can write

$$\mathcal{P}u = \frac{\langle u, \psi \rangle \psi}{\|\psi\|^2}$$

and the bifurcation equation (I.4.17) gets the form

$$G(\varepsilon, \mu, \alpha) := \frac{\langle F(\xi(\varepsilon, \mu, \alpha), \varepsilon, \mu, \alpha), \psi \rangle \psi}{\|\psi\|^2} = 0. \tag{I.4.18}$$

Since $\xi(0, \mu, \alpha) = x_0$ and x_0 is a fixed point of $\widetilde{P}(\cdot, 0, \mu, \alpha)$, $G(0, \mu, \alpha) = 0$ for all $(\mu, \alpha) \in \mathbb{R}^p \times \mathbb{R}$. Next, we want the periodic orbit to persist for all $\varepsilon \neq 0$ small, so we set $D_\varepsilon G(0, \mu, \alpha) = 0$, i.e.

$$D_\varepsilon G(0, \mu, \alpha) = \frac{\langle (D_\xi F(x_0, 0, \mu, \alpha) D_\varepsilon \xi(0, \mu, \alpha) + D_\varepsilon F(x_0, 0, \mu, \alpha)), \psi \rangle \psi}{\|\psi\|^2}$$

$$= \frac{\langle D_\varepsilon F(x_0, 0, \mu, \alpha), \psi \rangle \psi}{\|\psi\|^2} = -\frac{\langle D_\varepsilon \widetilde{P}(x_0, 0, \mu, \alpha), \psi \rangle \psi}{\|\psi\|^2} = 0.$$

We define the sliding Poincaré-Andronov-Melnikov function as

$$M^\mu(\alpha) = \int_0^T \langle g(\gamma(s), s + \alpha, \mu), A^*(s)\psi \rangle ds \tag{I.4.19}$$

where $g(x, t, \mu)$ is given by (I.4.10) and

$$A^*(t) = \begin{cases} X_1^{-1*}(t) X_1^*(t_1) S^* X_2^*(t_2) X_3^*(T) & \text{if } t \in [0, t_1), \\ X_2^{-1*}(t) X_2^*(t_2) X_3^*(T) & \text{if } t \in [t_1, t_2), \\ X_3^{-1*}(t) X_3^*(T) & \text{if } t \in [t_2, T]. \end{cases} \tag{I.4.20}$$

We note that $A^*(t)\psi = X_1^{-1*}(t)\psi$ for any $t \in [0, t_1)$ and $D_\varepsilon G(0, \mu, \alpha) = -\frac{M^\mu(\alpha)\psi}{\|\psi\|^2}$. Linearization of equations (I.4.1), (I.4.2) with $\varepsilon = 0$ along $\gamma(t)$ gives the variational equation

$$\begin{aligned} \dot{x}(t) &= Df_+(\gamma(t))x(t) \quad \text{if } t \in [0, t_1) \cup [t_2, T], \\ \dot{x}(t) &= Df_0(\gamma(t))x(t) \quad \text{if } t \in [t_1, t_2) \end{aligned} \tag{I.4.21}$$

with the impulsive condition

$$x(t_1^+) = S x(t_1^-) \tag{I.4.22}$$

and the periodic condition

$$B(x(0) - x(T)) = 0 \qquad (I.4.23)$$

where $B = \frac{\psi\psi^*}{\|\psi\|^2}$ is the orthogonal projection onto $Y^\perp$. Note that C^1-smoothness of $\gamma(t)$ at t_2 corresponds to the second "impulsive" condition $x(t_2^+) = \mathbb{I}x(t_2^-)$. Due to definitions of $X_1(t)$, $X_2(t)$, $X_3(t)$ in (I.4.13), it is obvious that

$$X(t) = \begin{cases} X_1(t) & \text{if } t \in [0, t_1), \\ X_2(t)S\,X_1(t_1) & \text{if } t \in [t_1, t_2), \\ X_3(t)X_2(t_2)S\,X_1(t_1) & \text{if } t \in [t_2, T] \end{cases}$$

satisfies the variational equation (I.4.21) together with conditions (I.4.22), (I.4.23). Using the classical result – Lemma I.1.5 – and our result – Lemma I.2.4 – we shall derive the adjoint variational equation to (I.4.1), (I.4.2) with $\varepsilon = 0$ along $\gamma(t)$. One can simply set $B_1 = S$, $B_2 = \mathbb{I}$, $B_3 = B$ in Lemma I.2.4 to see that the adjoint variational system is given by the following linear impulsive boundary value problem

$$\begin{aligned}
\dot{x}(t) &= -Df_+^*(\gamma(t))x(t) & \text{if } t \in [0, t_1) \cup [t_2, T], \\
\dot{x}(t) &= -Df_0^*(\gamma(t))x(t) & \text{if } t \in [t_1, t_2), \\
x(t_1^-) &= S^* x(t_1^+), \\
x(T) &= x(0) \in Y^\perp.
\end{aligned} \qquad (I.4.24)$$

From the definition of $A(t)$ in (I.4.11) it is easy to see that $A^{-1}(t)$ solves the variational equation (I.4.21) with the impulsive condition (I.4.22). Then Lemma I.1.5 yields that $A^*(t)\psi$ solves the adjoint variational equation with the corresponding impulsive condition. In fact, it satisfies the boundary condition as well. Indeed, from Lemma I.4.4 for any $\xi \in [f_+(x_0)]$,

$$0 = \langle (\mathbb{I} - A(0))\xi, \psi \rangle = \langle \xi, (\mathbb{I} - A^*(0))\psi \rangle$$

and the same holds in the orthogonal complement to $[f_+(x_0)]$, i.e. if $\xi \in [f_+(x_0)]^\perp$ then

$$0 = \langle D_\xi F(x_0, 0, \mu, \alpha)\xi, \psi \rangle = \langle (\mathbb{I} - A(0))\xi, \psi \rangle = \langle \xi, (\mathbb{I} - A^*(0))\psi \rangle.$$

Consequently, we can take in (I.4.19) any solution of the adjoint variational system (I.4.24) instead of $A^*(t)\psi$. In conclusion, we get the main result.

Theorem I.4.5. *Let Y, $M^\mu(\alpha)$, g, $A^*(t)$ be given by (I.4.14), (I.4.19), (I.4.10), (I.4.20),* *respectively, and $\psi \in Y^\perp$ be arbitrary and fixed. If α_0 is a simple root of M^{μ_0}, i.e.*

$$\int_0^T \langle g(\gamma(s), s + \alpha_0, \mu_0), A^*(s)\psi \rangle ds = 0,$$

$$\int_0^T \langle D_t g(\gamma(s), s + \alpha_0, \mu_0), A^*(s)\psi \rangle ds \neq 0,$$

then there exists a unique C^{r-2}-function $\alpha(\varepsilon, \mu)$ for $\varepsilon \sim 0$, $\mu \sim \mu_0$ such that $\alpha(0, \mu_0) = \alpha_0$, and there is a unique T-periodic solution $x(\varepsilon, \mu)(t)$ of equation (I.4.1) with parameters ε, μ and $\alpha = \alpha(\varepsilon, \mu)$, which solves equation (I.4.2) on Ω_0 and is orbitally close to $\gamma(t)$, i.e. $|x(\varepsilon, \mu)(t) - \gamma(t - \alpha(\varepsilon, \mu))| = O(\varepsilon)$ for any $t \in \mathbb{R}$.

I.4.2. Piecewise-linear application

Motivated by [22], in this section we shall consider the following three-dimensional piecewise-linear problem

$$
\begin{aligned}
\dot{x} &= -\delta_3 x + \varepsilon \cos \mu_1(t + \alpha) \\
\dot{y} &= \delta_2 y - \omega(z - \delta_1) + \varepsilon \sin \mu_2(t + \alpha) \qquad \text{if } z > 0, \\
\dot{z} &= \omega y + \delta_2(z - \delta_1)
\end{aligned}
$$

$$(I.4.25)_\varepsilon$$

$$
\begin{aligned}
\dot{x} &= -\delta_3 x + u \\
\dot{y} &= \delta_2(y + \delta) \qquad\qquad\qquad\quad \text{if } z < 0 \\
\dot{z} &= \omega(y + \delta)
\end{aligned}
$$

with constants $\delta_{1,2,3} > 0$, $\delta > -y_1$ (y_1 will be determined later; for this time we consider δ sufficiently large), $u \in \mathbb{R}$ and parameters $\alpha \in \mathbb{R}$, $\mu_1, \mu_2 > 0$, $\varepsilon \sim 0$. We shall shorten the notation $\mu = (\mu_1, \mu_2)$ for vector of parameters.

Clearly, we have $h(x, y, z) = z$, $\Omega_\pm = \{(x, y, z) \in \mathbb{R}^3 \mid \pm z > 0\}$, $\Omega_0 = \mathbb{R}^2 \times \{0\}$. Consequently, we obtain the equation on Ω_0

$$
\begin{aligned}
\dot{x} &= Ax + By + C + \varepsilon G \cos \mu_1(t + \alpha) \\
\dot{y} &= Dy + E + \varepsilon G \sin \mu_2(t + \alpha) \\
\dot{z} &= 0
\end{aligned}
$$

$$(I.4.26)_\varepsilon$$

where

$$A = -\delta_3, \qquad B = -\frac{\omega u}{\omega\delta + \delta_1\delta_2}, \qquad C = \frac{\delta_1\delta_2 u}{\omega\delta + \delta_1\delta_2},$$

$$D = \frac{\delta_1(\delta_2^2 + \omega^2)}{\omega\delta + \delta_1\delta_2}, \qquad E = \delta D, \qquad G = \frac{\omega(y + \delta)}{\omega\delta + \delta_1\delta_2}.$$

Since both equations – the first part of $(I.4.25)_0$ and $(I.4.26)_0$ – are linear, one can easily derive their flows in Ω_+ and Ω_0, respectively,

$$\varphi_+(x, y, z, t) = \left(xe^{-\delta_3 t}, e^{\delta_2 t}(y \cos \omega t - (z - \delta_1) \sin \omega t),\right.$$

$$e^{\delta_2 t}(y \sin \omega t + (z - \delta_1) \cos \omega t) + \delta_1\Big),$$

$$\varphi_0(x, y, 0, t) = \left(xe^{At} - \frac{(C - \delta B)(1 - e^{At})}{A} + \frac{B(y + \delta)(e^{Dt} - e^{At})}{D - A}, -\delta + (y + \delta)e^{Dt}, 0\right).$$

Now we need to find points $\bar{x}_i = (x_i, y_i, z_i)$, $i = 0, 1, 2$ and then we set

$$\gamma_1(t) = \varphi_+(\bar{x}_0, t), \qquad \gamma_2(t) = \varphi_0(\bar{x}_1, t - t_1), \qquad \gamma_3(t) = \varphi_+(\bar{x}_2, t - t_2). \qquad \text{(I.4.27)}$$

At the grazing point $\bar{x}_2$ two conditions are satisfied

$$z_2 = 0 \qquad \text{and} \qquad \omega y_2 + \delta_2(z_2 - \delta_1) = 0$$

due to the assumption H2). Thus we get $K = \mathbb{R} \times \left\{\frac{\delta_1 \delta_2}{\omega}\right\} \times \{0\}$ and $\bar{x}_2 = (x_2, \frac{\delta_1 \delta_2}{\omega}, 0)$. Following γ_3 we reach the point $\gamma_3(T) = \bar{x}_0$. If we set $\bar{x}_0 = (x_0, 0, z_0)$, we get an equation for the period T

$$e^{\delta_2(T - t_2)}\left(\frac{\delta_1 \delta_2}{\omega} \cos \omega(T - t_2) + \delta_1 \sin \omega(T - t_2)\right) = 0.$$

Denoting $c = \arctan \frac{\delta_2}{\omega} \in (0, \pi/2)$ we have $T = t_2 + (\pi - c)/\omega$ as the time of first intersection of $\{\gamma_3(t) \mid t > t_2\}$ with $\mathbb{R} \times \{0\} \times \mathbb{R}$. Then the other coordinates of $\bar{x}_0$ are

$$x_0 = x_2 e^{-\delta_3(\pi - c)/\omega} \qquad \text{and} \qquad z_0 = \frac{\delta_1}{\omega}e^{\delta_2(\pi - c)/\omega}\sqrt{\delta_2^2 + \omega^2} + \delta_1.$$

Note that $z_0 > 2\delta_1$. The relation $\gamma_1(t_1) = \bar{x}_1 \in \Omega_0$ yields the implicit equation for t_1:

$$e^{\delta_2 t_1} \cos \omega t_1 = -\frac{\delta_1}{z_0 - \delta_1} \qquad \text{(I.4.28)}$$

where we look for the smallest positive root. Note that the right-hand side of the last identity is from $(-1, 0)$. Therefore $t_1 \in (\pi/(2\omega), \pi/\omega)$. Next we obtain x_1, y_1 and, finally, we connect $\bar{x}_1$ with $\bar{x}_2$ via γ_2. In conclusion, we have the following lemma.

Lemma I.4.6. *The unperturbed system* (I.4.25)$_0$, (I.4.26)$_0$ *possesses a T-periodic sliding solution $\gamma(t)$ of* (I.4.3) *given by* (I.4.27). *Moreover,* $\bar{x}_i = (x_i, y_i, z_i)$, $i = 0, 1, 2$ *where*

$$x_0 = \frac{e^{\delta_3 t_2}}{e^{\delta_3(t_2 + (\pi - c)/\omega)} - 1}\left[\frac{B(y_1 + \delta)(e^{D(t_2 - t_1)} - e^{A(t_2 - t_1)})}{D - A} - \frac{(C - \delta B)(1 - e^{A(t_2 - t_1)})}{A}\right],$$

$$y_0 = 0, \qquad z_0 = \frac{\delta_1}{\omega}e^{\delta_2(\pi - c)/\omega}\sqrt{\delta_2^2 + \omega^2} + \delta_1,$$

$$\bar{x}_1 = \left(x_0 e^{-\delta_3 t_1}, -e^{\delta_2 t_1}(z_0 - \delta_1)\sin\omega t_1, 0\right), \quad \bar{x}_2 = \left(x_0 e^{\delta_3(\pi-c)/\omega}, \frac{\delta_1\delta_2}{\omega}, 0\right),$$

t_1 is given by (I.4.28) and

$$t_2 = t_1 + \frac{1}{D}\ln\frac{y_2 + \delta}{y_1 + \delta}, \qquad T = t_2 + (\pi - c)/\omega.$$

Fundamental matrices $X_1(t)$, $X_2(t)$, $X_3(t)$ of (I.4.13) have, respectively, the form

$$X_1(t) = \begin{pmatrix} e^{-\delta_3 t} & 0 & 0 \\ 0 & e^{\delta_2 t}\cos\omega t & -e^{\delta_2 t}\sin\omega t \\ 0 & e^{\delta_2 t}\sin\omega t & e^{\delta_2 t}\cos\omega t \end{pmatrix},$$

$$X_2(t) = \begin{pmatrix} e^{A(t-t_1)} & \frac{B(e^{D(t-t_1)}-e^{A(t-t_1)})}{D-A} & 0 \\ 0 & e^{D(t-t_1)} & 0 \\ 0 & 0 & 1 \end{pmatrix}, \qquad X_3(t) = X_1(t - t_2)$$

and the saltation matrix of (I.4.12) is

$$S = \begin{pmatrix} 1 & 0 & -\frac{u}{\omega\delta+\delta_1\delta_2} \\ 0 & 1 & \frac{\delta_1\omega-\delta\delta_2}{\omega\delta+\delta_1\delta_2} \\ 0 & 0 & 0 \end{pmatrix}.$$

Proof. The statement on the periodic trajectory and the mentioned points follows from the preceding discussion. The fundamental matrices are easily obtained due to the linearity of unperturbed systems (I.4.25)$_0$ and (I.4.26)$_0$, matrix S from its definition (I.4.12). □

Now, we verify the basic assumptions of this chapter.

Lemma I.4.7. System (I.4.25)$_0$, (I.4.26)$_0$ satisfies conditions H1), H2).

Proof. We have constructed $\gamma(t)$ such that the first condition would be satisfied. To verify the second one, we estimate

$$Dh(\bar{x})(f_-(\bar{x}) - f_+(\bar{x})) = \omega\delta + \delta_1\delta_2 > 0$$

for $\bar{x} \in \Omega_0$,

$$Dh(\gamma(t))f_+(\gamma(t)) = -\delta_1\delta_2 + \omega\left(-\delta + (y_1 + \delta)e^{D(t-t_1)}\right)$$
$$< -\delta_1\delta_2 + \omega\left(-\delta + (y_1 + \delta)e^{D(t_2-t_1)}\right) = 0$$

for $t \in [t_1, t_2)$, and

$$Dh(\bar{x}_2)f_+(\bar{x}_2) = \omega y_2 - \delta_1\delta_2 = 0.$$

Furthermore,

$$Dh(\gamma(t))f_-(\gamma(t)) = \omega(y_1 + \delta)e^{D(t-t_1)} \geq \omega(y_1 + \delta) > 0$$

for $t \in [t_1, t_2]$, and

$$D_s^2[h(\gamma(t_2 + s))]_{s=0^+} = Dh(\bar{x}_2)Df_+(\bar{x}_2)f_+(\bar{x}_2) = \omega(\delta_2 y_2 + \delta_1 \omega) > 0.$$

Moreover, note that for $t \in (t_2, T]$ we have

$$Dh(\gamma(t)) = \frac{\delta_1(\delta_2^2 + \omega^2)}{\omega}e^{\delta_2(t-t_2)}\sin\omega(t - t_2) > 0.$$

Hence $h(\gamma(t)) > 0$ for all $t \in (t_2, T]$. This completes the proof.   □

Since we cannot express t_1 explicitly from equation (I.4.28), we proceed numerically. From now on, we consider fixed values of $\delta_{1,2,3}$, δ and ω. We set

$$\delta_1 = \delta_3 = 1, \qquad \delta_2 = 1/2, \qquad \delta = 10, \qquad \omega = 1. \qquad (I.4.29)$$

From Lemma I.4.6 we get (see Figure I.4.2)

$$\begin{aligned}
\bar{x}_0 &\doteq (0.007u, 0, 5.265), & t_1 &\doteq 1.673, \\
\bar{x}_1 &\doteq (0.001u, -9.793, 0), & t_2 &\doteq 34.642, \\
\bar{x}_2 &\doteq (0.106u, 0.5, 0), & T &\doteq 37.320.
\end{aligned}$$

When parameters are fixed, we verify the third assumption.

Lemma I.4.8. *System* $(I.4.25)_\varepsilon$, $(I.4.26)_\varepsilon$ *with parameters* (I.4.29) *satisfies condition* H3) *if* $u \neq 0$.

Proof. From Lemma I.4.4, $\dim Z \geq 1$ for Z given by (I.4.14). Let us suppose that $\dim Z > 1$. Then there exists a vector $v \in Z$ such that $\langle v, f_+(\bar{x}_0)\rangle = 0$, i.e. there exist constants $a, b \in \mathbb{R}$ such that $|a| + |b| > 0$ and

$$v = a\begin{pmatrix} \delta_2(z_0 - \delta_1) \\ 0 \\ \delta_3 x_0 \end{pmatrix} + b\begin{pmatrix} -\omega(z_0 - \delta_1) \\ \delta_3 x_0 \\ 0 \end{pmatrix}$$

since $[f_+(\bar{x}_0)]^\perp = [(\delta_2(z_0 - \delta_1), 0, \delta_3 x_0)^*, (-\omega(z_0 - \delta_1), \delta_3 x_0, 0)^*]$. On substituting parameters (I.4.29) we obtain

$$v \doteq a\begin{pmatrix} 2.133 \\ 0 \\ 0.007u \end{pmatrix} + b\begin{pmatrix} -4.265 \\ 0.007u \\ 0 \end{pmatrix}.$$

Consequently,

$$(\mathbb{I} - A(0))v \doteq \begin{pmatrix} (-0.005a - 0.002b)u^2 + 2.133a - 4.265b \\ (-2.787a - 1.394b)u \\ (1.401a + 0.701b)u \end{pmatrix}$$

which is equal to zero for $u \neq 0$ if and only if $a = b = 0$, and we get a contradiction with the dimension of Z. $\qquad\qquad\qquad\qquad\qquad\qquad\qquad\qquad\qquad\qquad\qquad\square$

Note that in the case $u = 0$ the condition H3) is broken and the considered system possesses a degenerate sliding periodic solution which remains in the yz-plane for all time.

Using the Fredholm alternative we get

$$Y^{\perp} = \mathcal{R}(D_{\xi}F(\bar{x}_0, 0, \mu, \alpha))^{\perp} = \mathcal{N}(D_{\xi}F(\bar{x}_0, 0, \mu, \alpha))^* = [(0, 1, 1.98957)^*]$$

where the last equality was derived numerically. So we choose $\psi = (0, 1, 1.98957)^*$. Then the Poincaré-Andronov-Melnikov function defined by (I.4.19) has the form

$$
\begin{aligned}
M^{\mu}(\alpha) \doteq & \int_0^{t_1} (\cos s - 1.98957 \sin s) e^{-s/2} \sin \mu_2(s + \alpha) ds \\
& - \int_{t_1}^{t_2} 0.0178 \sin \mu_2(s + \alpha) ds \\
& + \int_{t_2}^{T} 10^7 (2.458 \cos s - 28.186 \sin s) e^{-s/2} \sin \mu_2(s + \alpha) ds.
\end{aligned}
\tag{I.4.30}
$$

This can be easily transformed to

$$M^{\mu}(\alpha) = K \sin \mu_2 \alpha + L \cos \mu_2 \alpha$$

where

$$
\begin{aligned}
K = & \int_0^{t_1} (\cos s - 1.98957 \sin s) e^{-s/2} \cos \mu_2 s\, ds - \int_{t_1}^{t_2} 0.0178 \cos \mu_2 s\, ds \\
& + \int_{t_2}^{T} 10^7 (2.458 \cos s - 28.186 \sin s) e^{-s/2} \cos \mu_2 s\, ds, \\
L = & \int_0^{t_1} (\cos s - 1.98957 \sin s) e^{-s/2} \sin \mu_2 s\, ds - \int_{t_1}^{t_2} 0.0178 \sin \mu_2 s\, ds \\
& + \int_{t_2}^{T} 10^7 (2.458 \cos s - 28.186 \sin s) e^{-s/2} \sin \mu_2 s\, ds.
\end{aligned}
\tag{I.4.31}
$$

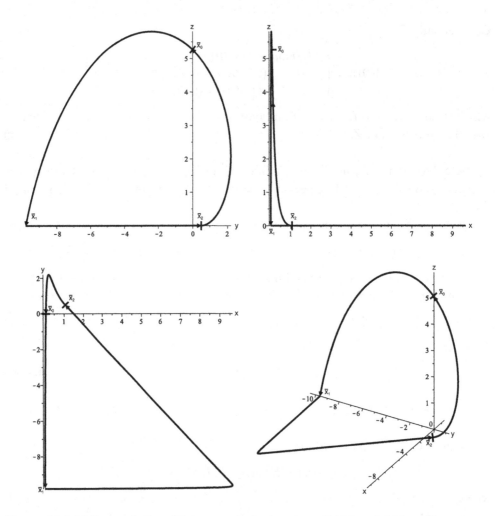

Figure I.4.2 Sliding periodic orbit in unperturbed system $(I.4.25)_0$, $(I.4.26)_0$ with parameters $(I.4.29)$ and $u = 10$ projected onto yz-, xz- and xy-plane, and the 3-dimensional overview

Clearly, $M^\mu(\alpha)$ has a simple root if and only if $K^2 + L^2 \neq 0$ or, alternatively, if function

$$\Phi(\mu_2) = \int_0^T \varphi(s)e^{\iota\mu_2 s}ds$$

with

$$\varphi(s) = \begin{cases} (\cos s - 1.98957 \sin s)e^{-s/2} & \text{if } s \in [0, t_1), \\ -0.0178 & \text{if } s \in [t_1, t_2), \\ 10^7(2.458 \cos s - 28.186 \sin s)e^{-s/2} & \text{if } s \in [t_2, T] \end{cases}$$

and $\iota = \sqrt{-1}$, is nonzero. After integrating we obtain

$$\Phi(\mu_2) = \frac{1}{\mu_2(4\mu_2^2 + 4\iota\mu_2 - 5)}\left[5.958\mu_2 + 4\iota\mu_2^2 + (0.089\iota - 3.106\mu_2 + 3.536\iota\mu_2^2)e^{\iota t_1\mu_2}\right.$$

$$\left. -(0.089\iota + 34.016\mu_2)e^{\iota t_2\mu_2} - (5.958\mu_2 + 4\iota\mu_2^2)e^{\iota T\mu_2}\right].$$

(I.4.32)

Since the denominator is nonzero for any $\mu_2 > 0$, function $\Phi(\mu_2)$ is well-defined and it is enough to investigate the numerator. Now we apply the condition on the period of the perturbation function, more precisely

$$(\cos\mu_1 T, \sin\mu_2 T, 0) = (1, 0, 0),$$

i.e. $\mu_1 = 2k_1\pi/T$, $\mu_2 = 2k_2\pi/T$ for some $k_1, k_2 \in \mathbb{N}$. For such μ_2 the numerator in (I.4.32) has the form

$$(0.089\iota - 3.106\mu_2 + 3.536\iota\mu_2^2)e^{\iota t_1\mu_2} - (0.089\iota + 34.016\mu_2)e^{\iota t_2\mu_2}.$$

For simplicity, we denote it by $\varphi(\mu_2)$. Then

$$|\varphi(\mu_2)| \geq 3.536\mu_2^2 - 37.123\mu_2 - 0.178$$

which is greater than zero for μ_2 positive if $\mu_2 > 11$. Hence the function $\Phi(\mu_2)$ is nonzero for such $\mu_2 \in (2\pi/T)\mathbb{N}$. Moreover, it can be numerically shown that $\varphi(\mu_2)$ is nonzero for $\mu_2 \in (0, 11]$ as well (cf. Figure I.4.3). Consequently, $\Phi(2k_2\pi/T)$ is nonzero for each $k_2 \in \mathbb{N}$ and Theorem I.4.5 can be applied.

Proposition I.4.9. *Let $u \neq 0$, $\mu_1 = 2k_1\pi/T$, $\mu_2 = 2k_2\pi/T$ for given $k_1, k_2 \in \mathbb{N}$. Then for each $k \in R$ where*

$$R = \{r \in \mathbb{Z} \mid r\pi - \lambda \in [0, 2k_2\pi)\}$$

and λ is such that

$$\cos\lambda = \frac{K}{\sqrt{K^2 + L^2}}, \qquad \sin\lambda = \frac{L}{\sqrt{K^2 + L^2}}$$

for K, L defined by (I.4.31), there exists a unique T-periodic sliding solution $x_k(\varepsilon)(t)$ of system (I.4.25)$_\varepsilon$ with $\varepsilon \neq 0$ sufficiently small and

$$\alpha = \alpha_k(\varepsilon) = \frac{k\pi - \lambda}{\mu_2} + O(\varepsilon)$$

such that

$$|x_k(\varepsilon)(t) - \gamma(t - \alpha)| = O(\varepsilon)$$

for any $t \in \mathbb{R}$. So for each $u \neq 0$, $k_1, k_2 \in \mathbb{N}$ there are at least as many different T-periodic sliding solutions as the number of elements of R.

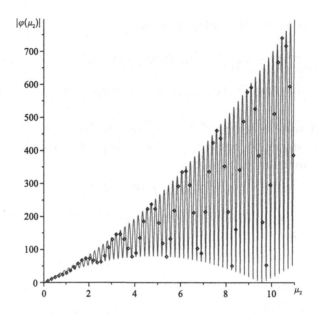

Figure I.4.3 The dependence of $|\varphi(\mu_2)|$ on μ_2. Diamonds depict the values at $2k_2\pi/T$ for $k_2 \in \{1, \ldots, 65\}$

Proof. First, the period matching condition $T = 2k_1\pi/\mu_1 = 2k_2\pi/\mu_2$ for $k_1, k_2 \in \mathbb{N}$ has to be satisfied. Next, for μ_2 such that $\Phi(\mu_2) \neq 0$, the root α_0 of Poincaré-Andronov-Melnikov function $M^\mu(\alpha)$ of (I.4.30) satisfies $\sin(\mu_2\alpha + \lambda) = 0$. Therefore, $\alpha_0 = (k\pi - \lambda)/\mu_2$ for $k \in \mathbb{Z}$. Moreover, the T-periodicity of functions g_+, g_- (and consequently of function g_0) in t means that only for $\alpha_0 \in [0, T)$, i.e. for $k \in R$, do we get the different solutions. The rest follows from Theorem I.4.5.   □

Sliding periodic solutions are also investigated in Chapter III.2.

CHAPTER I.5

Weakly coupled oscillators

I.5.1. Setting of the problem

In this chapter we consider the following weakly coupled and periodically forced systems of ordinary differential inclusions

$$x' \in \varepsilon f(x, y, t, \varepsilon)$$
$$y' \in g(x, y) + \varepsilon h(x, y, t, \varepsilon), \tag{I.5.1}$$

where $x \in \mathbb{R}^n$, $y \in \mathbb{R}^m$, $g \in C^3(\mathbb{R}^{n+m}, \mathbb{R}^m)$, $f: \mathbb{R}^{n+m+2} \to 2^{\mathbb{R}^n} \setminus \{\emptyset\}$ and $h: \mathbb{R}^{n+m+2} \to 2^{\mathbb{R}^m} \setminus \{\emptyset\}$ are upper semi-continuous multivalued mappings with compact and convex set values (see [39] for these notions), which are 1-periodic in $t \in \mathbb{R}$. For $\varepsilon = 0$, (I.5.1) becomes an autonomous ODE

$$y' = g(x, y) \tag{I.5.2}$$

parameterized by $x \in \mathbb{R}^n$. Throughout this chapter, we use $'$ to denote the partial derivative with respect to t.

We consider two cases:

1. for any $x \in U$, (I.5.2) has a single 1-periodic solution,
2. for any $x \in U$, (I.5.2) has a non-degenerate family of 1-periodic solutions

for an open subset $U \subset \mathbb{R}^n$. Our aim is to find topological degree bifurcation conditions under which (I.5.1) has forced 1-periodic solutions for $\varepsilon \neq 0$ small. For this purpose, a combination of the Lyapunov-Schmidt method together with the Leray-Schauder degree theory for multivalued mappings is applied [29] (see also A.1.3 and A.1.5). The main results of this chapter, Theorems I.5.2, I.5.3 and I.5.4, are extensions of similar ones for ODEs [29, 40–43] to ordinary differential inclusion (I.5.1). We also present Examples I.5.5 and I.5.8 to illustrate our theory. Averaging methods for symmetric ODEs are given in [26, 44, 45]. We note that this chapter is added to show that, when differential inclusion is studied, then a Poincaré-Andronov-Melnikov-like multifunction can still be derived. The method is performed by topological degree arguments in place of the implicit function theorem. So we do not follow the Poincaré mapping approach, but we note that the multivalued Poincaré mapping is used for instance in [39].

Poincaré-Andronov-Melnikov Analysis for Non-Smooth Systems.
http://dx.doi.org/10.1016/B978-0-12-804294-6.50007-2

I.5.2. Bifurcations from single periodic solutions

In this section, we consider the case 1. from Section I.5.1, which is related to Chapters I.1 and I.3 but here we consider nonautonomous differential inclusions. More precisely, we suppose the condition

H1) (I.5.2) has a 1-periodic solution $y = \varphi(t, x)$ for any $x \in U \subset \mathbb{R}^n$, where $\varphi \in C^1(\mathbb{R} \times U, \mathbb{R}^m)$ is 1-periodic in $t \in \mathbb{R}$.

Certainly the function $\varphi'(t, x)$ satisfies the variational equation

$$v' = g_y(x, \varphi(t, x))v. \tag{I.5.3}$$

We also consider the dual variational system

$$w' = -g_y^*(x, \varphi(t, x))w. \tag{I.5.4}$$

Next, we suppose

H2) There are smooth bases

$$\left\{v_0(t, x), v_1(t, x), \cdots, v_r(t, x)\right\} \quad \text{and} \quad \left\{w_0(t, x), w_1(t, x), \cdots, w_r(t, x)\right\}$$

of 1-periodic solutions of (I.5.3) and (I.5.4), respectively, for any $x \in U$. We assume that $v_0(t, x) = \varphi'(t, x)$.

For handling the problem, we consider the Banach spaces

$$X := \left\{x \in C(\mathbb{R}, \mathbb{R}^n) \mid x(t) \text{ is 1-periodic}\right\},$$
$$Y := \left\{y \in C(\mathbb{R}, \mathbb{R}^m) \mid y(t) \text{ is 1-periodic}\right\},$$
$$X_\infty := \left\{x \in L^\infty(\mathbb{R}, \mathbb{R}^n) \mid x(t) \text{ is 1-periodic}\right\},$$
$$Y_\infty := \left\{y \in L^\infty(\mathbb{R}, \mathbb{R}^m) \mid y(t) \text{ is 1-periodic}\right\},$$
$$W_n^{1,\infty} := \left\{x \in W^{1,\infty}(\mathbb{R}, \mathbb{R}^n) \mid x(t) \text{ is 1-periodic}\right\},$$
$$W_m^{1,\infty} := \left\{y \in W^{1,\infty}(\mathbb{R}, \mathbb{R}^m) \mid y(t) \text{ is 1-periodic}\right\}$$

with the usual sup-norms. Then we introduce the projections

$$P_1 : X \to X, \qquad P_x : Y \to Y$$

defined as follows

$$(P_1 x)(t) := x(t) - \int_0^1 x(s)\, ds,$$
$$(P_x y)(t) := y(t) - q_0 w_0(t, x) - q_1 w_1(t, x) - \cdots - q_r w_r(t, x),$$
$$(q_0, q_1, \cdots, q_r)^* := A(x)^{-1} \left(\int_0^1 \langle y(t), w_0(t, x)\rangle\, dt, \ldots, \int_0^1 \langle y(t), w_r(t, x)\rangle\, dt\right)^*$$

where $\langle \cdot, \cdot \rangle$ is the scalar product on $\mathbb{R}^m$, and $A(x): \mathbb{R}^{r+1} \to \mathbb{R}^{r+1}$ is the matrix given by

$$A(x) := \left(\int_0^1 \langle w_i(t, x), w_j(t, x) \rangle \, dt \right)_{i,j=0}^r .$$

The meaning of these projections is the following: The nonhomogeneous variational equation (I.5.3) along $\varphi(t, x)$ is given by

$$u' = \widetilde{h}_1,$$
$$v' = g_y(x, \varphi(t, x))v + \widetilde{h}_2. \tag{I.5.5}$$

From [14, Theorem 1.2, p. 411] we have the following result.

Lemma I.5.1. *System (I.5.5) has a 1-periodic solution in $W_n^{1,\infty} \times W_m^{1,\infty}$ for $\widetilde{h}_1 \in X_\infty$ and $\widetilde{h}_2 \in Y_\infty$ if and only if $P_1 \widetilde{h}_1 = \widetilde{h}_1$ and $P_x \widetilde{h}_2 = \widetilde{h}_2$. Moreover this solution is unique if $P_1 u = u$ and $\int_0^1 \langle v(t), v_i(t, x) \rangle \, dt = 0$, $i = 0, 1, \ldots, r$.*

Next for any $h_1 \in X_\infty$, $h_2 \in Y_\infty$ we set $\widetilde{h}_1 := P_1 h_1$, $\widetilde{h}_2 := P_x h_2$, and we denote by $(u, v) := \mathcal{K}_x(h_1, h_2)$ the solution of (I.5.5) from Lemma I.5.1. Then

$$\mathcal{K}_x : X_\infty \times Y_\infty \to X \times Y$$

is compact and linear, since $W_n^{1,\infty} \times W_m^{1,\infty} \subset\subset X \times Y$ is a compact embedding. Moreover, a mapping

$$\mathcal{K} : U \to L(X_\infty \times Y_\infty, X \times Y)$$

defined as $\mathcal{K}(x) := \mathcal{K}_x$ is continuous.

Now we shift $t \to t + \alpha$, $\alpha \in \mathbb{R}$, and then make in (I.5.1) the change of variables

$$\varepsilon \to \varepsilon^2, \quad x = \varepsilon^2 u + x_1, \quad u \in X, \quad P_1 u = u, \quad x_1 \in \mathbb{R}^n,$$

$$y = \varphi(t, \varepsilon^2 u + x_1) + \varepsilon^2 v + \varepsilon \sum_{i=1}^r \beta_i v_i(t, \varepsilon^2 u + x_1), \tag{I.5.6}$$

$$\int_0^1 \langle v(t), v_i(t, x_1) \rangle \, dt = 0, \quad i = 0, 1, \ldots, r$$

to derive

$$u' \in f\left(\varepsilon^2 u + x_1, \varphi(t, \varepsilon^2 u + x_1) + \varepsilon^2 v + \varepsilon \sum_{i=1}^r \beta_i v_i(t, \varepsilon^2 u + x_1), t + \alpha, \varepsilon^2 \right), \tag{I.5.7}$$

$$v' - g_y(x_1, \varphi(t, x_1))v \in H(u, v, x_1, \alpha, \beta, t, \varepsilon),$$

where $\beta := (\beta_1, \beta_2, \ldots, \beta_r)$ and

$$H(u, v, x_1, \alpha, \beta, t, \varepsilon) :=$$

$$\frac{1}{\varepsilon^2} \left\{ g\left(\varepsilon^2 u + x_1, \varphi(t, \varepsilon^2 u + x_1) + \varepsilon^2 v + \varepsilon \sum_{i=1}^{r} \beta_i v_i(t, \varepsilon^2 u + x_1) \right) \right.$$

$$-g(\varepsilon^2 u + x_1, \varphi(t, \varepsilon^2 u + x_1))$$

$$\left. -g_y(\varepsilon^2 u + x_1, \varphi(t, \varepsilon^2 u + x_1)) \left(\varepsilon^2 v + \varepsilon \sum_{i=1}^{r} \beta_i v_i(t, \varepsilon^2 u + x_1) \right) \right\}$$

$$+ \left(g_y(\varepsilon^2 u + x_1, \varphi(t, \varepsilon^2 u + x_1)) - g_y(x_1, \varphi(t, x_1)) \right) v$$

$$- \left(\varepsilon \sum_{i=1}^{r} \beta_i v_{ix}(t, \varepsilon^2 u + x_1) + \varphi_x(t, \varepsilon^2 u + x_1) \right) u'$$

$$+ h \left(\varepsilon^2 u + x_1, \varepsilon^2 v + \varepsilon \sum_{i=1}^{r} \beta_i v_i(t, \varepsilon^2 u + x_1) + \varphi(t, \varepsilon^2 u + x_1), t + \alpha, \varepsilon^2 \right).$$

Next we set the following mapping

$$G: X \times Y \times U \times \mathbb{R}^{r+2} \times [0, 1] \to 2^{X_\infty \times Y_\infty} \setminus \{\emptyset\}$$

given by

$$G(u, v, x_1, \alpha, \beta, \varepsilon, \lambda) :=$$

$$\left\{ (h_1, h_2) \in X_\infty \times Y_\infty \,\middle|\, h_1(t) \in f\left(\lambda \varepsilon^2 u(t) + x_1, \varphi(t, \lambda \varepsilon^2 u(t) + x_1) \right. \right.$$

$$\left. + \lambda \varepsilon^2 v(t) + \lambda \varepsilon \sum_{i=1}^{r} \beta_i v_i(t, \varepsilon^2 u(t) + x_1), t + \alpha, \lambda \varepsilon^2 \right),$$

$$h_2(t) + \left(\lambda \varepsilon \sum_{i=1}^{r} \beta_i v_{ix}(t, \varepsilon^2 u(t) + x_1) + \varphi_x(t, \lambda \varepsilon^2 u(t) + x_1) \right) h_1(t) \in$$

$$\frac{1}{\varepsilon^2} \left[g\left(\lambda \varepsilon^2 u(t) + x_1, \varphi(t, \lambda \varepsilon^2 u(t) + x_1) + \lambda \varepsilon^2 v(t) \right. \right.$$

$$\left. + \lambda \varepsilon \sum_{i=1}^{r} \beta_i v_i(t, \varepsilon^2 u(t) + x_1) \right) - g(\lambda \varepsilon^2 u(t) + x_1, \varphi(t, \lambda \varepsilon^2 u(t) + x_1))$$

$$- \lambda g_y(\lambda \varepsilon^2 u(t) + x_1, \varphi(t, \lambda \varepsilon^2 u(t) + x_1)) \left(\varepsilon^2 v(t) + \varepsilon \sum_{i=1}^{r} \beta_i v_i(t, \varepsilon^2 u(t) + x_1) \right)$$

$$-\frac{\lambda^2\varepsilon^2}{2}g_{yy}(\lambda\varepsilon^2 u(t) + x_1, \varphi(t, \lambda\varepsilon^2 u(t) + x_1))\left(\sum_{i=1}^{r}\beta_i v_i(t, \varepsilon^2 u(t) + x_1)\right)^2\Bigg]$$

$$+\frac{1}{2}g_{yy}(\lambda\varepsilon^2 u(t) + x_1, \varphi(t, \lambda\varepsilon^2 u(t) + x_1))\left(\sum_{i=1}^{r}\beta_i v_i(t, \lambda\varepsilon^2 u(t) + x_1)\right)^2$$

$$+\lambda\Big(g_y(\varepsilon^2 u(t) + x_1, \varphi(t, \varepsilon^2 u(t) + x_1)) - g_y(x_1, \varphi(t, x_1))\Big)v(t)$$

$$+h\Big(\lambda\varepsilon^2 u(t) + x_1, \lambda\varepsilon^2 v(t) + \lambda\varepsilon\sum_{i=1}^{r}\beta_i v_i(t, \varepsilon^2 u(t) + x_1)$$

$$+\varphi(t, \lambda\varepsilon^2 u(t) + x_1), t + \alpha, \lambda\varepsilon^2\Big) \text{ for almost each (f.a.e.) } t \in \mathbb{R}\Bigg\}$$

where the term in the brackets $[\cdots]$ is $O(\varepsilon^3)$ when $\varepsilon \to 0$. So the term $\frac{1}{\varepsilon^2}[\cdots]$ can be set to 0 for $\varepsilon = 0$. It can be shown that $G(u, v, x_1, \alpha, \beta, \varepsilon, \lambda)$ is nonempty [29].

Now, using the Lyapunov-Schmidt approach, we rewrite (I.5.7) as follows

$$(u, v, 0, 0) \in F(u, v, x_1, \alpha, \beta, \varepsilon, 1), \tag{I.5.8}$$

where the mapping

$$F: X \times Y \times U \times \mathbb{R}^{r+2} \times [0, 1] \to 2^{X \times Y \times \mathbb{R}^{n+r+1}} \setminus \{\emptyset\}$$

is defined by

$$F(u, v, x_1, \alpha, \beta, t, \varepsilon, \lambda) := \Bigg\{\left(\lambda\mathcal{K}_{x_1}(h_1, h_2), \int_0^1 h_1(t)\,dt,\right.$$

$$\left.\int_0^1 \langle h_2(t), w_0(t, x_1)\rangle\,dt, \dots, \int_0^1 \langle h_2(t), w_r(t, x_1)\rangle\,dt\right)$$

$$\Bigg| (h_1, h_2) \in G(u, v, x_1, \alpha, \beta, t, \varepsilon, \lambda)\Bigg\}.$$

It is standard to verify that F is upper semicontinuous with compact and convex set values [29]. Furthermore, we set

$$M(x_1, \alpha, \beta) :=$$

$$\left\{ \left(\int_0^1 h_1(t)\, dt, \int_0^1 \langle h_2(t), w_0(t, x_1) \rangle\, dt, \ldots, \int_0^1 \langle h_2(t), w_r(t, x_1) \rangle\, dt \right) \right.$$

$$\left| \, h_1(t) \in f(x_1, \varphi(t, x_1), t + \alpha, 0) \text{ f.a.e. } t \in \mathbb{R}, \right.$$

$$h_2(t) + \varphi_x(t, x_1) h_1(t) \in \sum_{i,j=1}^r \beta_i \beta_j a_{ijk}(t, x_1)$$

$$\left. + h(x_1, \varphi(t, x_1), t + \alpha, 0) \text{ f.a.e. } t \in \mathbb{R} \right\},$$

(I.5.9)

where

$$a_{ijk}(t, x_1) := \frac{1}{2} g_{yy}(x_1, \varphi(t, x_1))(v_i(t, x_1), v_j(t, x_1)).$$

Again the mapping

$$M : U \times \mathbb{R}^{r+1} \to 2^{\mathbb{R}^{n+r+1}} \setminus \{\emptyset\}$$

is upper semicontinuous with compact and convex set values. Summarizing, we arrive at the following result.

Theorem I.5.2. *Suppose* H1) *and* H2) *hold. If there is an open bounded subset* $\Omega \subset \overline{\Omega} \subset U \times \mathbb{R}^{r+1}$ *such that*
a) $0 \notin M(x_1, \beta, \alpha)$ *on the boundary* $\partial\Omega$,
b) $\deg(M, \Omega, 0) \neq 0$, *where* deg *is the Brouwer degree,*
then (I.5.1) *has a 1-periodic solution for* $\varepsilon > 0$ *small.*

Proof. To solve (I.5.1), we need to solve (I.5.8), which we put in the homotopy

$$(u, v, 0, 0) \in F(u, v, x_1, \alpha, \beta, \varepsilon, \lambda)$$

for $\lambda \in [0, 1]$. It is not difficult to find positive constants c_1, ε_0 such that

$$(u, v, 0, 0) \notin F(u, v, x_1, \alpha, \beta, \varepsilon, \lambda)$$

for any $(u, v, x_1, \alpha, \beta) \in \partial O$ and any $(\varepsilon, \lambda) \in (0, \varepsilon_0) \times [0, 1]$, where

$$O := B_{c_1} \times \Omega, \quad B_{c_1} := \{(u, v) \in X \times Y \mid \|u\| + \|v\| < c_1\}.$$

Hence

$$\deg\big((u,v,0,0) - F(u,v,x_1,\alpha,\beta,\varepsilon,1),O,0\big)$$

$$= \deg\big((u,v,0,0) - F(u,v,x_1,\alpha,\beta,\varepsilon,0),O,0\big)$$

$$= \deg\big((u,v,-M(x_1,\alpha,\beta)),O,0\big) = \deg(-M,\Omega,0) \neq 0.$$

So (I.5.8) is solvable for any $\varepsilon \in (0,\varepsilon_0)$. The proof is finished.   □

Taking $\varepsilon \to -\varepsilon^2$ in (I.5.6), we can repeat the previous arguments, but (I.5.9) is changed to

$$\overline{M}(x_1,\alpha,\beta) :=$$

$$\left\{ \left(\int_0^1 h_1(t)\,dt, \int_0^1 \langle h_2(t), w_0(t,x_1)\rangle\,dt, \ldots, \int_0^1 \langle h_2(t), w_r(t,x_1)\rangle\,dt \right) \right.$$

$$\left| \, h_1(t) \in -f(x_1,\varphi(t,x_1),t+\alpha,0) \text{ f.a.e. } t \in \mathbb{R}, \right.$$

$$h_2(t) + \varphi_x(t,x_1)h_1(t) \in \sum_{i,j=1}^r \beta_i\beta_j a_{ijk}(t,x_1)$$

$$\left. -h(x_1,\varphi(t,x_1),t+\alpha,0) \text{ f.a.e. } t \in \mathbb{R} \right\},$$

to get the following result.

Theorem I.5.3. *Suppose* H1) *and* H2) *hold. If there is an open bounded subset* $\Omega \subset \overline{\Omega} \subset U \times \mathbb{R}^{r+1}$ *such that*
a) $0 \notin \overline{M}(x_1,\beta,\alpha)$ *on the boundary* $\partial\Omega$,
b) $\deg(\overline{M},\Omega,0) \neq 0$,
then (I.5.1) *has a* 1-*periodic solution for* $\varepsilon < 0$ *small.*

I.5.3. Bifurcations from families of periodics

In this section, we consider the case 2. from Section I.5.1, which is related to Chapter I.2, but here we consider nonautonomous differential inclusions. More precisely, when the unperturbed equation (I.5.2) has some symmetry then, in place of condition H1), the following one may hold
C1) (I.5.2) has a smooth family $\varphi(t,x,\theta)$ of 1-periodic solutions for any $x \in U$ and $\theta \in \Gamma$, where $U \subset \mathbb{R}^n$, $\Gamma \subset \mathbb{R}^r$ are open bounded subsets.
Then we can repeat the previous procedure to (I.5.1) with the following modifications: First, (I.5.3) is replaced with

$$v' = g_y(x,\varphi(t,x,\theta))v. \tag{I.5.10}$$

Clearly $\varphi'(t, x, \theta)$, $\varphi_{\theta_i}(t, x, \theta)$, $i = 1, 2, \ldots, r$, $\theta = (\theta_1, \theta_2, \ldots, \theta_r)$ are 1-periodic solutions of (I.5.10). We suppose

C2) The family $\varphi(t, x, \theta)$ is *non-degenerate*, i.e. the functions $\widetilde{v}_0(t, x, \theta) := \varphi'(t, x, \theta)$, $\widetilde{v}_i(t, x, \theta) := \varphi_{\theta_i}(t, x, \theta)$, $i = 1, 2, \ldots, r$ form a basis of the space of 1-periodic solutions of (I.5.10).

From [14, Lemma 1.3, p. 410] we know that condition C2) implies the existence of a smooth basis $\widetilde{w}_j(t, x, \theta)$, $j = 0, 1, \ldots, r$ of the space of 1-periodic solutions of the adjoint system

$$w' = -g_y^*(x, \varphi(t, x, \theta))w$$

to (I.5.10).

Now, in the this procedure, we keep the projection P_1, but we replace P_x with $P_{x,\theta} : Y \to Y$ defined by

$$P_{x,\theta}y := y(t) - \widetilde{q}_0\widetilde{w}_0(t, x, \theta) - \widetilde{q}_1\widetilde{w}_1(t, x, \theta) - \cdots - \widetilde{q}_r\widetilde{w}_r(t, x, \theta),$$
$$(\widetilde{q}_0, \widetilde{q}_1, \ldots, \widetilde{q}_r)^* :=$$
$$\widetilde{A}(x, \theta)^{-1} \left(\int_0^1 \langle y(t), \widetilde{w}_0(t, x, \theta) \rangle \, dt, \ldots, \int_0^1 \langle y(t), \widetilde{w}_r(t, x, \theta) \rangle \, dt \right)^*$$

where

$$\widetilde{A}(x, \theta) := \left(\int_0^1 \left\langle \widetilde{w}_i(t, x, \theta), \widetilde{w}_j(t, x, \theta) \right\rangle \, dt \right)_{i,j=0}^r$$

is an $(r + 1) \times (r + 1)$ matrix. Then changing

$$x = \varepsilon u + x_1, \quad u \in X, \quad P_1 u = u, \quad x_1 \in \mathbb{R}^n,$$

$$y = \varepsilon v + \varphi(t, \varepsilon u + x_1, \theta), \quad \int_0^1 \langle v(t), \widetilde{v}_i(t, x_1, \theta) \rangle \, dt = 0, \quad i = 0, 1, \ldots, r$$

in (I.5.1), we derive, as above,

$$u' \in f(\varepsilon u + x_1, \varepsilon v + \varphi(t, \varepsilon u + x_1, \theta), t + \alpha, \varepsilon)$$
$$v' - g_y(x_1, \varphi(t, x_1, \theta))v \in \widetilde{H}(u, v, \varepsilon, \alpha, \theta, t),$$

$$\text{(I.5.11)}$$

where

$$\widetilde{H}(u, v, \varepsilon, \alpha, \theta, t) := \frac{1}{\varepsilon}\Big(g(\varepsilon u + x_1, \varepsilon v + \varphi(t, \varepsilon u + x_1, \theta))$$
$$-g(\varepsilon u + x_1, \varphi(t, \varepsilon u + x_1, \theta))\Big) - g_y(x_1, \varphi(t, x_1, \theta))v$$
$$+h(\varepsilon u + x_1, \varepsilon v + \varphi(t, \varepsilon u + x_1, \theta), t + \alpha, \varepsilon) - \varphi_x(t, \varepsilon u + x_1, \theta)u'.$$

Furthermore, we set

$$\widetilde{M}(x_1, \alpha, \theta) :=$$

$$\left\{ \left(\int_0^1 h_1(t)\,dt, \int_0^1 \langle h_2(t), w_0(t, x_1) \rangle\,dt, \dots, \int_0^1 \langle h_2(t), w_r(t, x_1) \rangle\,dt \right) \right.$$

$$\left| h_1(t) \in f(x_1, \varphi(t, x_1, \theta), t + \alpha, 0) \text{ f.a.e. } t \in \mathbb{R}, \right. \qquad (I.5.12)$$

$$\left. h_2(t) + \varphi_x(t, x_1, \theta) h_1(t) \in h(x_1, \varphi(t, x_1, \theta), t + \alpha, 0) \text{ f.a.e. } t \in \mathbb{R} \right\}.$$

The mapping

$$\widetilde{M} : U \times \mathbb{R} \times \Gamma \to 2^{\mathbb{R}^{n+r+1}} \setminus \{\emptyset\}$$

is upper semicontinuous with compact and convex set values. Consequently, we can directly modify the method of Section I.5.2 to derive the following result.

Theorem I.5.4. *Suppose* C1) *and* C2). *If there is an open bounded subset* $\Omega \subset \overline{\Omega} \subset U \times \mathbb{R} \times \Gamma$ *such that*
a) $0 \notin M(x_1, \alpha, \theta)$ *on the boundary* $\partial\Omega$,
b) $\deg(M, \Omega, 0) \neq 0$,
then (I.5.1) *has a 1-periodic solution for* $\varepsilon \neq 0$ *small*.

I.5.4. Examples

We present two examples of weakly coupled discontinuous nonlinear oscillators by applying Theorems I.5.2 and I.5.4, respectively.

Example I.5.5. We first apply Theorem I.5.2 to the system

$$\begin{aligned}
y_1' &\in y_1 - y_2 - x^2(y_1^2 + y_2^2)y_1 + \varepsilon\mu_1 \operatorname{Sgn} y_2 \\
y_2' &\in y_1 + y_2 - x^2(y_1^2 + y_2^2)y_2 + \varepsilon\mu_2 \operatorname{Sgn} y_1 + \varepsilon\mu_3 \cos t \qquad (I.5.13) \\
x' &\in \varepsilon(y_1 \cos t + y_2 \sin t + \operatorname{Sgn} x),
\end{aligned}$$

where $\mu_{1,2,3}$ are positive parameters and $\operatorname{Sgn} : \mathbb{R} \to 2^{\mathbb{R}} \setminus \{\emptyset\}$ is given by

$$\operatorname{Sgn} y := \begin{cases} \operatorname{sgn} y, & y \neq 0, \\ [-1, 1], & y = 0, \end{cases}$$

where $\operatorname{sgn} y := y/|y|$ for $y \neq 0$ and $\operatorname{sgn} 0 = 0$. We need to verify conditions H1) and

H2) for the unperturbed system (I.5.13) of the form

$$y_1' = y_1 - y_2 - x^2(y_1^2 + y_2^2)y_1$$
$$y_2' = y_1 + y_2 - x^2(y_1^2 + y_2^2)y_2 \tag{I.5.14}$$

possessing a smooth family of 2π-periodic solutions

$$\varphi(t, x) = \frac{1}{x}(\cos t, \sin t) \tag{I.5.15}$$

for $x \neq 0$. The linearization of (I.5.14) along (I.5.15) is as follows

$$v_1' = -(1 + \cos 2t)v_1 - (1 + \sin 2t)v_2$$
$$v_2' = (1 - \sin 2t)v_1 - (1 - \cos 2t)v_2, \tag{I.5.16}$$

and its adjoint system is given by

$$w_1' = (1 + \cos 2t)w_1 - (1 - \sin 2t)w_2$$
$$w_2' = (1 + \sin 2t)w_1 + (1 - \cos 2t)w_2. \tag{I.5.17}$$

One readily verifies that (I.5.16) has solutions

$$v_0(t, x) = (-\sin t, \cos t),$$
$$\widetilde{v}(x, t) = \left(e^{-2t}\cos t, e^{-2t}\sin t\right).$$

Hence $v_0(x, t)$ is a basis of 2π-periodic solutions of (I.5.16). Furthermore, the functions

$$w_0(t, x) = (\sin t, -\cos t),$$
$$\widetilde{w}(x, t) = e^{2t}(\cos t, \sin t)$$

are solutions of (I.5.17), so $w_0(t, x)$ is a basis of 2π-periodic solutions of (I.5.17). Consequently, now we do not have parameters β. After some computations, the function M of (I.5.9), for this case (I.5.13), has the form

$$M(x_1, \alpha) = \left(\frac{2\pi}{x_1}(\cos \alpha + |x_1|), 4 \operatorname{sgn} x_1(\mu_1 - \mu_2) - \pi\mu_3 \cos \alpha\right). \tag{I.5.18}$$

We immediately see that (I.5.18) has a simple root

$$\widetilde{x}_1 = \frac{4(\mu_2 - \mu_1)}{\pi\mu_3}, \qquad \widetilde{\alpha} = \arccos\left[-\frac{4|\mu_1 - \mu_2|}{\pi\mu_3}\right], \tag{I.5.19}$$

provided

$$0 < 4|\mu_1 - \mu_2| < \pi\mu_3. \tag{I.5.20}$$

Taking a small neighborhood $\Omega \subset \mathbb{R}^2$ of the point $(\widetilde{x}_1, \widetilde{\alpha})$, Theorem I.5.2 gives the following result.

Theorem I.5.6. *If positive parameters $\mu_{1,2,3}$ satisfy assumption (I.5.20), then (I.5.13) has a 2π-periodic solution for any $\varepsilon \neq 0$ small, which is located in an $O(|\varepsilon|)$-neighborhood of the vector function*

$$\left(\frac{1}{\widetilde{x}_1} \cos(t - \widetilde{\alpha}), \frac{1}{\widetilde{x}_1} \sin(t - \widetilde{\alpha}), \widetilde{x}_1 \right),$$

where $\widetilde{x}_1$ and $\widetilde{\alpha}$ are given by (I.5.19).

To visualize the set given by (I.5.20), see Figure I.5.1, where its section for $\mu_1 = 1$ and $\mu_2 \in (0, 2]$ is sketched. A similar figure is for the section for $\mu_2 = 1$ and $\mu_1 \in (0, 2]$.

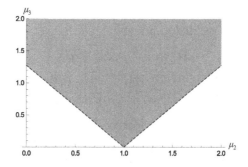

Figure I.5.1 Set of (μ_2, μ_3) satisfying (I.5.20) with $\mu_1 = 1$

Note if (μ_1, μ_2, μ_3) satisfies (I.5.20), then also $\xi(\mu_1, \mu_2, \mu_3)$ satisfies it for any $\xi > 0$.

Remark I.5.7. We can repeat the previous arguments for an example where (I.5.14) is replaced with (see [29, Example 3.5.3] or [42]),

$$y_1' = (x^2 + 1)(y_1^2 + y_2^2)y_2$$
$$y_2' = -(x^2 + 1)(y_1^2 + y_2^2)y_1$$

possessing a smooth family of 1-periodic solutions

$$\varphi(t, x) = \sqrt{\frac{2\pi k}{x^2 + 1}} \left(\sin 2\pi kt, \cos 2\pi kt \right)$$

for $k \in \mathbb{N}$. Then

$$v_0(t, x) = (\cos 2\pi kt, -\sin 2\pi kt),$$
$$w_0(t, x) = (\sin 2\pi kt, \cos 2\pi kt).$$

We do not perform further computations in this book.

Example I.5.8. Finally, we consider the system

$$y_1' \in -y_2 + \varepsilon \frac{\mu_1}{y_2^2 + 1} \operatorname{Sgn} y_1 + \varepsilon \mu_2 \cos t$$

$$y_2' \in y_1 + \varepsilon \frac{\mu_3}{y_1^2 + 1} \operatorname{Sgn} y_2 + \varepsilon \mu_4 \cos t \qquad (I.5.21)$$

$$x' \in \varepsilon \left(\frac{\operatorname{Sgn} x}{x^2 + 1} + y_1 \cos t + y_2 \sin t \right),$$

where $\mu_{1,2,3,4}$ are positive parameters. We verify assumptions C1) and C2) for its unperturbed system

$$y_1' = -y_2, \qquad y_2' = y_1 \qquad (I.5.22)$$

which is just the harmonic oscillator. So now

$$\varphi(t, x, \theta) = \theta(\cos t, \sin t),$$

$$\widetilde{v}_0(t, x, \theta) = \theta(-\sin t, \cos t), \qquad \widetilde{v}_1(t, x, \theta) = (\cos t, \sin t),$$

$$\widetilde{w}_0(t, x, \theta) = (-\sin t, \cos t), \qquad \widetilde{w}_1(t, x, \theta) = (\cos t, \sin t)$$

for $\theta \neq 0$. After some computations, we derive (I.5.12) of the form

$$\widetilde{M}(x_1, \alpha, \theta) = \left(2\pi \left(\frac{\operatorname{sgn} x_1}{x_1^2 + 1} + \theta \cos \alpha \right), \pi \left(\mu_2 \sin \alpha + \mu_4 \cos \alpha \right), \right.$$

$$\left. \pi \left(\mu_2 \cos \alpha - \mu_4 \sin \alpha \right) + 4 \frac{\arctan \theta}{|\theta|} \left(\mu_1 + \mu_3 \right) \right). \qquad (I.5.23)$$

Now we need the following obvious result [29].

Lemma I.5.9. *Let $F_1 \in C^1(\Omega_1 \times \Omega_2, \mathbb{R}^n)$, $F_2 \in C^1(\Omega_1 \times \Omega_2, \mathbb{R}^m)$, and $\Omega_1 \subset \mathbb{R}^n$, $\Omega_2 \subset \mathbb{R}^m$ be open subsets. Suppose that for any $y \in \Omega_2$ there is an $x := f(y) \in \Omega_1$ such that $F_1(f(y), y) = 0$, and $D_x F_1(f(y), y) \colon \mathbb{R}^n \to \mathbb{R}^n$ is regular, i.e. $F_1(x, y) = 0$ has a simple root $x = f(y)$ in Ω_1 for any $y \in \Omega_2$. Assume that $G(y) := F_2(f(y), y) = 0$ has a simple root $y_0 \in \Omega_2$, i.e. $G(y_0) = 0$ and $DG(y_0)$ is regular. Then (x_0, y_0), $x_0 := f(y_0)$ is a simple root of $F = (F_1, F_2)^*$, i.e. $F(x_0, y_0) = 0$ and $DF(x_0, y_0)$ is regular. Note a local uniqueness of simple roots and their smooth dependence on parameters follow from the implicit function theorem, so we suppose that $f \in C^1(\Omega_2, \Omega_1)$.*

Applying Lemma I.5.9 to (I.5.23), we solve the system

$$\frac{\operatorname{sgn} \widetilde{x}_1}{\widetilde{x}_1^2 + 1} + \widetilde{\theta} \cos \widetilde{\alpha} = 0 \qquad (I.5.24)$$

$$\mu_2 \sin \widetilde{\alpha} + \mu_4 \cos \widetilde{\alpha} = 0 \qquad (I.5.25)$$

$$\pi (\mu_2 \cos \widetilde{\alpha} - \mu_4 \sin \widetilde{\alpha}) + 4 \frac{\arctan \widetilde{\theta}}{|\widetilde{\theta}|} (\mu_1 + \mu_3) = 0. \qquad (I.5.26)$$

Clearly (I.5.24) and (I.5.25) give

$$\widetilde{\theta} = -\frac{\operatorname{sgn} \widetilde{x}_1}{\cos \widetilde{\alpha} (\widetilde{x}_1^2 + 1)}, \qquad \tan \widetilde{\alpha} = -\frac{\mu_4}{\mu_2}. \qquad (I.5.27)$$

Then for $\widetilde{\alpha} \in (-\pi/2, \pi/2)$, by inserting (I.5.27) into (I.5.26) we obtain

$$\pi \frac{\mu_2^2 + \mu_4^2}{4\mu_2(\mu_1 + \mu_3)} - \operatorname{sgn} \widetilde{x}_1 (\widetilde{x}_1^2 + 1) \arctan \frac{\sqrt{\mu_2^2 + \mu_4^2}}{\mu_2(\widetilde{x}_1^2 + 1)} = 0. \qquad (I.5.28)$$

First, (I.5.28) implies $\widetilde{x}_1 > 0$. Next, the function

$$z \mapsto (z^2 + 1) \arctan \frac{\sqrt{\mu_2^2 + \mu_4^2}}{\mu_2(z^2 + 1)}$$

is strictly increasing on $[0, \infty)$ from $\arctan \frac{\sqrt{\mu_2^2 + \mu_4^2}}{\mu_2}$ to $\frac{\sqrt{\mu_2^2 + \mu_4^2}}{\mu_2}$. Consequently, if

$$\arctan \frac{\sqrt{\mu_2^2 + \mu_4^2}}{\mu_2} < \frac{\pi}{4} \frac{\mu_2^2 + \mu_4^2}{\mu_2(\mu_1 + \mu_3)} < \frac{\sqrt{\mu_2^2 + \mu_4^2}}{\mu_2},$$

which is equivalent to

$$\frac{\pi}{4} \sqrt{\mu_2^2 + \mu_4^2} < \mu_1 + \mu_3 < \frac{\pi}{4} \frac{\mu_2^2 + \mu_4^2}{\mu_2 \arctan \frac{\sqrt{\mu_2^2 + \mu_4^2}}{\mu_2}}, \qquad (I.5.29)$$

then (I.5.28) possesses a unique simple zero $\widetilde{x}_1$ on $(0, \infty)$. Summarizing, by Lemma I.5.9 and Theorem I.5.4, we obtain the following result.

Theorem I.5.10. *If positive parameters $\mu_{1,2,3,4}$ satisfy assumption (I.5.29), then (I.5.21) has a 2π-periodic solution for any $\varepsilon \neq 0$ small, which is located in an $O(|\varepsilon|)$-neighborhood of the vector function*

$$\left(\widetilde{\theta} \cos(t - \widetilde{\alpha}), \widetilde{\theta} \sin(t - \widetilde{\alpha}), \widetilde{x}_1 \right),$$

where $\widetilde{\alpha} \in (-\pi/2, \pi/2)$, $\widetilde{\theta}$ and $\widetilde{x}_1$ are given by (I.5.27) and (I.5.28).

To visualize the set given by (I.5.29), first we consider its section for $\mu_2 = 1$ and $\mu_4 \in (0, 1]$ in Figure I.5.2. Then see Figure I.5.3 for $\mu_4 = 1$ and $\mu_2 \in (0, 1]$. Again, note

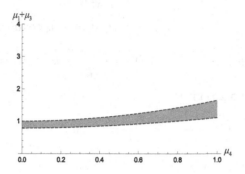

Figure I.5.2 Set of $(\mu_4, \mu_1 + \mu_3)$ given by (I.5.29) with $\mu_2 = 1$, $\mu_4 \in (0, 1]$

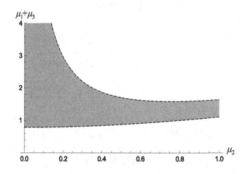

Figure I.5.3 Set of $(\mu_2, \mu_1 + \mu_3)$ given by (I.5.29) with $\mu_4 = 1$, $\mu_2 \in (0, 1]$

if $(\mu_1, \mu_2, \mu_3, \mu_4)$ satisfies (I.5.29) then also $\xi(\mu_1, \mu_2, \mu_3, \mu_4)$ satisfies it for any $\xi > 0$.

Remark I.5.11. Finally, we can consider a system more complicated than (I.5.21), when the unperturbed one has the form

$$
\begin{aligned}
y_1' &= y_2, & y_2' &= -y_1 - (x^2 + 1)(y_1^2 + y_3^2)y_1 \\
y_3' &= y_4, & y_4' &= -y_3 - (x^2 + 1)(y_1^2 + y_3^2)y_3.
\end{aligned}
\tag{I.5.30}
$$

We note that (I.5.30) has the form

$$
\ddot{w} + (1 + (x^2 + 1)\|w\|^2)w = 0
\tag{I.5.31}
$$

for $w = (y_1, y_3)$ and $\|w\| = \sqrt{y_1^2 + y_3^2}$. For $\Gamma(\theta) = \left(\begin{smallmatrix} \cos\theta & -\sin\theta \\ \sin\theta & \cos\theta \end{smallmatrix}\right)$ we see that if $w(t)$ solves (I.5.31) then $\Gamma(\theta)w(t)$ is also its solution. We know [46] that

$$y_1(t) = v(t, x, k) = \frac{\sqrt{2}k}{\sqrt{(1 - 2k^2)(x^2 + 1)}} \, \text{cn} \, \frac{t}{\sqrt{1 - 2k^2}}$$

solves $y_1' = y_2$, $y_2' = -y_1 - (x^2 + 1)y_1^3$, where cn is the Jacobi elliptic function and k is the elliptic modulus [46]. Consequently, (I.5.30) has a smooth family of periodic solutions

$$y(t, x, \theta, k) = \Big(\cos\theta \, v(t, x, k), \cos\theta \, v'(t, x, k), \sin\theta \, v(t, x, k), \sin\theta \, v'(t, x, k)\Big). \quad (\text{I.5.32})$$

The function $y(t, x, \theta, k)$ has the period $T(k) = 4K(k)\sqrt{1 - 2k^2}$ for the complete elliptic integral $K(k)$ of the first kind. We note $T(0) = 2\pi$ and $T(\sqrt{2}/2) = 0$. Numerically solving the equation $T(k) = 1$, we find its unique solution $k_0 \doteq 0.700595$ with $T(k_0)' \neq 0$. So we fix $k = k_0$ and take

$$\varphi(t, x, \theta) = y(t, x, \theta, k_0)$$

to satisfy condition C1). Condition C2) is verified for this case in (see [29, Example 3.5.4] or [42]). Again we do not carry out more computations for this example.

REFERENCE

[1] J. Awrejcewicz, M. M. Holicke, *Smooth and Nonsmooth High Dimensional Chaos and the Melnikov-Type Methods*, World Scientific Publishing Company 2007.

[2] M. di Bernardo, C. J. Budd, A. R. Champneys, P. Kowalczyk, *Piecewise-smooth Dynamical Systems: Theory and Applications*, Applied Mathematical Sciences 163, Springer-Verlag 2008.

[3] B. Brogliato, *Nonsmooth Impact Mechanics*, Lecture Notes in Control and Information Sciences 220, Springer 1996.

[4] A. Fidlin, *Nonlinear Oscillations in Mechanical Engineering*, Springer 2006.

[5] A. F. Filippov, *Differential Equations with Discontinuous Righthand Sides*, Mathematics and Its Applications 18, Kluwer Academic 1988.

[6] M. Kunze, *Non-smooth Dynamical Systems*, Lecture Notes in Mathematics 1744, Springer 2000.

[7] Yu. A. Kuznetsov, S. Rinaldi, A. Gragnani, One-parametric bifurcations in planar Filippov systems, *Internat. J. Bifur. Chaos Appl. Sci. Engrg.* **13** (2003) 2157–2188.

[8] R. I. Leine, H. Nijmeijer, *Dynamics and Bifurcations of Non-smooth Mechanical Systems*, Lecture Notes in Applied and Computational Mechanics 18, Springer-Verlag 2004.

[9] A. A. Andronov, A. A. Vitt, S. E. Khaikin, *Theory of Oscillators*, Pergamon Press 1966.

[10] K. Deimling, *Nonlinear Functional Analysis*, Springer-Verlag 1985.

[11] P. Kukučka, Jumps of the fundamental solution matrix in discontinuous systems and applications, *Nonlinear Anal.* **66** (2007) 2529–2546.

[12] P. Kukučka, Melnikov method for discontinuous planar systems, *Nonlinear Anal.* **66** (2007) 2698–2719.

[13] R. I. Leine, D. H. van Campen, B. L. van de Vrande, Bifurcations in nonlinear discontinuous systems, *Nonlinear Dynam.* **23** (2000) 105–164.

[14] P. Hartman, *Ordinary Differential Equations*, John Wiley & Sons, Inc. 1964.

[15] M. Bonnin, F. Corinto, M. Gilli, Diliberto's theorem in higher dimension, *Internat. J. Bifur. Chaos Appl. Sci. Engrg.* **19** (2009) 629–637.

[16] C. Chicone, *Ordinary Differential Equations with Applications*, Texts in Applied Mathematics 34, Springer 2006.

[17] M. Medved', *Dynamické systémy*, Comenius University in Bratislava 2000, in Slovak.

[18] F. Battelli, M. Fečkan, Homoclinic trajectories in discontinuous systems, *J. Dynam. Differential Equations* **20** (2008) 337–376.

[19] W. Rudin, *Real and Complex Analysis*, McGraw-Hill, Inc. 1974.

[20] F. Battelli, M. Fečkan, Some remarks on the Melnikov function, *Electron. J. Differential Equations* **2002** (13) (2002) 1–29.

[21] V. Acary, O. Bonnefon, B. Brogliato, *Nonsmooth Modeling and Simulation for Switched Circuits*, Springer 2011.

[22] F. Giannakopoulos, K. Pliete, Planar systems of piecewise linear differential equations with a line of discontinuity, *Nonlinearity* **14** (2001) 1611–1632.

[23] M. U. Akhmet, On the smoothness of solutions of differential equations with a discontinuous right-hand side, *Ukrainian Math. J.* **45** (1993) 1785–1792.

[24] M. Golubitsky, V. Guillemin, *Stable Mappings and Their Singularities*, Kluwer Academic Publishers 1999.

[25] S. N. Chow, J. K. Hale, *Methods of Bifurcation Theory*, Texts in Applied Mathematics 34, Springer-Verlag 1982.

[26] N. Dilna, M. Fečkan, On the uniqueness, stability and hyperbolicity of symmetric and periodic solutions of weakly nonlinear ordinary differential equations, *Miskolc Math. Notes* **10** (1) (2009) 11–40.

[27] J. Murdock, C. Robinson, Qualitative dynamics from asymptotic expansions: local theory, *J.*

Differential Equations **36** (3) (1980) 425–441.

[28] J. Moser, Regularization of Kepler's problem and the averaging method on a manifold, *Comm. Pure Appl. Math.* **23** (1970) 609–636.

[29] M. Fečkan, *Topological Degree Approach to Bifurcation Problems*, Springer Netherlands 2008.

[30] F. R. Gantmacher, *Applications of the Theory of Matrices*, Interscience 1959.

[31] M. Farkas, *Periodic Motions*, Springer-Verlag 1994.

[32] J. Guckenheimer, P. Holmes, *Nonlinear Oscillations, Dynamical Systems and Bifurcations of Vector Fields*, Springer-Verlag 1983.

[33] J. K. Hale, H. Koçak, *Dynamics and Bifurcations*, Springer-Verlag 1991.

[34] J. Awrejcewicz, J. Delfs, Dynamics of a self-excited stick-slip oscillator with two degrees of freedom, Part I: Investigation of equilibria, *Eur. J. Mech. A Solids* **9** (4) (1990) 269–282.

[35] J. Awrejcewicz, J. Delfs, Dynamics of a self-excited stick-slip oscillator with two degrees of freedom, Part II: Slip-stick, slip-slip, stick-slip transitions, periodic and chaotic orbits, *Eur. J. Mech. A Solids* **9** (5) (1990) 397–418.

[36] J. Awrejcewicz, P. Olejnik, Stick-slip dynamics of a two-degree-of-freedom system, *Internat. J. Bifur. Chaos Appl. Sci. Engrg.* **13** (4) (2003) 843–861.

[37] M. Kunze, T. Küpper, Qualitative bifurcation analysis of a non-smooth friction-oscillator model, *Z. Angew. Math. Phys.* **48** (1997) 87–101.

[38] P. Olejnik, J. Awrejcewicz, M. Fečkan, An approximation method for the numerical solution of planar discontinuous dynamical systems with stick-slip friction, *Appl. Math. Sci. (Ruse)* **8** (145) (2014) 7213–7238.

[39] J. Andres, L. Górniewicz, *Topological Fixed Point Principles for Boundary Value Problems*, Kluwer 2003.

[40] M.U. Akhmet, C. Buyukadali, T. Ergenc, Periodic solutions of the hybrid system with small parameter, *Nonlinear Anal. Hybrid Syst.* **2** (2008) 532–543.

[41] A. Buica, J. Llibre, Averaging methods for finding periodic orbits via Brouwer degree, *Bull. Sci. Math.* **128** (2004) 7–22.

[42] M. Fečkan, R. Ma, B. Thompson, Weakly coupled oscillators and topological degree, *Bull. Sci. Math.* **131** (2007) 559–571.

[43] J. Llibre, O. Makarenkov, Asymptotic stability of periodic solutions for nonsmooth differential equations with application to the nonsmooth van der Pol oscillator, *SIAM J. Math. Anal.* **40** (2009) 2478–2495.

[44] N. Dilna, M. Fečkan, About the uniqueness and stability of symmetric and periodic solutions of weakly nonlinear ordinary differential equations, *Dopovidi Nac. Acad. Nauk Ukraini (Reports of the National Academy of Sciences of Ukraine)* **5** (2009) 22–28.

[45] N. Dilna, M. Fečkan, Weakly non-linear and symmetric periodic systems at resonance, *Nonlinear Stud.* **16** (2009) 23–44.

[46] D. F. Lawden, *Elliptic Functions and Applications*, Appl. Math. Sciences 80, Springer-Verlag 1989.

PART II

Forced hybrid systems

Introduction

The combination of differential and difference equations is known as a hybrid system (cf. [1–9]). In this system, a mapping given by the difference equation is applied on a solution $x(t)$ of the differential equation at appropriate times, which leads to a time-switching system, or impacting hybrid system (hard-impact oscillator), where the switching depends on the position $x(t)$, not on time t. In this part we study the latter case. For better visualizing of the problem, we introduce a motivating example.

Consider a motion of a single particle in one spatial dimension described by the position $x(t)$ and the velocity $\dot{x}(t)$. We suppose that it is moving under a linear spring, it is weakly damped and forced. So its position satisfies the ordinary differential equation

$$\ddot{x} + \varepsilon\zeta\dot{x} + x = \varepsilon\mu\cos\omega t \quad \text{if} \quad x(t) < \sigma, \tag{II.0.1}$$

where $\varepsilon\zeta$ measures the viscous damping, $\varepsilon\mu$ is the magnitude of forcing, ε is a small parameter, and we suppose that the motion is free to move in the region $x < \sigma$ for $\sigma > 0$, until some time $t = t_0$ at which $x = \sigma$ where there is an impact with a rigid obstacle. Then, at $t = t_0$, we suppose that $(x(t_0^-), \dot{x}(t_0^-))$ is mapped in zero time, i.e. immediately via an impact law to

$$x(t_0^+) = x(t_0^-) \quad \text{and} \quad \dot{x}(t_0^+) = -(1 + \varepsilon r)\dot{x}(t_0^-),$$

where $x(t^\pm) = \lim_{s \to t^\pm} x(s)$, $\dot{x}(t^\pm) = \lim_{s \to t^\pm} \dot{x}(s)$ and $0 < 1 + \varepsilon r \leq 1$ is the Newton coefficient of restriction [2]. Another well-known piecewise-linear impact system is a weakly damped and forced inverted pendulum given by equations [10]

$$\ddot{x} + \varepsilon\zeta\dot{x} - x = \varepsilon\mu\cos\omega t \quad \text{if} \quad |x(t)| < \sigma,$$

$$x(t^+) = x(t^-) \quad \text{and} \quad \dot{x}(t^+) = -(1+\varepsilon r)\dot{x}(t^-) \quad \text{if} \quad |x(t^-)| = \sigma. \tag{II.0.2}$$

We can study coupled (II.0.1) and (II.0.2) to get a higher dimensional impact system.

So a system for a weakly forced impact oscillator consists of an ordinary differential equation

$$\ddot{x} = f_1(x, \dot{x}) + \varepsilon g_1(x, \dot{x}, t, \varepsilon, \mu) \tag{II.0.3}$$

and an impact condition

$$\dot{x}(t^+) = f_2(\dot{x}(t^-)) + \varepsilon g_2(x(t^-), \dot{x}(t^-), t, \varepsilon, \mu) \quad \text{if} \quad h(x(t^-)) = 0. \tag{II.0.4}$$

This equation rewritten as an evolution system has a form

$$\dot{x}_1 = x_2$$
$$\dot{x}_2 = f_1(x_1, x_2) + \varepsilon g_1(x_1, x_2, t, \varepsilon, \mu),$$

$$x_2(t^+) = f_2(x_2(t^-)) + \varepsilon g_2(x_1(t^-), x_2(t^-), t, \varepsilon, \mu) \quad \text{if} \quad h(x_1(t^-)) = 0.$$

In fact, we shall investigate a more general case (see (II.1.1), (II.1.2)).

CHAPTER II.1

Periodically forced impact systems

II.1.1. Setting of the problem and main results

In this chapter we investigate the persistence of a single T-periodic orbit of an autonomous system with impact under nonautonomous perturbation, and derive a sufficient condition.

Let $\Omega \subset \mathbb{R}^n$ be an open set in $\mathbb{R}^n$ and $h(x)$ be a C^r-function on $\overline{\Omega}$, with $r \geq 3$. We set $\Omega_0 := \{x \in \Omega \mid h(x) = 0\}$, $\Omega_1 := \Omega \backslash \Omega_0$. Let $f_1 \in C_b^r(\overline{\Omega})$, $f_2 \in C_b^r(\Omega_0, \Omega_0)$, $g_1 \in C_b^r(\overline{\Omega} \times \mathbb{R} \times \mathbb{R} \times \mathbb{R}^p)$, $g_2 \in C_b^r(\Omega_0 \times \mathbb{R} \times \mathbb{R} \times \mathbb{R}^p)$ and $h \in C_b^r(\overline{\Omega}, \mathbb{R})$. Furthermore, we suppose that $g_{1,2}$ are T-periodic in $t \in \mathbb{R}$ and 0 is a regular value of h. Let $\varepsilon, \alpha \in \mathbb{R}$ and $\mu \in \mathbb{R}^p$, $p \geq 1$ be parameters.

Definition II.1.1. We say that a function $x(t)$ is a solution of an impact system

$$\dot{x} = f_1(x) + \varepsilon g_1(x, t, \varepsilon, \mu), \quad x \in \Omega_1, \tag{II.1.1}$$

$$x(t^+) = f_2(x(t^-)) + \varepsilon g_2(x(t^-), t, \varepsilon, \mu) \quad \text{if} \quad h(x(t^\pm)) = 0, \tag{II.1.2}$$

if it is piecewise C^1-smooth satisfying equation (II.1.1) on Ω_1, equation (II.1.2) on Ω_0 and, moreover, the following holds: if for some t_0 we have $x(t_0) \in \Omega_0$, then there exists $\rho > 0$ such that for any $t \in (t_0 - \rho, t_0)$, $s \in (t_0, t_0 + \rho)$ we have $h(x(t))h(x(s)) > 0$.

Furthermore, we always naturally suppose that the problem (II.1.1), (II.1.2) is consistent in the sense that $h(f_2(x) + \varepsilon g_2(x, t, \varepsilon, \mu)) = 0$ whenever $h(x) = 0$.

For modelling the problem given by (II.1.1), (II.1.2) we assume
H1) The unperturbed equation

$$\dot{x} = f_1(x) \tag{II.1.3}$$

has a T-periodic orbit $\gamma(t)$ which is discontinuous at $t = t_1 \in (0, T)$ where it satisfies the impact condition

$$x(t^+) = f_2(x(t^-)) \quad \text{if} \quad h(x(t^-)) = 0. \tag{II.1.4}$$

Poincaré-Andronov-Melnikov Analysis for Non-Smooth Systems.
http://dx.doi.org/10.1016/B978-0-12-804294-6.50010-2

The orbit is given by its initial point $x_0 \in \Omega_1$, and consists of two branches

$$\gamma(t) = \begin{cases} \gamma_1(t) & \text{if } t \in [0, t_1), \\ \{x_1, x_2\} & \text{if } t = t_1, \\ \gamma_2(t) & \text{if } t \in (t_1, T], \end{cases} \qquad \text{(II.1.5)}$$

where $0 < t_1 < T$, $\gamma(t) \in \Omega_1$ for $t \in [0, t_1) \cup (t_1, T]$, $\gamma(t_1) \subset \Omega_0$, and

$$x_1 := \gamma_1(t_1^-) \in \Omega_0,$$
$$x_2 := \gamma_2(t_1^+) \in \Omega_0, \qquad \text{(II.1.6)}$$
$$x_0 := \gamma_2(T) = \gamma_1(0) \in \Omega_1.$$

H2) Moreover, we also assume that

$$Dh(x_1)f_1(x_1)Dh(x_2)f_1(x_2) < 0.$$

The geometric meaning of assumption H2) is that the impact periodic solution $\gamma(t)$ from H1) transversally hits and leaves the impact surface Ω_0 at t_1^- and t_1^+, respectively.

Next, since the impact system (II.1.3), (II.1.4) is autonomous, $\gamma(t - \alpha)$ is also its solution for any $\alpha \in \mathbb{R}$. So we are looking for a forced T-periodic solution $x(t)$ of the perturbed impact system (II.1.1), (II.1.2) which is orbitally close to γ, i.e. $x(t) \sim \gamma(t - \alpha)$ for some α depending on $\varepsilon \neq 0$ small. For this reason, by shifting the time, we study a shifted (II.1.1), (II.1.2) of the form

$$\dot{x} = f_1(x) + \varepsilon g_1(x, t + \alpha, \varepsilon, \mu), \quad x \in \Omega_1, \qquad \text{(II.1.7)}$$
$$x(t^+) = f_2(x(t^-)) + \varepsilon g_2(x(t^-), t + \alpha, \varepsilon, \mu) \quad \text{if} \quad h(x(t^-)) = 0 \qquad \text{(II.1.8)}$$

with additional parameter $\alpha \in \mathbb{R}$.

Let $x(\tau, \xi)(t, \varepsilon, \mu, \alpha)$ denote the solution of the initial value problem

$$\dot{x} = f_1(x) + \varepsilon g_1(x, t + \alpha, \varepsilon, \mu)$$
$$x(\tau) = \xi. \qquad \text{(II.1.9)}$$

First, we modify Lemma I.1.2 for impact system (II.1.7), (II.1.8).

Lemma II.1.2. *Assume H1) and H2). Then there exist $\varepsilon_0, r_0 > 0$ and a Poincaré mapping (cf. Figure II.1.1)*

$$P(\cdot, \varepsilon, \mu, \alpha) \colon B(x_0, r_0) \to \Sigma$$

for all fixed $\varepsilon \in (-\varepsilon_0, \varepsilon_0)$, $\mu \in \mathbb{R}^p$, $\alpha \in \mathbb{R}$, where $B(x, r)$ is a ball in $\mathbb{R}^n$ with center at x and radius r, and

$$\Sigma = \{y \in \mathbb{R}^n \mid \langle y - x_0, f_1(x_0) \rangle = 0\}.$$

Moreover, $P \colon B(x_0, r_0) \times (-\varepsilon_0, \varepsilon_0) \times \mathbb{R}^p \times \mathbb{R} \to \mathbb{R}^n$ is C^r-smooth in all arguments and

$B(x_0, r_0) \subset \Omega_1$.

Proof. The lemma follows from the implicit function theorem (IFT) [11]. We obtain the existence of positive constants $\tau_1, r_1, \delta_1, \varepsilon_1$ and C^r-function

$$t_1(\cdot, \cdot, \cdot, \cdot, \cdot) \colon (-\tau_1, \tau_1) \times B(x_0, r_1) \times (-\varepsilon_1, \varepsilon_1) \times \mathbb{R}^p \times \mathbb{R} \to (t_1 - \delta_1, t_1 + \delta_1)$$

such that $h(x(\tau, \xi)(t, \varepsilon, \mu, \alpha)) = 0$ for $\tau \in (-\tau_1, \tau_1)$, $\xi \in B(x_0, r_1) \subset \Omega_1$, $\varepsilon \in (-\varepsilon_1, \varepsilon_1)$, $\mu \in \mathbb{R}^p$, $\alpha \in \mathbb{R}$ and $t \in (t_1 - \delta_1, t_1 + \delta_1)$ if and only if $t = t_1(\tau, \xi, \varepsilon, \mu, \alpha)$. Moreover, $t_1(0, x_0, 0, \mu, \alpha) = t_1$.

Analogously, we derive function t_2 satisfying

$$\langle x(t_1(\tau, \xi, \varepsilon, \mu, \alpha), f_2(x(\tau, \xi)(t_1(\tau, \xi, \varepsilon, \mu, \alpha), \varepsilon, \mu, \alpha))$$
$$+ \varepsilon g_2(x(\tau, \xi)(t_1(\tau, \xi, \varepsilon, \mu, \alpha), \varepsilon, \mu, \alpha), t_1(\tau, \xi, \varepsilon, \mu, \alpha) + \alpha, \varepsilon, \mu))$$
$$(t_2(\tau, \xi, \varepsilon, \mu, \alpha), \varepsilon, \mu, \alpha) - x_0, f_1(x_0)\rangle = 0.$$

Moreover, we have $t_2(0, x_0, 0, \mu, \alpha) = T$. Poincaré mapping is then defined as

$$P(\xi, \varepsilon, \mu, \alpha) = x(t_1(0, \xi, \varepsilon, \mu, \alpha), f_2(x(0, \xi)(t_1(0, \xi, \varepsilon, \mu, \alpha), \varepsilon, \mu, \alpha))$$
$$+ \varepsilon g_2(x(0, \xi)(t_1(0, \xi, \varepsilon, \mu, \alpha), \varepsilon, \mu, \alpha), t_1(0, \xi, \varepsilon, \mu, \alpha) + \alpha, \varepsilon, \mu))(t_2(0, \xi, \varepsilon, \mu, \alpha), \varepsilon, \mu, \alpha).$$
$$\text{(II.1.10)}$$

The proof is finished.   □

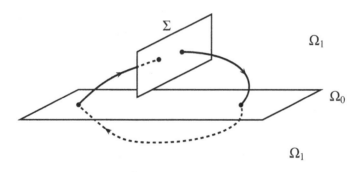

Figure II.1.1 Impact Poincaré mapping

Since we are looking for a persisting periodic orbit with fixed period T, we have to solve a couple of equations

$$P(\xi, \varepsilon, \mu, \alpha) = \xi,$$
$$t_2(0, \xi, \varepsilon, \mu, \alpha) = T.$$

Therefore, we define the stroboscopic Poincaré mapping (cf. (I.1.12), (I.4.7) in Part I)

$$\widetilde{P}(\xi, \varepsilon, \mu, \alpha) = x(t_1(0, \xi, \varepsilon, \mu, \alpha), f_2(x(0, \xi)(t_1(0, \xi, \varepsilon, \mu, \alpha), \varepsilon, \mu, \alpha))$$
$$+\varepsilon g_2(x(0, \xi)(t_1(0, \xi, \varepsilon, \mu, \alpha), \varepsilon, \mu, \alpha), t_1(0, \xi, \varepsilon, \mu, \alpha) + \alpha, \varepsilon, \mu))(T, \varepsilon, \mu, \alpha) \qquad \text{(II.1.11)}$$

and solve a single equation

$$F(\xi, \varepsilon, \mu, \alpha) := \xi - \widetilde{P}(\xi, \varepsilon, \mu, \alpha) = 0 \quad \text{for} \quad \xi \in \Sigma. \qquad \text{(II.1.12)}$$

Now, we calculate the linearization of $\widetilde{P}$ at $(\xi, \varepsilon) = (x_0, 0)$. We obtain the next result.

Lemma II.1.3. *Let $\widetilde{P}(\xi, \varepsilon, \mu, \alpha)$ be defined by (II.1.11). Then for all $\mu \in \mathbb{R}^p$, $\alpha \in \mathbb{R}$,*

$$\widetilde{P}_\xi(x_0, 0, \mu, \alpha) = A(0), \qquad \text{(II.1.13)}$$

$$\widetilde{P}_\varepsilon(x_0, 0, \mu, \alpha) = \int_0^T A(s)g_1(\gamma(s), s + \alpha, 0, \mu)ds + X_2(T)g_2(x_1, t_1 + \alpha, 0, \mu), \qquad \text{(II.1.14)}$$

where $\widetilde{P}_\xi$, $\widetilde{P}_\varepsilon$ are partial derivatives of $\widetilde{P}$ with respect to ξ, ε, respectively. Here $A(t)$ is given by

$$A(t) = \begin{cases} X_2(T)S X_1(t_1)X_1^{-1}(t) & \text{if } t \in [0, t_1), \\ X_2(T)X_2^{-1}(t) & \text{if } t \in [t_1, T] \end{cases} \qquad \text{(II.1.15)}$$

with impact saltation matrix

$$S = Df_2(x_1) + \frac{(f_1(x_2) - Df_2(x_1)f_1(x_1))Dh(x_1)}{Dh(x_1)f_1(x_1)} \qquad \text{(II.1.16)}$$

and fundamental matrix solutions $X_1(t)$, $X_2(t)$ satisfying, respectively,

$$\dot{X}_1(t) = Df_1(\gamma(t))X_1(t) \qquad \dot{X}_2(t) = Df_1(\gamma(t))X_2(t)$$
$$X_1(0) = \mathbb{I}, \qquad\qquad X_2(t_1) = \mathbb{I}. \qquad \text{(II.1.17)}$$

In addition, $\widetilde{P}_\xi(x_0, 0, \mu, \alpha)$ has an eigenvalue 1 with corresponding eigenvector $f_1(x_0)$, i.e.

$$\widetilde{P}_\xi(x_0, 0, \mu, \alpha)f_1(x_0) = f_1(x_0).$$

Proof. The statement on the derivatives of $\widetilde{P}$ follows from its definition with the aid of the following identities

$$D_\xi x(0, x_0)(t, 0, \mu, \alpha) = X_1(t),$$

$$D_\varepsilon x(0, x_0)(t, 0, \mu, \alpha) = \int_0^t X_1(t)X_1^{-1}(s)g_1(\gamma(s), s + \alpha, 0, \mu)ds$$

for $t \in [0, t_1]$,

$$D_\xi x(t_1, x_2)(t, 0, \mu, \alpha) = X_2(t), \qquad D_\tau x(t_1, x_2)(t, 0, \mu, \alpha) = -X_2(t)f_1(x_2),$$

$$D_\varepsilon x(t_1, x_2)(t, 0, \mu, \alpha) = \int_{t_1}^{t} X_2(t)X_2^{-1}(s)g_1(\gamma(s), s + \alpha, 0, \mu)ds$$

for $t \in [t_1, T]$, and

$$D_\xi t_1(0, x_0, 0, \mu, \alpha) = -\frac{Dh(x_1)X_1(t_1)}{Dh(x_1)f_1(x_1)},$$

$$D_\varepsilon t_1(0, x_0, 0, \mu, \alpha) = -\frac{Dh(x_1)\int_0^{t_1} X_1(t_1)X_1^{-1}(s)g_1(\gamma(s), s + \alpha, 0, \mu)ds}{Dh(x_1)f_1(x_1)}.$$

To proceed with the proof, we note $\varepsilon = 0$ results in $x(\tau, \xi)(t, 0, \mu, \alpha) = x(\tau, \xi)(t)$, the solution of $\dot{x} = f_1(x)$, $x(\tau) = \xi$, and it is independent of (μ, α). Analogously, $t_1(\tau, \xi, 0, \mu, \alpha) = t_1(\tau, \xi)$, $t_2(\tau, \xi, 0, \mu, \alpha) = t_2(\tau, \xi)$, $P(\xi, 0, \mu, \alpha) = P(\xi)$ and $\widetilde{P}(\xi, 0, \mu, \alpha) = \widetilde{P}(\xi)$.

Now, since for all $t > 0$ small, it holds that

$$x(0, \gamma_1(t))(t_1(0, \gamma_1(t))) = x(0, x_0)(t + t_1(0, \gamma_1(t))) = \gamma_1(t + t_1(0, \gamma_1(t)))$$

is an element of Ω_0 as well as of $\{\gamma(t) \mid t \in \mathbb{R}\}$, we have

$$t + t_1(0, \gamma_1(t)) = t_1.$$

Consequently,

$$x(0, \gamma_1(t))(t_1(0, \gamma_1(t))) = x_1.$$

Then we obtain

$$\widetilde{P}(\gamma_1(t)) = x(t_1(0, \gamma_1(t)), f_2(x(0, \gamma_1(t))(t_1(0, \gamma_1(t)))))(T)$$
$$= x(t_1(0, \gamma_1(t)), f_2(x_1))(T) = x(t_1 - t, x_2)(T) = x(t_1, x_2)(T + t) = \gamma_1(t).$$

Hence,

$$\widetilde{P}_\xi(x_0, 0, \mu, \alpha)f_1(x_0) = D_t[\widetilde{P}(\gamma_1(t))]_{t=0^+} = D_t[\gamma_1(t)]_{t=0^+} = f_1(x_0).$$

The proof is finished.   □

We solve equation (II.1.12) for $(\xi, \alpha) \in \Sigma \times \mathbb{R}$ with parameters ε, μ using the Lyapunov-Schmidt reduction method. Obviously, $F(x_0, 0, \mu, \alpha) = 0$ for all $(\mu, \alpha) \in \mathbb{R}^p \times \mathbb{R}$. Let us denote

$$Z = \mathcal{N}D_\xi F(x_0, 0, \mu, \alpha), \qquad Y = \mathcal{R}D_\xi F(x_0, 0, \mu, \alpha) \tag{II.1.18}$$

the null space and the range of the corresponding operator and

$$Q: \mathbb{R}^n \to Y, \qquad \mathcal{P}: \mathbb{R}^n \to Y^\perp \tag{II.1.19}$$

orthogonal projections onto Y and $Y^\perp$, respectively, where $Y^\perp$ is the orthogonal complement to Y in $\mathbb{R}^n$. Here we take the third assumption

H3) $\mathcal{N}D_\xi F(x_0, 0, \mu, \alpha) = [f_1(x_0)]$.

Equation (II.1.12) is split into the couple of equations

$$\begin{aligned} \mathcal{Q}F(\xi, \varepsilon, \mu, \alpha) &= 0, \\ \mathcal{P}F(\xi, \varepsilon, \mu, \alpha) &= 0 \end{aligned} \tag{II.1.20}$$

where the first one can be solved using IFT, since

$$\mathcal{Q}F(x_0, 0, \mu, \alpha) = 0$$

and $\mathcal{Q}D_\xi F(x_0, 0, \mu, \alpha)$ is an isomorphism from $[f_1(x_0)]^\perp$ onto Y for all $(\mu, \alpha) \in \mathbb{R}^p \times \mathbb{R}$. Thus we get the existence of a C^r-function $\xi = \xi(\varepsilon, \mu, \alpha)$ for ε close to 0 and $(\mu, \alpha) \in \mathbb{R}^p \times \mathbb{R}$ such that $\mathcal{Q}F(\xi(\varepsilon, \mu, \alpha), \varepsilon, \mu, \alpha) = 0$ for all such $(\varepsilon, \mu, \alpha)$ and $\xi(0, \mu, \alpha) = x_0$. The second equation is the so-called persistence equation for $\alpha \in \mathbb{R}$,

$$\mathcal{P}F(\xi(\varepsilon, \mu, \alpha), \varepsilon, \mu, \alpha) = 0. \tag{II.1.21}$$

Let $\psi \in Y^\perp$ be arbitrary and fixed. Then we can write

$$\mathcal{P}u = \frac{\langle u, \psi \rangle \psi}{\|\psi\|^2},$$

and the persistence equation (II.1.21) has the form

$$G(\varepsilon, \mu, \alpha) := \frac{\langle F(\xi(\varepsilon, \mu, \alpha), \varepsilon, \mu, \alpha), \psi \rangle \psi}{\|\psi\|^2} = 0. \tag{II.1.22}$$

Clearly, $G(0, \mu, \alpha) = 0$ for all $(\mu, \alpha) \in \mathbb{R}^p \times \mathbb{R}$. Moreover, we want the periodic orbit to persist, so we need to solve $G(\varepsilon, \mu, \alpha) = 0$ for $\varepsilon \neq 0$ small. But since $G(\varepsilon, \mu, \alpha) = D_\varepsilon G(0, \mu, \alpha)\varepsilon + o(\varepsilon)$, the equality $D_\varepsilon G(0, \mu_0, \alpha_0) = 0$ is a necessary condition for a point $(0, \mu_0, \alpha_0)$ to be a starting persistence value. This means if there is a sequence $\{(\varepsilon_n, \mu_n, \alpha_n)\}_{n \in \mathbb{N}}$ such that $\varepsilon_n \neq 0$, $(\varepsilon_n, \mu_n, \alpha_n) \to (0, \mu_0, \alpha_0)$ for $n \to \infty$ and $G(\varepsilon_n, \mu_n, \alpha_n) = 0$, then $D_\varepsilon G(0, \mu_0, \alpha_0) = 0$. So we derive

$$\begin{aligned} D_\varepsilon G(0, \mu, \alpha) &= \frac{\langle (D_\xi F(x_0, 0, \mu, \alpha)D_\varepsilon \xi(0, \mu, \alpha) + D_\varepsilon F(x_0, 0, \mu, \alpha)), \psi \rangle \psi}{\|\psi\|^2} \\ &= \frac{\langle D_\varepsilon F(x_0, 0, \mu, \alpha), \psi \rangle \psi}{\|\psi\|^2} = -\frac{\langle D_\varepsilon \widetilde{P}(x_0, 0, \mu, \alpha), \psi \rangle \psi}{\|\psi\|^2}. \end{aligned}$$

We denote

$$M^\mu(\alpha) = \int_0^T \langle g_1(\gamma(s), s + \alpha, 0, \mu), A^*(s)\psi \rangle ds + \langle X_2(T)g_2(x_1, t_1 + \alpha, 0, \mu), \psi \rangle \tag{II.1.23}$$

the impact Poincaré-Andronov-Melnikov function, where

$$A^*(t) = \begin{cases} X_1^{-1*}(t)X_1^*(t_1)S^*X_2^*(T) & \text{if } t \in [0, t_1), \\ X_2^{-1*}(t)X_2^*(T) & \text{if } t \in [t_1, T]. \end{cases} \tag{II.1.24}$$

Note that $D_\varepsilon G(0, \mu, \alpha) = -\frac{M^\mu(\alpha)\psi}{\|\psi\|^2}$.

Linearization of the unperturbed impact system (II.1.3), (II.1.4) along the T-periodic solution $\gamma(t)$ gives the variational equation

$$\dot{x}(t) = Df_1(\gamma(t))x(t) \tag{II.1.25}$$

with impulsive condition

$$x(t_1^+) = S x(t_1^-) \tag{II.1.26}$$

and periodic condition

$$B(x(0) - x(T)) = 0 \tag{II.1.27}$$

where $B = \frac{\psi\psi^*}{\|\psi\|^2}$ is the orthogonal projection onto $Y^\perp$. From the definition of $X_1(t), X_2(t)$,

$$X(t) = \begin{cases} X_1(t) & \text{if } t \in [0, t_1), \\ X_2(t)S X_1(t_1) & \text{if } t \in [t_1, T] \end{cases}$$

solves the variational equation (II.1.25) and the conditions (II.1.26), (II.1.27).

Now, on letting $B_1 = S$, $B_2 = \mathbb{I}$, $B_3 = B$ in Lemma I.2.4 one can see that the adjoint variational system of (II.1.3) and impact condition (II.1.4) (i.e. adjoint system of (II.1.25), (II.1.26), (II.1.27)) is given by the following linear impulsive boundary value problem

$$\begin{aligned} \dot{x}(t) &= -Df_1^*(\gamma(t))x(t), & t \in [0, T], \\ x(t_1^-) &= S^* x(t_1^+), \\ x(T) &= x(0) \in Y^\perp. \end{aligned} \tag{II.1.28}$$

From (II.1.24) we know that $A^*(t)\psi$ solves the adjoint variational equation with impulsive condition. To see that it satisfies the boundary condition as well, we consider

$$0 = \langle D_\xi F(x_0, 0, \mu, \alpha)\xi, \psi \rangle = \langle ((\mathbb{I} - A(0))\xi, \psi \rangle = \langle \xi, (\mathbb{I} - A^*(0))\psi \rangle$$

for all $\xi \in [f_1(x_0)]^\perp$, and if $\xi \in [f_1(x_0)]$, Lemma II.1.3 yields

$$0 = \langle \xi - \xi, \psi \rangle = \langle (\mathbb{I} - A(0))\xi, \psi \rangle = \langle \xi, (\mathbb{I} - A^*(0))\psi \rangle.$$

As a consequence, we can take in (II.1.23) any solution of the adjoint variational system (II.1.28). Summarizing, we get the main result.

Theorem II.1.4. *Let $\psi \in Y^\perp$ be arbitrary and fixed, Y be given by (II.1.18) and $A^*(t)$,*

$M^{\mu}(\alpha)$ be defined by (II.1.24), (II.1.23), respectively. If α_0 is a simple root of function $M^{\mu_0}(\alpha)$, i.e.

$$\int_0^T \langle g_1(\gamma(s), s + \alpha_0, 0, \mu_0), A^*(s)\psi \rangle \, ds + \langle X_2(T)g_2(x_1, t_1 + \alpha_0, 0, \mu_0), \psi \rangle = 0,$$

$$\int_0^T \langle D_t g_1(\gamma(s), s + \alpha_0, 0, \mu_0), A^*(s)\psi \rangle \, ds + \langle X_2(T)D_t g_2(x_1, t_1 + \alpha_0, 0, \mu_0), \psi \rangle \neq 0,$$

then there exists a unique C^{r-1}-function $\alpha(\varepsilon, \mu)$ for $\varepsilon \sim 0$ small and $\mu \sim \mu_0$ such that $\alpha(0, \mu_0) = \alpha_0$, and there is a unique T-periodic solution $x_{\varepsilon,\mu}(t)$ of equation (II.1.1) with parameters $\varepsilon \neq 0$ sufficiently small, μ close to μ_0 and $\alpha = \alpha(\varepsilon, \mu)$, which satisfies condition (II.1.2) and $|x_{\varepsilon,\mu}(t) - \gamma(t - \alpha(\varepsilon, \mu))| = O(\varepsilon)$.

Proof. We set

$$H(\varepsilon, \mu, \alpha) = \begin{cases} D_\varepsilon G(0, \mu, \alpha) & \text{for } \varepsilon = 0, \\ G(\varepsilon, \mu, \alpha)/\varepsilon & \text{for } \varepsilon \neq 0. \end{cases}$$

Then H is C^{r-1}-smooth. Assumptions of our theorem imply $H(0, \mu_0, \alpha_0) = 0$ and $D_\alpha H(0, \mu_0, \alpha_0) \neq 0$. By IFT there exists a unique C^{r-1}-function $\alpha(\varepsilon, \mu)$ for $\varepsilon \sim 0$ small and $\mu \sim \mu_0$ such that $\alpha(0, \mu_0) = \alpha_0$ and $H(\varepsilon, \mu, \alpha(\varepsilon, \mu)) = 0$. But this means that $x(0, \xi(\varepsilon, \mu, \alpha(\varepsilon, \mu)))(t, \varepsilon, \mu, \alpha(\varepsilon, \mu))$ is a solution of (II.1.7), (II.1.8) with $\alpha = \alpha(\varepsilon, \mu)$ and it satisfies

$$|x(0, \xi(\varepsilon, \mu, \alpha(\varepsilon, \mu)))(t, \varepsilon, \mu, \alpha(\varepsilon, \mu)) - \gamma(t)| = O(\varepsilon).$$

Then

$$x_{\varepsilon,\mu}(t) = x(0, \xi(\varepsilon, \mu, \alpha(\varepsilon, \mu)))(t - \alpha(\varepsilon, \mu), \varepsilon, \mu, \alpha(\varepsilon, \mu))$$

is the desired solution of (II.1.1), (II.1.2). The proof is complete.   □

II.1.2. Pendulum hitting moving obstacle

Here we provide an application of derived theory to the problem of a mathematical pendulum which impacts an oscillating wall (see Figure II.1.2). The horizontal distance between the wall and the center of the pendulum is $\delta + \varepsilon z(t, \varepsilon, \mu)$ where z is periodic in t and δ is a positive constant. We denote x the angle and l the length of the massless cord.

Then x satisfies the dimensionless equation

$$\ddot{x} = -\omega^2 x$$

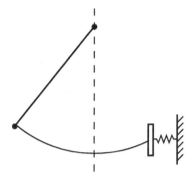

Figure II.1.2 Impacting pendulum

with a given frequency $\omega > 0$, and impact condition

$$\dot{x}(t^+) = -\dot{x}(t^-) + \varepsilon\dot{z}(t^-, \varepsilon, \mu)\frac{\sqrt{l^2 - (\delta + \varepsilon z(t^-, \varepsilon, \mu))^2}}{l}$$

whenever

$$x(t^-) - \arcsin\frac{\delta + \varepsilon z(t^-, \varepsilon, \mu)}{l} = 0,$$

which follows from the actual position of the wall and its speed projected onto a tangent line to the trajectory of the bob. Writing as a system we get

$$\dot{x} = \omega y$$
$$\dot{y} = -\omega x$$

and

$$x(t^+) = x(t^-)$$

$$y(t^+) = -y(t^-) + \varepsilon\dot{z}(t^-, \varepsilon, \mu)\frac{\sqrt{l^2 - (\delta + \varepsilon z(t^-, \varepsilon, \mu))^2}}{\omega l}$$

if

$$x(t^-) - \arcsin\frac{\delta + \varepsilon z(t^-, \varepsilon, \mu)}{l} = 0.$$

To obtain a problem in the form of (II.1.7), (II.1.8), we introduce parameter α and transform the variables

$$u = x - \arcsin\frac{\delta + \varepsilon z(t + \alpha, \varepsilon, \mu)}{l} + \arcsin\frac{\delta}{l}, \qquad v = y.$$

So we get

$$
\dot{u}(t) = \omega v(t) - \varepsilon \frac{\dot{z}(t + \alpha, \varepsilon, \mu)}{\sqrt{l^2 - (\delta + \varepsilon z(t + \alpha, \varepsilon, \mu))^2}}
$$

$$
\dot{v}(t) = -\omega u(t) - \varepsilon \frac{\omega z(t + \alpha, 0, \mu)}{\sqrt{l^2 - \delta^2}} + O(\varepsilon^2)
$$

$(\text{II.1.29})_\varepsilon$

with impact condition

$$
u(t^+) = u(t^-)
$$

$$
v(t^+) = -v(t^-) + \varepsilon \dot{z}(t^- + \alpha, \varepsilon, \mu) \frac{\sqrt{l^2 - (\delta + \varepsilon z(t^- + \alpha, \varepsilon, \mu))^2}}{\omega l}
$$

$(\text{II.1.30})_\varepsilon$

$$
\text{if} \quad h(u(t^-), v(t^-)) = 0
$$

where

$$
h(u, v) = u - \arcsin \frac{\delta}{l}.
$$

Note the dependence on ε in the notation, i.e. $(\text{II.1.29})_0$, $(\text{II.1.30})_0$ denotes the unperturbed system and the unperturbed impact condition, respectively. Moreover, we denote $\hat{u} = \arcsin \frac{\delta}{l}$.

In this case we have $\Omega_0 = \{(u, v) \in \mathbb{R}^2 \mid u = \hat{u}\}$, $\Sigma = \{(u, 0) \in \mathbb{R}^2 \mid u < -\hat{u}\}$ and the following lemma describing the unperturbed problem.

Lemma II.1.5. *System* $(\text{II.1.29})_0$, $(\text{II.1.30})_0$ *possesses a family of periodic orbits* $\gamma^u(t)$ *parametrized by* $u < -\hat{u}$ *such that*

$$
\gamma^u(t) = \begin{cases} (u \cos \omega t, -u \sin \omega t) & \text{if } t \in [0, t_1), \\ \{(u_1, v_1), (u_2, v_2)\} & \text{if } t = t_1, \\ (u \cos \omega(T - t), u \sin \omega(T - t)) & \text{if } t \in (t_1, T] \end{cases}
$$

where

$$
t_1 = \frac{1}{\omega} \arccos \frac{\hat{u}}{u}, \qquad (u_1, v_1) = (u \cos \omega t_1, -u \sin \omega t_1) = \left(\hat{u}, \sqrt{u^2 - \hat{u}^2} \right),
$$

$$
T = 2t_1, \qquad (u_2, v_2) = (u_1, -v_1) = \left(\hat{u}, -\sqrt{u^2 - \hat{u}^2} \right).
$$

The fundamental matrices defined by (II.1.17) *and the impact saltation matrix of* (II.1.16) *have the form*

$$
X_1(t) = \begin{pmatrix} \cos \omega t & \sin \omega t \\ -\sin \omega t & \cos \omega t \end{pmatrix}, \qquad X_2(t) = X_1(t - t_1), \qquad S = \begin{pmatrix} -1 & 0 \\ -\frac{2u_1}{v_1} & -1 \end{pmatrix},
$$

respectively.

Proof. Due to the linearity of (II.1.29)$_0$, taking $(u, 0) \in \Sigma$ as an initial point of $\gamma^u(t)$ one can easily compute $\gamma_1^u(t)$ (see (II.1.5)). Time of impact t_1 is the first intersection point of $\{\gamma_1^u(t) \mid t > 0\}$ with Ω_0 and is obtained from the identity $h(\gamma_1^u(t_1)) = 0$. Accordingly, we get $(u_1, v_1) = \gamma_1^u(t_1)$ and $(u_2, v_2) = f_2(u_1, v_1)$ where $f_2(u, v) = (u, -v)$ is the right-hand side of (II.1.30)$_0$. Similarly to $\gamma_1^u(t)$ we get

$$\gamma_2^u(t) = (u_2 \cos \omega(t - t_1) + v_2 \sin \omega(t - t_1), -u_2 \sin \omega(t - t_1) + v_2 \cos \omega(t - t_1)).$$

From the periodicity of $\gamma^u(t)$ we get T as a solution of the equation

$$\gamma_2^u(T) = (u, 0)$$

or equivalently of a couple of equations

$$u_2 \cos \omega(T - t_1) + v_2 \sin \omega(T - t_1) = u$$
$$-u_2 \sin \omega(T - t_1) + v_2 \cos \omega(T - t_1) = 0.$$

We have

$$T = t_1 + \frac{1}{\omega} \operatorname{arccot} \frac{u_2}{v_2} = 2t_1.$$

Therefore, using trigonometric sum identities,

$$\gamma_2^u(t) = (u \cos \omega(T - t), u \sin \omega(T - t)).$$

Matrices $X_1(t)$, $X_2(t)$ and S are obtained directly from their definitions, since (II.1.29)$_0$ is linear.   □

Now we verify the basic assumptions.

Lemma II.1.6. *System* (II.1.29)$_0$, (II.1.30)$_0$ *satisfies conditions* H1), H2) *and* H3).

Proof. From construction of $\gamma^u(t)$, H1) is immediately verified. So is H2) since

$$Dh(u_1, v_1) f_1(u_1, v_1) = \omega v_1 > 0, \qquad Dh(u_2, v_2) f_1(u_2, v_2) = \omega v_2 < 0.$$

From Lemma II.1.5 we get

$$D_\xi F(u, 0, 0, \mu, \alpha) = \mathbb{I} - X_2(T) S X_1(t_1) = \begin{pmatrix} 0 & 0 \\ \frac{2u_1}{v_1} & 0 \end{pmatrix}.$$

Hence, it is easy to see that $f_1(u, 0) = (0, -\omega u) \in Z$ for Z of (II.1.18). Suppose that $\dim Z > 1$. Then there exists $w \in Z$ such that $\langle w, f_1(u, 0) \rangle = 0$, i.e. $w = (\zeta, 0)$. Next

$$D_\xi F(u, 0, 0, \mu, \alpha) w = \begin{pmatrix} 0 \\ \frac{2\zeta u_1}{v_1} \end{pmatrix}.$$

Thus $\zeta = 0$ and the verification of the last condition is complete.   □

According to (II.1.15), by multiplying the corresponding matrices from Lemma II.1.5 we derive

$$
A(t) = \begin{cases}
\begin{pmatrix} \cos \omega t & -\sin \omega t \\ -\frac{2u_1}{v_1}\cos \omega t + \sin \omega t & \frac{2u_1}{v_1}\sin \omega t + \cos \omega t \end{pmatrix} & \text{if } t \in [0, t_1), \\[2ex]
\begin{pmatrix} \cos\left(\omega t - 2\arccos \frac{\hat{u}}{u}\right) & -\sin\left(\omega t - 2\arccos \frac{\hat{u}}{u}\right) \\ \sin\left(\omega t - 2\arccos \frac{\hat{u}}{u}\right) & \cos\left(\omega t - 2\arccos \frac{\hat{u}}{u}\right) \end{pmatrix} & \text{if } t \in [t_1, T].
\end{cases}
$$

The Fredholm alternative yields

$$
\mathcal{R}(\mathbb{I} - A(0))^{\perp} = \mathcal{N}(\mathbb{I} - A^*(0)) = \mathcal{N}\begin{pmatrix} 0 & \frac{2u_1}{v_1} \\ 0 & 0 \end{pmatrix} = [(1,0)],
$$

thus we take $\psi = (1,0)^*$.

Let us consider $z(t, \varepsilon, \mu) = \sin \mu t$. It is sufficient to assume $\mu > 0$ (the case $\mu < 0$ is covered by parameter $\alpha \in \mathbb{R}$). Then the Poincaré-Andronov-Melnikov function defined by (II.1.23) has the form

$$
M^{\mu}(\alpha) = \frac{1}{2\sqrt{l^2 - \delta^2}} \int_0^{t_1} (\omega - \mu)\cos(\omega t - \mu(t + \alpha))
$$

$$
-(\omega + \mu)\cos(\omega t + \mu(t + \alpha))dt
$$

$$
+ \frac{1}{2\sqrt{l^2 - \delta^2}} \int_{t_1}^{T} (\omega - \mu)\cos\left(\omega t - 2\arccos \frac{\hat{u}}{u} - \mu(t + \alpha)\right)
$$

$$
-(\omega + \mu)\cos\left(\omega t - 2\arccos \frac{\hat{u}}{u} + \mu(t + \alpha)\right)dt - \frac{\sqrt{l^2 - \delta^2}}{\omega l}\frac{\mu v_1}{u}\cos \mu(t_1 + \alpha).
$$

After some algebra we get

$$
M^{\mu}(\alpha) = \frac{v(u)\cos\mu(t_1 + \alpha)}{\sqrt{l^2 - \delta^2}}
$$

where

$$
v(u) = \frac{\mu(l^2 - \delta^2)}{\omega l}\sqrt{1 - \left(\frac{\hat{u}}{u}\right)^2} - 2\sin\mu t_1.
$$

Function $M^{\mu}(\alpha)$ can be easily differentiated with respect to α and one can apply Theorem II.1.4.

Proposition II.1.7. *Let $0 < \omega$, $0 < \mu$ and $k \in \mathbb{N}$ be such that $k\omega < \mu < 2k\omega$. Then for each $r \in \{0, 1, \cdots, 2k - 1\}$, there exists a unique $2k\pi/\mu$-periodic solution $x_{k,r,\varepsilon}(t)$ of* *system* (II.1.29)$_\varepsilon$, (II.1.30)$_\varepsilon$ *with* $\varepsilon \ne 0$ *sufficiently small and*

$$\alpha = \alpha_{k,r}(\varepsilon) = \frac{\pi(2r+1)}{2\mu} + O(\varepsilon)$$

such that

$$|x_{k,r,\varepsilon}(t) - \gamma^u(t - \alpha)| = O(\varepsilon)$$

for any $t \in \mathbb{R}$ *and* $u = u(k) = \frac{\hat{u}}{\cos \frac{k\omega\pi}{\mu}}$. *So there are at least* $2\sum_{k \in (\frac{\mu}{2\omega}, \frac{\mu}{\omega}) \cap \mathbb{N}} k$ *different impact periodic solutions.*

Proof. Since the forcing $\sin \mu t$ has periods $2k\pi/\mu$, $k \in \mathbb{Z}$, we need the period matching condition $T = 2k\pi/\mu$ for some $k \in \mathbb{N}$. This gives

$$2k\pi/\mu = T = 2t_1 = 2\frac{1}{\omega} \arccos \frac{\hat{u}}{u}.$$

Since $\arccos \frac{\hat{u}}{u} \in (\pi/2, \pi)$ we get the assumption $k\omega < \mu < 2k\omega$. Then

$$u = u(k) = \frac{\hat{u}}{\cos \frac{k\omega\pi}{\mu}} < -\hat{u}.$$

Hence

$$v(u(k)) = \frac{\mu(l^2 - \delta^2)}{\omega l} \sin \frac{k\omega\pi}{\mu} > 0.$$

So clearly, $M^\mu(\alpha_0) = 0$ if and only if

$$\alpha_0 = \frac{\pi(2s+1)}{2\mu} - t_1 = \frac{\pi(2(s-k)+1)}{2\mu}$$

for $s \in \mathbb{Z}$. In the period interval $[0, T] = [0, 2k\pi/\mu]$, we have $2k$ different

$$\alpha_0 \in \left\{ \frac{(2r+1)\pi}{2\mu}, \quad r \in \{0, 1, \cdots, 2k-1\} \right\}.$$

Obviously, each α_0 is a simple root of $M^\mu(\alpha)$. The rest follows from Theorem II.1.4.
□

Related problems are also studied in [12, 13].

II.1.3. Forced reflection pendulum

In this section, we study (0.8) in more detail. In spite of the fact that (II.1.1), (II.1.2) is simpler than (0.8), we can follow the previous computations. Now we take any C^∞-function $h(x, y) = h(x)$ which coincides with $h(x) = x$ and $h(x) = x - l$ in neighborhoods of 0 and l, respectively. In general we obtain rather messy formulas, so we pass to the concrete equation

$$\dot{x} = y, \quad \dot{y} = 2 - \varepsilon\eta_4 y + \varepsilon\eta_3 \cos t \quad \text{for} \quad x < 0,$$
$$x(\tilde{t}_1^+) = x(\tilde{t}_1^-) = 0, \quad y(\tilde{t}_1^+) = (1 + \varepsilon\eta_1)y(\tilde{t}_1^-),$$
$$\dot{x} = y, \quad \dot{y} = -\varepsilon\eta_4 y + \varepsilon\eta_3 \cos t \quad \text{for} \quad 0 < x < 1, \qquad \text{(II.1.31)}$$
$$x(\tilde{t}_2^+) = x(\tilde{t}_2^-) = 1, \quad y(\tilde{t}_2^+) = (1 + \varepsilon\eta_2)y(\tilde{t}_2^-),$$
$$\dot{x} = y, \quad \dot{y} = -1 - \varepsilon\eta_4 y + \varepsilon\eta_3 \cos t \quad \text{for} \quad 1 < x,$$

where $\tilde{t}_1, \tilde{t}_2$ are hitting times. Then $T_0 = 4\sqrt{\frac{3}{2}}$, i.e. (0.10) holds for $k \in \mathbb{N}$. We carry on computations for $k = 1$, the other cases being analogous. Condition (0.7) gives $T(\xi) = 2\left(3 - \frac{1}{2\xi}\right)\sqrt{-\xi}$, and (0.9) has solutions

$$\xi_{1,2} = \frac{1}{18}\left(3 - \pi^2 \mp \pi\sqrt{\pi^2 - 6}\right) \doteq -0.724974, -0.0383156.$$

First, we consider the case ξ_1. The unperturbed (II.1.31) is just (see (0.2))

$$\dot{x} = y, \quad \dot{y} = 2 \quad \text{for} \quad x < 0,$$
$$x(\tilde{t}_1^+) = x(\tilde{t}_1^-) = 0, \quad y(\tilde{t}_1^+) = y(\tilde{t}_1^-),$$
$$\dot{x} = y, \quad \dot{y} = 0 \quad \text{for} \quad 0 < x < 1, \qquad \text{(II.1.32)}$$
$$x(\tilde{t}_2^+) = x(\tilde{t}_2^-) = 1, \quad y(\tilde{t}_2^+) = y(\tilde{t}_2^-),$$
$$\dot{x} = y, \quad \dot{y} = -1 \quad \text{for} \quad 1 < x,$$

so there are no impact times. We also note that we need just $\neq 0$ in H2) instead of < 0, which is satisfied for (II.1.32). The solution of (II.1.32) is given by (see (0.5))

$$x(t) = \xi_1 + t^2, \quad y(t) = 2t \quad \text{for} \quad t \in [0, t_1),$$
$$x(t) = 2t_1(t - t_1), \quad y(t) = 2t_1 \quad \text{for} \quad t \in [t_1, t_2),$$
$$x(t) = 1 + 2t_1(t - t_2) - \frac{(t - t_2)^2}{2}, \quad y(t) = 2t_1 - (t - t_2) \quad \text{for} \quad t \in [t_2, \bar{t}_2), \quad \text{(II.1.33)}$$
$$x(t) = 1 - 2t_1(t - \bar{t}_2), \quad y(t) = -2t_1 \quad \text{for} \quad t \in [\bar{t}_2, \bar{t}_1),$$
$$x(t) = -2t_1(t - \bar{t}_1) + (t - \bar{t}_1)^2, \quad y(t) = -2t_1 + 2(t - \bar{t}_1) \quad \text{for} \quad t \in [\bar{t}_1, T],$$

where

$$t_1 \doteq 0.851454, \quad t_2 \doteq 1.43868, \quad \bar{t}_2 \doteq 4.8445, \quad \bar{t}_1 \doteq 5.43173, \quad T = 2\pi.$$

The linearization of (0.2) is $\dot{x} = y$ and $\dot{y} = 0$. So we have $X_1(t) = \begin{pmatrix} 1 & t \\ 0 & 1 \end{pmatrix}$, $X_2(t) = X_1(t - t_1)$, $X_3(t) = X_1(t - t_2)$, $X_4(t) = X_1(t - \bar{t}_2)$ and $X_5(t) = X_1(t - \bar{t}_1)$. Saltation matrices are computed by (I.1.16), subsequently at $(x(t_1), y(t_1))$, $(x(t_2), y(t_2))$, $(x(\bar{t}_2), y(\bar{t}_2))$,

$(x(\bar{t}_1), y(\bar{t}_2))$,

$$S_1 = S_4 = \begin{pmatrix} 1 & 0 \\ -\sqrt{-\frac{1}{\xi_1}} & 1 \end{pmatrix} \doteq \begin{pmatrix} 1 & 0 \\ -1.17446 & 1 \end{pmatrix},$$

$$S_2 = S_3 \doteq \begin{pmatrix} 1 & 0 \\ -\frac{1}{2}\sqrt{-\frac{1}{\xi_1}} & 1 \end{pmatrix} = \begin{pmatrix} 1 & 0 \\ -0.587231 & 1 \end{pmatrix}.$$

Then $A(t)$ is given by (I.1.21),

$$A(t) = \begin{cases} X_1(T - \bar{t}_1)S_4 X_1(\bar{t}_1 - \bar{t}_2)S_3 X_1(\bar{t}_2 - t_2)S_2 X_1(t_2 - t_1) \\ \times S_1 X_1(t_1 - t) & \text{if } t \in [0, t_1), \\ X_1(T - \bar{t}_1)S_4 X_1(\bar{t}_1 - \bar{t}_2)S_3 X_1(\bar{t}_2 - t_2)S_2 X_1(t_2 - t) & \text{if } t \in [t_1, t_2), \\ X_1(T - \bar{t}_1)S_4 X_1(\bar{t}_1 - \bar{t}_2)S_3 X_1(\bar{t}_2 - t) & \text{if } t \in [t_2, \bar{t}_2), \\ X_1(T - \bar{t}_1)S_4 X_1(\bar{t}_1 - t) & \text{if } t \in [\bar{t}_2, \bar{t}_1), \\ X_1(T - t) & \text{if } t \in [\bar{t}_1, T]. \end{cases}$$

Thus we get $A(0) \doteq \begin{pmatrix} 1 & 0 \\ 5.42676 & 1 \end{pmatrix}$. Then (see (II.1.18)) $Z = \mathcal{N}(\mathbb{I} - A(0)) = [(0, 2)^*]$, $Y = \mathcal{R}(\mathbb{I} - A(0)) = \{x = 0\}$ and $H3$ is satisfied along with $Y^{\perp} = [(1, 0)^*]$. So we set $\psi = (1, 0)^*$. Summarizing, we compute the Poincaré-Andronov-Melnikov function (II.1.23) for (II.1.31), where the last term is naturally extended to

$$\sum_{i=1}^{k} \langle A(t_i)g_2(x_i, t_i + \alpha, 0, \mu), \psi \rangle$$

for k impact points x_i with impact times t_i, $i = 1, 2, \ldots, k$. So using

$$g_1(x, y, t, 0, \mu) = (0, -\eta_4 y + \eta_3 \cos t)^*, \quad g_2(x, y, t, 0, \mu) = (0, \eta_{1,2} y)^*,$$

we derive

$$M_1^{\mu}(\alpha) \doteq 2.49576\eta_3 \sin \alpha + 4.17204\eta_4 - 2.8999(\eta_1 + \eta_2). \tag{II.1.34}$$

For completeness, we finish with the case ξ_2. Then (II.1.33) is changed to

$$x(t) = \xi_2 + t^2, \quad y(t) = 2t \quad \text{for} \quad t \in [0, s_1),$$
$$x(t) = 2s_1(t - s_1), \quad y(t) = 2t_1 \quad \text{for} \quad t \in [s_1, s_2),$$
$$x(t) = 1 + 2s_1(t - s_2) - \frac{(t - s_2)^2}{2}, \quad y(t) = 2s_1 - (t - s_2) \quad \text{for} \quad t \in [s_2, \bar{s}_2),$$
$$x(t) = 1 - 2s_1(t - \bar{s}_2), \quad y(t) = -2s_1 \quad \text{for} \quad t \in [\bar{s}_2, \bar{s}_1),$$
$$x(t) = -2s_1(t - \bar{s}_1) + (t - \bar{s}_1)^2, \quad y(t) = -2s_1 + 2(t - \bar{s}_1) \quad \text{for} \quad t \in [\bar{s}_1, T],$$
$$\tag{II.1.35}$$

where

$$s_1 \doteq 0.195744, \quad s_2 \doteq 2.75011, \quad \bar{s}_2 \doteq 3.53308, \quad \bar{s}_1 \doteq 6.08744, \quad T = 2\pi.$$

Matrix $X_1(t)$ remains, but now $X_2(t) = X_1(t - s_1)$, $X_3(t) = X_1(t - s_2)$, $X_4(t) = X_1(t - \bar{s}_2)$, $X_5(t) = X_1(t - \bar{s}_1)$, and saltation matrices are changed to

$$S_1 = S_4 = \begin{pmatrix} 1 & 0 \\ -\sqrt{-\frac{1}{\xi_2}} & 1 \end{pmatrix} \doteq \begin{pmatrix} 1 & 0 \\ -5.10872 & 1 \end{pmatrix},$$

$$S_2 = S_3 = \begin{pmatrix} 1 & 0 \\ -\frac{1}{2}\sqrt{-\frac{1}{\xi_2}} & 1 \end{pmatrix} \doteq \begin{pmatrix} 1 & 0 \\ -2.55436 & 1 \end{pmatrix}.$$

The above formula for $A(t)$ also remains, so now we derive

$$A(0) \doteq \begin{pmatrix} 1 & 0 \\ -102.68053 & 1 \end{pmatrix}.$$

Then again $Z = \mathcal{N}(\mathbb{I} - A(0)) = [(0, 2)^*]$, $Y = \mathcal{R}(\mathbb{I} - A(0)) = \{x = 0\}$, and H3) is satisfied along with $Y^\perp = [(1, 0)^*]$. Hence we set $\psi = (1, 0)^*$. Repeating the previous arguments, we compute the Poincaré-Andronov-Melnikov function (II.1.23) for ξ_2,

$$M_2^\mu(\alpha) \doteq 0.770556\eta_3 \sin\alpha + 0.421487\eta_4 - 0.153262(\eta_1 + \eta_2). \tag{II.1.36}$$

Applying Theorem II.1.4, we arrive at the following result.

Theorem II.1.8. *Let* $\eta_3 \neq 0$ *and set*

$$\Upsilon_1 = \left| 1.161927\frac{\eta_1 + \eta_2}{\eta_3} - 1.671646\frac{\eta_4}{\eta_3} \right|,$$

$$\Upsilon_2 = \left| 0.198898\frac{\eta_1 + \eta_2}{\eta_3} - 0.546991\frac{\eta_4}{\eta_3} \right|.$$

If $\Upsilon_1 < 1$ *(*$\Upsilon_2 < 1$*) (see Figure II.1.3) then (II.1.31) has a 2π-periodic solution for any* $\varepsilon \neq 0$ *small orbitally located near to (II.1.33) ((II.1.35)).*

Proof. Since $\eta_3 \neq 0$, equations $M_1^\mu(\alpha) = 0$ and $M_2^\mu(\alpha) = 0$ are equivalent to

$$\sin\alpha \doteq 1.161927\frac{\eta_1 + \eta_2}{\eta_3} - 1.671646\frac{\eta_4}{\eta_3},$$

$$\sin\alpha \doteq 0.198898\frac{\eta_1 + \eta_2}{\eta_3} - 0.546991\frac{\eta_4}{\eta_3},$$

respectively. Clearly the assumptions $\Upsilon_1 < 1$ or $\Upsilon_2 < 1$ ensure that M_1^μ or M_2^μ has a simple zero. The proof is finished. $\quad\square$

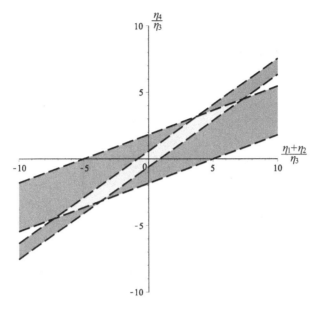

Figure II.1.3 Regions $\Upsilon_1 < 1$ (steeper – green, yellow) and $\Upsilon_2 < 1$ (blue, yellow) in $\frac{\eta_1+\eta_2}{\eta_3} \times \frac{\eta_4}{\eta_3}$ coordinates

CHAPTER II.2

Bifurcation from family of periodic orbits in forced billiards

II.2.1. Setting of the problem and main results

In this chapter, we investigate the persistence of a single periodic orbit from a family of periodic orbits in an autonomous system with $N \geq 1$ impacts under autonomous perturbation, and derive a sufficient condition. So, it is assumed that the persisting orbit will be of period close to the period of the original trajectory. For brevity, we omit the proofs that can be derived from Chapters I.2 and II.1.

Let $\Omega \subset \mathbb{R}^n$ be an open set in $\mathbb{R}^n$. We set $\Omega_0 := \{x \in \Omega \mid h(x) = 0\}$ and $\Omega_1 := \Omega \backslash \Omega_0$ for C^r-function $h(x)$ defined on $\overline{\Omega}$, with $r \geq 3$. Let $f_1 \in C_b^r(\overline{\Omega})$, $f_2 \in C_b^r(\Omega_0, \Omega_0)$, $g_1 \in C_b^r(\overline{\Omega} \times \mathbb{R} \times \mathbb{R} \times \mathbb{R}^p)$, $g_2 \in C_b^r(\Omega_0 \times \mathbb{R} \times \mathbb{R} \times \mathbb{R}^p)$ and $h \in C_b^r(\overline{\Omega}, \mathbb{R})$. Furthermore, we suppose that 0 is a regular value of h. Let $\varepsilon, \alpha \in \mathbb{R}$ and $\mu \in \mathbb{R}^p$, $p \geq 1$ be parameters. Now we modify Definition II.1.1 to the autonomous case.

Definition II.2.1. We say that a function $x(t)$ is a solution of an impact system

$$\dot{x} = f_1(x) + \varepsilon g_1(x, \varepsilon, \mu), \quad x \in \Omega_1, \tag{II.2.1}$$

$$x(t^+) = f_2(x(t^-)) + \varepsilon g_2(x(t^-), \varepsilon, \mu) \quad \text{if} \quad h(x(t^{\pm})) = 0, \tag{II.2.2}$$

if it is piecewise C^1-smooth satisfying equation (II.2.1) on Ω_1, equation (II.2.2) on Ω_0 and, moreover, the following holds: if for some t_0 we have $x(t_0) \in \Omega_0$, then there exists $\rho > 0$ such that for any $t \in (t_0 - \rho, t_0)$, $s \in (t_0, t_0 + \rho)$ we have $h(x(t))h(x(s)) > 0$.

Again, we always naturally suppose that the problem (II.2.1), (II.2.2) is consistent in the sense that $h(f_2(x) + \varepsilon g_2(x, \varepsilon, \mu)) = 0$ whenever $h(x) = 0$.

Let us assume

H1) For $\varepsilon = 0$ equation (II.2.1) has a family of T^β-periodic orbits $\{\gamma(\beta, t)\}$ smooth in β, discontinuous in t at $t \in \{t_i^\beta\}_{i=1}^N$, $0 < t_1^\beta < \cdots < t_N^\beta < T^\beta$, and parametrized by $\beta \in V \subset \mathbb{R}^k$, $0 < k < n$ where V is an open set in $\mathbb{R}^k$. Each of the orbits is uniquely determined by its initial point $x_0(\beta) := \gamma(\beta, 0) \in \Omega_1$, $x_0 \in C_b^r$, and consists of the branches

$$\gamma(\beta, t) = \begin{cases} \gamma_1(\beta, t) & \text{if } t \in [0, t_1^\beta), \\ \gamma_i(\beta, t) & \text{if } t \in (t_{i-1}^\beta, t_i^\beta], \; i = 2, 3, \dots, N, \\ \gamma_{N+1}(\beta, t) & \text{if } t \in (t_N^\beta, T^\beta], \\ \{x_{2i-1}^\beta, x_{2i}^\beta\} & \text{if } t = t_i^\beta, \; i = 1, 2, \dots, N, \end{cases}$$

(II.2.3)

where $\gamma(\beta, t) \in \Omega_1$ for $t \in [0, T^\beta] \backslash \{t_i^\beta\}_{i=1}^N$, $\gamma(\beta, t_i^\beta) \subset \Omega_0$ for $i = 1, 2, \dots, N$, and

$$x_{2i-1}(\beta) := \gamma_i(\beta, t_i^\beta) \in \Omega_0, \quad i = 1, 2, \dots, N,$$

$$x_{2i}(\beta) := \gamma_{i+1}(\beta, t_i^{\beta+}) \in \Omega_0, \quad i = 1, 2, \dots, N,$$

(II.2.4)

$$x_0(\beta) := \gamma_{N+1}(\beta, T^\beta) = \gamma_1(\beta, 0) \in \Omega_1.$$

Moreover, equation (II.2.2) holds at $t \in \{t_i^\beta\}_{i=1}^N$ with $\varepsilon = 0$, and we suppose in addition that vectors

$$\frac{\partial x_0(\beta)}{\partial \beta_1}, \dots, \frac{\partial x_0(\beta)}{\partial \beta_k}, f_1(x_0(\beta))$$

are linearly independent whenever $\beta \in V$.

H2) Furthermore, we assume that

$$Dh(x_{2i-1}(\beta))f_1(x_{2i-1}(\beta))Dh(x_{2i}(\beta))f_1(x_{2i}(\beta)) < 0, \quad i = 1, 2, \dots, N.$$

The geometric meaning of assumption H2) is that each of the impact periodic solutions $\gamma(\beta, t)$ from H1) transversally hits and leaves the impact surface Ω_0 aiming back to one of Ω_1^+ and Ω_1^- for $\Omega_1^\pm := \{x \in \Omega \mid \pm h(x) > 0\}$, at $\{t_i^{\beta-}\}_{i=1}^N$ and $\{t_i^{\beta+}\}_{i=1}^N$, respectively. Note that the assumption H2) with $\neq$ instead of $<$ would be sufficient for our purposes.

Note that by H1), H2) and the implicit function theorem (IFT) it can be shown that $\{t_i^\beta\}_{i=1}^N$ and T^β are C_b^r-functions of β [14].

We study local bifurcations for $\gamma(\beta, t)$, hence we fix $\beta_0 \in V$ and denote $x_0^0 = x_0(\beta_0)$, $t_1^0 = t_1^{\beta_0}$, etc. Note that by H1), $x_0(V)$ is an immersed C^r-submanifold of $\mathbb{R}^n$.

Let $x(\tau, \xi)(t, \varepsilon, \mu)$ denote a solution of the initial value problem

$$\dot{x} = f_1(x) + \varepsilon g_1(x, \varepsilon, \mu)$$

$$x(\tau) = \xi.$$

(II.2.5)

First, we have the existence of a Poincaré mapping (see Lemmas I.1.2 and II.1.2).

Lemma II.2.2. *Assume* H1) *and* H2). *Then there exist* $\varepsilon_1, r_1 > 0$, *a neighborhood* $W \subset V$ *of* β_0 *in* $\mathbb{R}^k$ *and a Poincaré mapping*

$$P(\cdot, \beta, \varepsilon, \mu) \colon B(x_0^0, r_1) \to \Sigma_\beta$$

for all fixed $\beta \in W$, $\varepsilon \in (-\varepsilon_1, \varepsilon_1)$, $\mu \in \mathbb{R}^p$, where

$$\Sigma_\beta = \{y \in \mathbb{R}^n \mid \langle y - x_0(\beta), f_1(x_0(\beta)) \rangle = 0\}.$$

The Poincaré mapping is given by

$$P(\xi, \beta, \varepsilon, \mu) = P_N(0, \xi, \beta, \varepsilon, \mu) \tag{II.2.6}$$

where

$$P_N(\tau, \xi, \beta, \varepsilon, \mu) = x(t_N(\tau, \xi, \varepsilon, \mu), F_2(P_{N-1}(\tau, \xi, \varepsilon, \mu), \varepsilon, \mu))$$
$$(t_{N+1}(\tau, \xi, \beta, \varepsilon, \mu), \varepsilon, \mu),$$
$$P_i(\tau, \xi, \varepsilon, \mu) = x(t_i(\tau, \xi, \varepsilon, \mu), F_2(P_{i-1}(\tau, \xi, \varepsilon, \mu), \varepsilon, \mu))$$
$$(t_{i+1}(\tau, \xi, \varepsilon, \mu), \varepsilon, \mu), \quad i = 1, 2, \ldots, N - 1,$$
$$P_0(\tau, \xi, \varepsilon, \mu) = x(\tau, \xi)(t_1(\tau, \xi, \varepsilon, \mu), \varepsilon, \mu),$$
$$F_2(\xi, \varepsilon, \mu) = f_2(\xi) + \varepsilon g_2(\xi, \varepsilon, \mu),$$

$t_i(\tau, \xi, \varepsilon, \mu)$ is the unique solution of $h(P_{i-1}(\tau, \xi, \varepsilon, \mu)) = 0$, $i = 1, 2, \ldots, N$ for $\tau \sim 0$, $\xi \in B(x_0^0, r_1)$, $\varepsilon \in (-\varepsilon_1, \varepsilon_1)$, $\mu \in \mathbb{R}^p$ such that $t_i(0, x_0(\beta), 0, \mu) = t_i^\beta$, $i = 1, 2, \ldots, N$, and $t_{N+1}(\tau, \xi, \beta, \varepsilon, \mu)$ is the unique solution of $P_N(\tau, \xi, \beta, \varepsilon, \mu) \in \Sigma_\beta$ for $\tau \sim 0$, $\xi \in B(x_0^0, r_1)$, $\beta \in W$, $\varepsilon \in (-\varepsilon_1, \varepsilon_1)$, $\mu \in \mathbb{R}^p$ such that $t_{N+1}(0, x_0(\beta), \beta, 0, \mu) = T^\beta$.

Moreover, $P: B(x_0^0, r_1) \times W \times (-\varepsilon_1, \varepsilon_1) \times \mathbb{R}^p \to \mathbb{R}^n$ is C^r-smooth in all arguments and $x_0(W) \subset B(x_0^0, r_1) \subset \Omega_1$.

The next result describes some properties of the Poincaré mapping (see Lemma II.1.3).

Lemma II.2.3. *Let $P(\xi, \beta, \varepsilon, \mu)$ be defined by (II.2.6). Then*

$$P_\xi(x_0(\beta), \beta, 0, \mu) = (\mathbb{I} - S_\beta)A(\beta, 0), \tag{II.2.7}$$

$$P_\beta(x_0(\beta), \beta, 0, \mu) = S_\beta Dx_0(\beta), \tag{II.2.8}$$

$$P_\varepsilon(x_0(\beta), \beta, 0, \mu) = (\mathbb{I} - S_\beta)\Bigg(\int_0^{T^\beta} A(\beta, s) g_1(\gamma(\beta, s), 0, \mu) ds$$
$$+ \sum_{i=1}^N A(\beta, t_i^\beta) g_2(x_{2i-1}(\beta), 0, \mu) \Bigg), \tag{II.2.9}$$

where P_ξ, P_β, P_ε are partial derivatives of P with respect to ξ, β, ε, respectively.

Here S_β is the orthogonal projection onto the 1-dimensional space $[f_1(x_0(\beta))]$ defined by

$$S_\beta u = \frac{\langle u, f_1(x_0(\beta)) \rangle f_1(x_0(\beta))}{\|f_1(x_0(\beta))\|^2} \tag{II.2.10}$$

and $A(\beta, t)$ is given by

$$
A(\beta, t) = \begin{cases}
X_{N+1}(\beta, T^{\beta}) S_N(\beta) X_N(\beta, t_N^{\beta}) S_{N-1} X_{N-1}(\beta, t_{N-1}^{\beta}) \\
\quad \times \cdots \times S_1(\beta) X_1(\beta, t_1^{\beta}) X_1^{-1}(\beta, t), & \text{if } t \in [0, t_1^{\beta}), \\
\ldots, & \ldots, \\
X_{N+1}(\beta, T^{\beta}) S_N(\beta) X_N(\beta, t_N^{\beta}) X_N^{-1}(\beta, t), & \text{if } t \in [t_{N-1}^{\beta}, t_N^{\beta}), \\
X_{N+1}(\beta, T^{\beta}) X_{N+1}^{-1}(\beta, t), & \text{if } t \in [t_N^{\beta}, T^{\beta}]
\end{cases}
\tag{II.2.11}
$$

with impact saltation matrices

$$
S_i(\beta) = Df_2(x_{2i-1}(\beta)) + \frac{(f_1(x_{2i}(\beta)) - Df_2(x_{2i-1}(\beta)) f_1(x_{2i-1}(\beta))) Dh(x_{2i-1}(\beta))}{Dh(x_{2i-1}(\beta)) f_1(x_{2i-1}(\beta))}
\tag{II.2.12}
$$

and fundamental matrix solutions $\{X_i(\beta, t)\}_{i=1}^{N+1}$ satisfying

$$
\begin{aligned}
\dot{X}_1(\beta, t) &= Df_1(\gamma(\beta, t)) X_1(\beta, t) & \dot{X}_i(\beta, t) &= Df_1(\gamma(\beta, t)) X_i(\beta, t) \\
X_1(\beta, 0) &= \mathbb{I}, & X_i(\beta, t_{i-1}^{\beta}) &= \mathbb{I}
\end{aligned}
\tag{II.2.13}
$$

for $i = 2, 3, \ldots, N + 1$, respectively.

We have the next result on an eigenvalue of P_{ξ}.

Lemma II.2.4. *For any $\xi \in B(x_0^0, r_1)$, $\beta \in W$, $\varepsilon \in (-\varepsilon_1, \varepsilon_1)$ and $\mu \in \mathbb{R}^p$, $P_{\xi}(\xi, \beta, \varepsilon, \mu)$ has eigenvalue 0 with the corresponding eigenvector $f_1(\xi) + \varepsilon g_1(\xi, \varepsilon, \mu)$, i.e.*

$$
P_{\xi}(\xi, \beta, \varepsilon, \mu)[f_1(\xi) + \varepsilon g_1(\xi, \varepsilon, \mu)] = 0.
$$

The initial point $\xi \in \Sigma_{\beta}$ is searched for as a solution of the equation

$$
F(\xi, \beta, \varepsilon, \mu) = 0, \qquad \xi \in \Sigma_{\beta}
\tag{II.2.14}
$$

for $F(\xi, \beta, \varepsilon, \mu) := \xi - S_{\beta}(\xi - x_0(\beta)) - P(\xi, \beta, \varepsilon, \mu)$, where

$$
S_{\beta} = \frac{f_1(x_0(\beta))(f_1(x_0(\beta)))^*}{\|f_1(x_0(\beta))\|^2}
\tag{II.2.15}
$$

is the symmetric matrix representation of S_{β} given by (II.2.10). Now, we take the third assumption
H3) The set

$$
\left\{ \frac{\partial x_0(\beta)}{\partial \beta_1}, \ldots, \frac{\partial x_0(\beta)}{\partial \beta_k}, f_1(x_0(\beta)) \right\}
$$

spans the null space of the operator $F_{\xi}(x_0(\beta), \beta, 0, \mu)$.
and solve the problem (II.2.14) using the Lyapunov-Schmidt reduction. The following

theorem can be proved like Theorem II.1.4.

Theorem II.2.5. *Let the conditions* H1), H2), H3) *be satisfied,* $\{\psi_1(\beta), \ldots, \psi_k(\beta)\}$ *be an arbitrary fixed basis of* $\left[\mathcal{R}F_\xi(x_0(\beta), \beta, 0, \mu)\right]^\perp \cap [f_1(x_0(\beta))]^\perp$, *and*

$$M^\mu(\beta) := (M_1^\mu(\beta), \ldots, M_k^\mu(\beta)),$$

$$M_i^\mu(\beta) = \int_0^{T^\beta} \langle g_1(\gamma(\beta, t), 0, \mu), A^*(\beta, t)\psi_i(\beta)\rangle \, dt \tag{II.2.16}$$

$$+ \sum_{j=1}^N \left\langle g_2(x_{2j-1}(\beta), 0, \mu), A^*(\beta, t_j^\beta)\psi_i(\beta)\right\rangle, \quad i = 1, 2, \ldots, k$$

with $A^*(\beta, t)$ *being the transpose of* $A(\beta, t)$ *given by* (II.2.11). *If* β_0 *is a simple root of the function* M^{μ_0}, *i.e.*

$$M^{\mu_0}(\beta_0) = 0, \qquad \det M^{\mu_0}(\beta_0) \neq 0,$$

then there exists a neighborhood U *of the point* $(0, \mu_0) \in \mathbb{R} \times \mathbb{R}^p$, *and a unique* C^{r-2}-*function* $\beta(\varepsilon, \mu)$ *for* $(\varepsilon, \mu) \in U$ *with* $\beta(0, \mu_0) = \beta_0$ *such that the perturbed impact system* (II.2.1), (II.2.2) *possesses a unique persisting closed trajectory.*

Moreover, the persisting trajectory is $T^{\beta(\varepsilon, \mu)} + O(\varepsilon)$-*periodic and contains the point*

$$x^*(\varepsilon, \mu) := x_0(\beta(\varepsilon, \mu)) + \xi(\beta(\varepsilon, \mu), \varepsilon, \mu) \in \Sigma_{\beta(\varepsilon, \mu)},$$

where

$$\xi(\cdot, \cdot, \cdot) \colon B(\beta_0, r_2) \times (-\varepsilon_2, \varepsilon_2) \times \mathbb{R}^p \to \left[\mathcal{N}F_\xi(x_0(\beta), \beta, 0, \mu) \cap [f_1(x_0(\beta))]^\perp\right]^\perp$$

for some constants $r_2, \varepsilon_2 > 0$, *is a unique* C^{r-1}-*solution of* $Q_\beta F(x_0(\beta) + \xi, \beta, \varepsilon, \mu) = 0$ *for* $\beta \in B(\beta_0, r_2)$, $\varepsilon \in (-\varepsilon_2, \varepsilon_2)$, $\mu \in \mathbb{R}^p$ *such that* $\xi(\beta, 0, \mu) = 0$, *where* $Q_\beta \colon \Sigma_\beta \to \mathcal{R}F_\xi(x_0(\beta), \beta, 0, \mu)$ *is the orthogonal projection.*

II.2.2. Application to a billiard in a circle

Here we consider a planar billiard of one ball moving in a circle with weak friction and a weakly nonlinear gravitational field. That means, writing as a system of the form (II.2.1), (II.2.2), we consider

$$\begin{aligned}
\dot{x}_1 &= x_2 \\
\dot{x}_2 &= -\varepsilon x_2 \\
\dot{y}_1 &= y_2 \\
\dot{y}_2 &= -\varepsilon y_2 - \varepsilon(\mu_0 + \mu_1 x_1 + \mu_2 x_1^2)
\end{aligned} \tag{II.2.17}$$

if $h(x_1, x_2, y_1, y_2) := 1 - x_1^2 - y_1^2 > 0$ with the impact condition

$$x_1(t^+) = x_1(t^-)$$
$$x_2(t^+) = (1 - 2x_1^2(t^-))x_2(t^-) - 2x_1(t^-)y_1(t^-)y_2(t^-)$$
$$y_1(t^+) = y_1(t^-)$$
$$y_2(t^+) = -2x_1(t^-)x_2(t^-)y_1(t^-) + (1 - 2y_1^2(t^-))y_2(t^-)$$

(II.2.18)

if $h(x_1(t^-), x_2(t^-), y_1(t^-), y_2(t^-)) = 0$. Here $\mu = (\mu_0, \mu_1, \mu_2) \in \mathbb{R}^3$ are parameters. We set $\Sigma = \{0\} \times \mathbb{R}^3$. Note that (II.2.17) rewritten in Newton mechanical form

$$\ddot{x}_1 + \varepsilon \dot{x}_1 = 0$$
$$\ddot{y}_1 + \varepsilon \dot{y}_1 = -\varepsilon(\mu_0 + \mu_1 x_1 + \mu_2 x_1^2)$$

is really weakly damped with gravitational field invariant in the y_1 axis, i.e. depending only on the x_1 variable.

The theory of flat billiards is by now classic and very well developed. We refer the reader to [15–17] for more details and references. However, other kinds of billiards are also studied. According to [18], for example, a billiard in a broad sense is the geodesic flow on a Riemannian manifold with boundary. Non-flat billiards are studied in [19]. Related problems to (II.2.17), (II.2.18) are studied in [20, 21].

The following statement can be proved for the unperturbed system.

Lemma II.2.6. *System (II.2.17), (II.2.18) with $\varepsilon = 0$ has a two-parameter family of periodic orbits $\gamma(\beta, t)$, $\beta = (\beta_1, \beta_2) \in (-\frac{\pi}{3}, \frac{\pi}{3}) \times (0, \infty)$ given by (II.2.3) with*

$$\gamma_1(\beta, t) = \gamma_0 \left(0, \beta_2 \cos\beta_1, \frac{1}{2\cos\beta_1}, \beta_2 \sin\beta_1, t \right),$$

$$\gamma_2(\beta, t) = \gamma_0(\bar{x}_2, t - t_1^\beta), \quad \gamma_3(\beta, t) = \gamma_0(\bar{x}_4, t - t_2^\beta), \quad \gamma_4(\beta, t) = \gamma_0(\bar{x}_6, t - t_3^\beta)$$

where $\gamma_0(\bar{x}, t) = (x_2 t + x_1, x_2, y_2 t + y_1, y_2)$, $\bar{x} = (x_1, x_2, y_1, y_2)$, $\bar{x}_i = (x_{i1}, x_{i2}, y_{i1}, y_{i2})$ for

$i = 1, 2, \ldots, 6$,

$$\bar{x}_1 = \left(\sin\left(\frac{\pi}{3} - \beta_1\right), \beta_2 \cos\beta_1, \cos\left(\frac{\pi}{3} - \beta_1\right), \beta_2 \sin\beta_1\right),$$

$$\bar{x}_2 = \left(\sin\left(\frac{\pi}{3} - \beta_1\right), -\beta_2 \cos\left(\frac{\pi}{3} + \beta_1\right), \cos\left(\frac{\pi}{3} - \beta_1\right), -\beta_2 \sin\left(\frac{\pi}{3} + \beta_1\right)\right),$$

$$\bar{x}_3 = \left(\sin\beta_1, -\beta_2 \cos\left(\frac{\pi}{3} + \beta_1\right), -\cos\beta_1, -\beta_2 \sin\left(\frac{\pi}{3} + \beta_1\right)\right),$$

$$\bar{x}_4 = \left(\sin\beta_1, -\beta_2 \cos\left(\frac{\pi}{3} - \beta_1\right), -\cos\beta_1, \beta_2 \sin\left(\frac{\pi}{3} - \beta_1\right)\right),$$

$$\bar{x}_5 = \left(-\sin\left(\frac{\pi}{3} + \beta_1\right), -\beta_2 \cos\left(\frac{\pi}{3} - \beta_1\right), \cos\left(\frac{\pi}{3} + \beta_1\right), \beta_2 \sin\left(\frac{\pi}{3} - \beta_1\right)\right),$$

$$\bar{x}_6 = \left(-\sin\left(\frac{\pi}{3} + \beta_1\right), \beta_2 \cos\beta_1, \cos\left(\frac{\pi}{3} + \beta_1\right), \beta_2 \sin\beta_1\right)$$

and $t_1^\beta = \frac{\sqrt{3} - \tan\beta_1}{2\beta_2}$, $t_2^\beta = \frac{\sqrt{3}}{\beta_2} + t_1^\beta$, $t_3^\beta = \frac{\sqrt{3}}{\beta_2} + t_2^\beta$, $T^\beta = \frac{3\sqrt{3}}{\beta_2}$.

Hence we have $\bar{x}_0(\beta) = \left(0, \beta_2 \cos\beta_1, \frac{1}{2\cos\beta_1}, \beta_2 \sin\beta_1\right)^*$. One can easily verify that assumptions H1) and H2) are fulfilled. To check H3), we need the next result.

Lemma II.2.7. *Fundamental matrices of* (II.2.13) *are of the forms*

$$X_1(\beta, t) = \begin{pmatrix} 1 & t & 0 & 0 \\ 0 & 1 & 0 & 0 \\ 0 & 0 & 1 & t \\ 0 & 0 & 0 & 1 \end{pmatrix}, \qquad \begin{aligned} X_2(\beta, t) &= X_1(\beta, t - t_1^\beta), \\ X_3(\beta, t) &= X_1(\beta, t - t_2^\beta), \\ X_4(\beta, t) &= X_1(\beta, t - t_3^\beta), \end{aligned}$$

and saltation matrices of (II.2.12) *are* $S_1(\beta) = S(\beta, 2\beta_1 + \frac{\pi}{3})$, $S_2(\beta) = S(\beta, 2\beta_1 + \pi)$ *and* $S_3(\beta) = S(\beta, 2\beta_1 - \frac{\pi}{3})$ *where*

$$S(\beta, t) = \begin{pmatrix} -\cos t & 0 & -\sin t & 0 \\ -\frac{\sqrt{3}}{3}\beta_2(1 - 2\cos t) & -\cos t & \frac{\sqrt{3}}{3}\beta_2\left(\sqrt{3} + 2\sin t\right) & -\sin t \\ -\sin t & 0 & \cos t & 0 \\ -\frac{\sqrt{3}}{3}\beta_2\left(\sqrt{3} - 2\sin t\right) & -\sin t & -\frac{\sqrt{3}}{3}\beta_2(1 + 2\cos t) & \cos t \end{pmatrix}.$$

Now, Lemma II.2.3 yields

$$F_\xi(\bar{x}_0(\beta), \beta, 0, \mu) = (\mathbb{I} - S_\beta)(\mathbb{I} - A(\beta, 0))$$

for S_β given by (II.2.15), and $A(\beta, t)$ by (II.2.11). Therefore, by Lemma II.2.7, one can show

$$F_\xi(\bar{x}_0(\beta), \beta, 0, \mu) = 2\sqrt{3}\cos^2\beta_1 v w^* \qquad (\text{II.2.19})$$

for $v = \left(-\frac{\tan^2 \beta_1}{2\beta_2}, -\tan\beta_1, \frac{\tan\beta_1}{2\beta_2}, 1\right)^*$, $w = (2\beta_2 \tan\beta_1, -\tan^2\beta_1, -2\beta_2, \tan\beta_1)^*$. Clearly,

$$\mathcal{N}F_\xi(\bar{x}_0(\beta), \beta, 0, \mu) = [w]^\perp$$
$$= [(1, 0, \tan\beta_1, 0)^*, (0, 1, 0, \tan\beta_1)^*, (0, 0, \tan\beta_1, 2\beta_2)^*].$$

So now, it is easy to verify that H3) is satisfied. Moreover, from (II.2.19) we know that $\mathcal{R}F_\xi(\bar{x}_0(\beta), \beta, 0, \mu) = [v]$. Thus

$$\left[\mathcal{R}F_\xi(x_0(\beta), \beta, 0, \mu)\right]^\perp \cap [f_1(x_0(\beta))]^\perp$$
$$= \left[(0, 1, 0, \tan\beta_1)^*, \left(-\tan\beta_1, \frac{\tan^2\beta_1}{2\beta_2}, 1, -\frac{\tan\beta_1}{2\beta_2}\right)^*\right].$$

For $\psi_1(\beta) = (0, 1, 0, \tan\beta_1)^*$, $\psi_2 = \left(-\tan\beta_1, \frac{\tan^2\beta_1}{2\beta_2}, 1, -\frac{\tan\beta_1}{2\beta_2}\right)^*$ the Poincaré-Andronov-Melnikov function of (II.2.16) is of the form

$$M^\mu(\beta) = \left(\frac{3\sqrt{3}(\mu_1 - 4\beta_2)}{4\beta_2 \cos\beta_1}, \frac{3\sqrt{3}(2\mu_1 - \mu_2 \sin 3\beta_1)}{16\beta_2^2 \cos\beta_1}\right). \tag{II.2.20}$$

Applying Theorem II.2.5 gives the following result.

Proposition II.2.8. *Let* $\mu_0 \in \mathbb{R}$, $\mu_1 > 0$, $0 \neq \mu_2 \in \mathbb{R}$ *satisfy* $\left|\frac{2\mu_1}{\mu_2}\right| < 1$. *Then the perturbed system* (II.2.17), (II.2.18) *possesses one persisting periodic orbit for* $\varepsilon \neq 0$, *which is close to* $\gamma(\beta_0, t)$ *with* $\beta_0 = \left(\frac{1}{3}\arcsin\frac{2\mu_1}{\mu_2}, \frac{\mu_1}{4}\right)$.

Proof. Clearly, β_0 is a zero of M^μ given by (II.2.20). It can be shown that

$$DM^\mu(\beta_0) = \begin{pmatrix} 0 & -\frac{12\sqrt{3}}{\mu_1 \cos\left(\frac{1}{3}\arcsin\frac{2\mu_1}{\mu_2}\right)} \\ -\frac{9\,\text{sgn}\,\mu_2\,\sqrt{3\mu_2^2 - 12\mu_1^2}}{\mu_1^2 \cos\left(\frac{1}{3}\arcsin\frac{2\mu_1}{\mu_2}\right)} & 0 \end{pmatrix}$$

which has a nonzero determinant due to the assumptions.   □

Figure II.2.1 illustrates the persisting orbit. The preceding computations could be extended to study persistence of periodic orbits for perturbed billiards on a triangle motivated by [22, 23], but we do not go into detail.

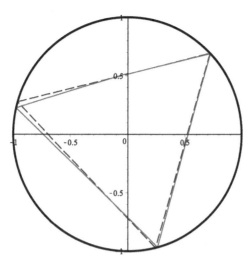

Figure II.2.1 Numerically computed solution of (II.2.17), (II.2.18) with $\varepsilon = 10^{-5}$, $\mu = (0, 0.1, 0.3)$ (solid red) persisting from the solution of the unperturbed system (dashed blue)

REFERENCE

[1] D. Bainov, P. S. Simeonov, *Impulsive Differential Equations: Asymptotic Properties of the Solutions*, World Scientific Publishing Company 1995.

[2] M. di Bernardo, C. J. Budd, A. R. Champneys, P. Kowalczyk, *Piecewise-smooth Dynamical Systems: Theory and Applications*, Applied Mathematical Sciences 163, Springer-Verlag 2008.

[3] D. R. J. Chillingworth, Discontinuous geometry for an impact oscillator, *Dyn. Syst.* **17** (2002) 389–420.

[4] B. Brogliato, *Nonsmooth Impact Mechanics*, Lecture Notes in Control and Information Sciences 220, Springer 1996.

[5] Z. Du, W. Zhang, Melnikov method for homoclinic bifurcation in nonlinear impact oscillators, *Comput. Math. Appl.* **50** (2005) 445–458.

[6] A. Halanay, D. Wexler, *Qualitative Theory of Impulsive Systems*, Editura Academiei Republicii Socialiste Romania 1968.

[7] A. Kovaleva, The Melnikov criterion of instability for random rocking dynamics of a rigid block with an attached secondary structure, *Nonlinear Anal. Real World Appl.* **11** (2010) 472–479.

[8] V. Lakshmikantham, D. Bainov, P.S. Simeonov, *Theory of Impulsive Differential Equation*, World Scientific Publishing Company 1989.

[9] A.M. Samoilenko, N.A. Perestyuk, *Impulsive Differential Equations*, World Scientific Publishing Company 1995.

[10] A. Fidlin, *Nonlinear Oscillations in Mechanical Engineering*, Springer 2006.

[11] K. Deimling, *Nonlinear Functional Analysis*, Springer-Verlag 1985.

[12] J. Awrejcewicz, G. Kudra, Modeling, numerical analysis and application of triple physical pendulum with rigid limiters of motion, *Arch. Appl. Mech.* **74** (11) (2005) 746–753.

[13] J. Awrejcewicz, G. Kudra, C.-H. Lamarque, Dynamics investigation of three coupled rods with a horizontal barrier, *Meccanica* **38** (6) (2003) 687–698.

[14] M. U. Akhmet, On the smoothness of solutions of differential equations with a discontinuous right-hand side, *Ukrainian Math. J.* **45** (1993) 1785–1792.

[15] N. Chernov, R. Markarian, *Chaotic Billiards*, Mathematical Surveys and Monographs 127, Amer. Math. Soc. 2006.

[16] V. V. Kozlov, D. V. Treshvhev, *Billiards: A Genetic Introduction to the Dynamics of Systems with Impacts*, Translations of Mathematical Monographs 89, American Mathematical Society 1991.

[17] M. Rychlik, Periodic points of the billiard ball map in a convex domain, *J. Differential Geom.* **30** (1989) 191–205.

[18] E. Gutkin, Billiard dynamics: a survey with the emphasis on open problems, *Regul. Chaotic Dyn.* **8** (2003) 1–13.

[19] F. Battelli, M. Fečkan, On the chaotic behavior of non-flat billiards, *Commun. Nonlinear Sci. Numer. Simul.* **19** (2014) 1442–1464.

[20] H.-J. Jodl H. J. Korsch, T. Hartmann, *Chaos: A Program Collection for the PC*, 3rd edition, Springer-Verlag Berlin 2008.

[21] S. Tabachnikov, *Geometry and Billiards*, Student Mathematical Library 30, American Mathematical Society 2005.

[22] A. M. Baxter, R. Umble, Periodic orbits of billiards on an equilateral triangle, *Amer. Math. Monthly* **115** (2008) 479–491.

[23] T. Ruijgrok, Periodic orbits in triangular billiards, *Acta Phys. Polon. B* **22** (1991) 955–981.

PART III

Continuous approximations of non-smooth systems

Introduction

One way of investigating the discontinuous vector field is an approximation by a one-parametric family of continuous vector fields which should keep certain dynamical properties of the original one. Continuous approximation is mostly used for differential equations with non-smooth nonlinearities, such as a dry friction nonlinearity (see a survey paper [1]). This approximation process is closely related to a geometric singular perturbation theory [2, 3]. On the other hand, it seems by [4] that an impact oscillator modeling a bouncing ball is better described by a harmonic oscillator with a jumping nonlinearity with the force field nearly infinite in one side, than by its limit version. This approach is used also in [5] when an impact oscillator is approximated by a one-parametric family of singularly perturbed differential equations, but the geometric singular perturbation theory does not apply. In this part, we develop suitable continuous approximation methods for transversal orbits studied in Chapters I.1, I.2 and I.3, then for sliding orbits investigated in Chapter I.4 and, finally, for impact orbits dealt within Chapter II.1. In the last chapter of this part, Chapter III.4, we study the relationship between dynamics and approximation by using the Lyapunov function method. The equivalence between differentiable and non-differentiable dynamics in $\mathbb{R}^n$ is studied also in [6].

CHAPTER III.1

Transversal periodic orbits

III.1.1. Setting of the problem and main result

In this chapter, we continuously approximate system (I.1.1) from Chapter I.1. We rewrite (I.1.1) in the form

$$\dot{x} = F_{\pm}(x, t, \alpha, \varepsilon, \mu), \quad x \in \overline{\Omega}_{\pm}, \tag{III.1.1}$$

where $F_{\pm}(x, t, \alpha, \varepsilon, \mu) = f_{\pm}(x) + \varepsilon g(x, t + \alpha, \varepsilon, \mu)$ keeping the notations of that chapter. Let us consider a C^{∞}-smooth cutoff function

$$\theta(r) = \begin{cases} -1 & \text{if } r \in (-\infty, -1], \\ \tanh \frac{2r}{1-r^2} & \text{if } r \in (-1, 1), \\ 1 & \text{if } r \in [1, \infty) \end{cases} \tag{III.1.2}$$

with the graph shown in Figure III.1.1.

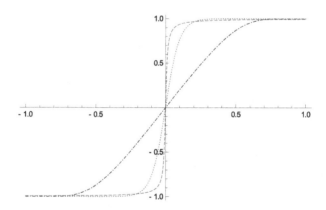

Figure III.1.1 Graphs of the functions $\frac{2}{\pi} \arctan 100r$ (dashed), $\tanh 10r$ (dotted) and (III.1.2) (dash-dotted)

Remark III.1.1. The functions $\frac{2}{\pi} \arctan 100r$ and $\tanh 10r$ and their modifications are also often used instead of (III.1.2). We now prefer the last one for easier computations. Nevertheless, the analysis in this section can be straightforwardly extended for the

Poincaré-Andronov-Melnikov Analysis for Non-Smooth Systems.
http://dx.doi.org/10.1016/B978-0-12-804294-6.50014-X

others.

Now we consider a smooth approximation of (III.1.1) given by [1]

$$\dot{x} = \frac{1 + \theta(h(x)/\eta)}{2} F_+(x, t, \alpha, \varepsilon, \mu) + \frac{1 - \theta(h(x)/\eta)}{2} F_-(x, t, \alpha, \varepsilon, \mu) \qquad \text{(III.1.3)}$$

for any $\eta > 0$. We denote $\Omega_{\pm\eta} := \{x \in \Omega \mid \pm h(x) > \eta\}$ and $\Omega_\eta := \{x \in \Omega \mid |h(x)| < \eta\}$. Clearly, (III.1.3) coincides with (III.1.1) on $\Omega_{+\eta} \cup \Omega_{-\eta}$. Let $x_\eta(\tau, \xi)(t, \varepsilon, \mu, \alpha)$ be a solution of (III.1.3) with the initial value condition $x(\tau) = \xi \in B(x_0, r_0)$ for $r_0 > 0$ sufficiently small. We assume conditions H1) and H2) of Section I.1.1. Then the implicit function theorem (IFT) yields the existence of $\tau_1, r_1, \delta_{1,2,3}, \varepsilon_1, \eta_1 > 0$ and C^r-functions

$$\widetilde{t}_1(\cdot, \cdot, \cdot, \cdot, \cdot) \colon B(x_0, r_1) \times (-\varepsilon_1, \varepsilon_1) \times [0, \eta_1) \times \mathbb{R}^p \times \mathbb{R} \to (t_1 - \delta_1, t_1 + \delta_1),$$

$$\widetilde{t}_2(\cdot, \cdot, \cdot, \cdot, \cdot, \cdot) \colon (t_1 - \tau_1, t_1 + \tau_1) \times B(x_1, r_1) \times (-\varepsilon_1, \varepsilon_1) \times [0, \eta_1) \times \mathbb{R}^p \times \mathbb{R}$$
$$\to (t_2 - \delta_2, t_2 + \delta_2),$$

$$\widetilde{t}_3(\cdot, \cdot, \cdot, \cdot, \cdot) \colon (t_2 - \tau_1, t_2 + \tau_1) \times B(x_2, r_1) \times (-\varepsilon_1, \varepsilon_1) \times \mathbb{R}^p \times \mathbb{R}$$
$$\to (T - \delta_3, T + \delta_3)$$

such that

$$h(x_+(0, \xi_1)(\bar{t}_1, \varepsilon, \mu, \alpha)) = \eta,$$
$$h(x_-(\tau, \xi_2)(\bar{t}_2, \varepsilon, \mu, \alpha)) = -\eta,$$
$$x_+(\bar{\tau}, \xi_3)(\bar{t}_3, \varepsilon, \mu, \alpha) \in \Sigma$$

for $\tau \in (t_1 - \tau_1, t_1 + \tau_1)$, $\bar{\tau} \in (t_2 - \tau_1, t_2 + \tau_1)$, $\xi_i \in B(x_{i-1}, r_1)$, $i = 1, 2, 3$, $\varepsilon \in (-\varepsilon_1, \varepsilon_1)$, $\eta \in [0, \eta_1)$, $\mu \in \mathbb{R}^p$, $\alpha \in \mathbb{R}$ and $\bar{t}_i \in (t_i - \delta_i, t_i + \delta_i)$, $i = 1, 2$, $\bar{t}_3 \in (T - \delta_3, T + \delta_3)$ if and only if $\bar{t}_1 = \widetilde{t}_1(\xi_1, \varepsilon, \eta, \mu, \alpha)$, $\bar{t}_2 = \widetilde{t}_2(\tau, \xi_2, \varepsilon, \eta, \mu, \alpha)$, $\bar{t}_3 = \widetilde{t}_3(\bar{\tau}, \xi_3, \varepsilon, \mu, \alpha)$. In this way, we get solutions of (III.1.3) on the intervals $[0, \widetilde{t}_1(\xi_1, \varepsilon, \eta, \mu, \alpha)]$, $[\tau, \widetilde{t}_2(\tau, \xi_2, \varepsilon, \eta, \mu, \alpha)]$ and $[\bar{\tau}, \widetilde{t}_3(\bar{\tau}, \xi_3, \varepsilon, \mu, \alpha)]$ with $x(0) = \xi_1$, $x(\tau) = \xi_2$ for $h(\xi_2) = -\eta$, and $x(\bar{\tau}) = \xi_3$ for $h(\xi_3) = \eta$.

Since the first part of assumption H2) also holds in a bounded neighborhood O_1 of x_1, we fix it such that $B(x_1, r_1) \subset O_1$. Then there are positive constants c_1, c_2 and c_3 such that

$$Dh(x)f_\pm(x) < -c_1, \quad \|F_\pm(x, t, \alpha, \varepsilon, \mu)\| \le c_2, \quad \|Dh(x)\| \le c_3$$

for any $(x, t, \alpha, \varepsilon, \mu) \in B(x_1, r_1) \times \mathbb{R}^2 \times (-\varepsilon_1, \varepsilon_1) \times \mathbb{R}^p$. Now, consider $\xi \in B(x_1, r_1/2)$ with $h(\xi) = \eta$ (we can take η_1 sufficiently small). Then

$$\|x_\eta(\tau, \xi)(t, \varepsilon, \mu, \alpha) - x_1\| \le \|\xi - x_1\| + c_2(t - \tau) \le r_1$$

for any $t \in [\tau, \tau + \frac{r_1}{2c_2}]$, and thus

$$-c_2 c_3 \leq \frac{d}{dt} h(x_\eta(\tau, \xi)(t, \varepsilon, \mu, \alpha)) < -c_1. \qquad \text{(III.1.4)}$$

Hence, integrating the above inequality by t from τ to $\tau + \frac{r_1}{2c_2}$,

$$-\frac{c_3 r_1}{2} \leq h\left(x_\eta(\tau, \xi)\left(\tau + \frac{r_1}{2c_2}, \varepsilon, \mu, \alpha\right)\right) - \eta < -\frac{c_1 r_1}{2c_2}.$$

Taking $\eta_1 < \frac{c_1 r_1}{4c_2}$, the IFT yields the existence of $\bar{\tau}_1, \bar{\varepsilon}_1 > 0$ and C^r-function

$$t_{1\eta}(\cdot, \cdot, \cdot, \cdot, \cdot) \colon (t_1 - \bar{\tau}_1, t_1 + \bar{\tau}_1) \times B_\eta(x_1, r_1/2) \times (-\bar{\varepsilon}_1, \bar{\varepsilon}_1) \times \mathbb{R}^p \times \mathbb{R} \to \mathbb{R}$$

such that

$$h(x_\eta(\tau, \xi)(t, \varepsilon, \mu, \alpha)) = -\eta$$

for $\tau \in (t_1 - \bar{\tau}_1, t_1 + \bar{\tau}_1), \xi \in B_\eta(x_1, r_1/2) := B(x_1, r_1/2) \cap h^{-1}(\eta), \varepsilon \in (-\bar{\varepsilon}_1, \bar{\varepsilon}_1), \mu \in \mathbb{R}^p,$ $\alpha \in \mathbb{R}$ and $t \in [\tau, \tau + \frac{r_1}{2c_2}]$ if and only if $t = t_{1\eta}(\tau, \xi, \varepsilon, \mu, \alpha)$. Integrating (III.1.4) by t from τ to $t_{1\eta}(\tau, \xi, \varepsilon, \mu, \alpha)$, we also derive

$$\frac{2\eta}{c_2 c_3} \leq t_{1\eta}(\tau, \xi, \varepsilon, \mu, \alpha) - \tau < \frac{2\eta}{c_1},$$

which implies

$$\|x_\eta(\tau, \xi)(t, \varepsilon, \mu, \alpha) - \xi\| \leq \frac{2c_2 \eta}{c_1}$$

for any $t \in [\tau, t_{1\eta}(\tau, \xi, \varepsilon, \mu, \alpha)]$. The same analysis can be done near x_2 to get a C^r-function $t_{2\eta}(\tau, \xi, \varepsilon, \mu, \alpha)$ for ξ near x_2 with $h(\xi) = -\eta$ and

$$h(x_\eta(\tau, \xi)(t_{2\eta}(\tau, \xi, \varepsilon, \mu, \alpha), \varepsilon, \mu, \alpha)) = \eta.$$

Summarizing, the solution $x_\eta(0,\xi)(t,\varepsilon,\mu,\alpha)$ of (III.1.3) is given by (see Figure III.1.2)

$$t \in [0, \widetilde{t}_1(\xi,\varepsilon,\eta,\mu,\alpha)]: \quad x_+(0,\xi)(t,\varepsilon,\mu,\alpha),$$

$$t \in [\widetilde{t}_1(\xi,\varepsilon,\eta,\mu,\alpha), t_{1\eta}(\widetilde{\tau}_1,\widetilde{\xi}_1,\varepsilon,\mu,\alpha)]: \quad x_\eta(\widetilde{\tau}_1,\widetilde{\xi}_1)(t,\varepsilon,\mu,\alpha)$$

$$\text{for } \widetilde{\tau}_1 = \widetilde{t}_1(\xi,\varepsilon,\eta,\mu,\alpha), \quad \widetilde{\xi}_1 = x_+(0,\xi)(\widetilde{\tau}_1,\varepsilon,\mu,\alpha),$$

$$t \in [t_{1\eta}(\widetilde{\tau}_1,\widetilde{\xi}_1,\varepsilon,\mu,\alpha), \widetilde{t}_2(\bar{\tau}_1,\bar{\xi}_1,\varepsilon,\eta,\mu,\alpha)]: \quad x_-(\bar{\tau}_1,\bar{\xi}_1)(t,\varepsilon,\mu,\alpha)$$

$$\text{for } \bar{\tau}_1 = t_{1\eta}(\widetilde{\tau}_1,\widetilde{\xi}_1,\varepsilon,\mu,\alpha), \quad \bar{\xi}_1 = x_\eta(\widetilde{\tau}_1,\widetilde{\xi}_1)(\widetilde{\tau}_1,\varepsilon,\mu,\alpha), \qquad \text{(III.1.5)}$$

$$t \in [\widetilde{t}_2(\bar{\tau}_1,\bar{\xi}_1,\varepsilon,\eta,\mu,\alpha), t_{2\eta}(\widetilde{\tau}_2,\widetilde{\xi}_2,\varepsilon,\mu,\alpha)]: \quad x_\eta(\widetilde{\tau}_2,\widetilde{\xi}_2)(t,\varepsilon,\mu,\alpha)$$

$$\text{for } \widetilde{\tau}_2 = \widetilde{t}_2(\bar{\tau}_1,\bar{\xi}_1,\varepsilon,\eta,\mu,\alpha), \quad \widetilde{\xi}_2 = x_-(\bar{\tau}_1,\bar{\xi}_1)(\widetilde{\tau}_2,\varepsilon,\mu,\alpha),$$

$$t \in [\bar{\tau}_2, T] \text{ or } t \in [\bar{\tau}_2, \widetilde{t}_3(\widetilde{\tau}_2,\widetilde{\xi}_2,\varepsilon,\mu,\alpha)]: \quad x_+(\bar{\tau}_2,\bar{\xi}_2)(t,\varepsilon,\mu,\alpha)$$

$$\text{for } \bar{\tau}_2 = t_{2\eta}(\widetilde{\tau}_2,\widetilde{\xi}_2,\varepsilon,\mu,\alpha), \quad \bar{\xi}_2 = x_\eta(\widetilde{\tau}_2,\widetilde{\xi}_2)(\bar{\tau}_2,\varepsilon,\mu,\alpha).$$

Finally, we get approximated stroboscopic Poincaré mapping

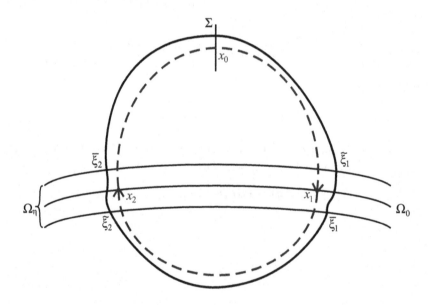

Figure III.1.2 Notation of points on the approximating orbit (III.1.5)

$$\widetilde{P}_\eta(\xi,\varepsilon,\mu,\alpha) = x_\eta(0,\xi)(T,\varepsilon,\mu,\alpha),$$

and approximated Poincaré mapping

$$P_\eta(\xi,\varepsilon,\mu,\alpha) = x_\eta(0,\xi)(\widetilde{t}_3(\widetilde{\tau}_2,\widetilde{\xi}_2,\varepsilon,\mu,\alpha),\varepsilon,\mu,\alpha).$$

The above decomposition of $x_\eta(0,\xi)(t,\varepsilon,\mu,\alpha)$ implies

$$\|\widetilde{P}_\eta(\xi,\varepsilon,\mu,\alpha) - \widetilde{P}(\xi,\varepsilon,\mu,\alpha)\| \to 0, \quad \|P_\eta(\xi,\varepsilon,\mu,\alpha) - P(\xi,\varepsilon,\mu,\alpha)\| \to 0$$

as $\eta \to 0$ uniformly with respect to other parameters, for P and $\widetilde{P}$ given by (I.1.10) and (I.1.12), respectively. This gives just C-approximations of Poincaré mappings useful for topological stable results such as the following.

Theorem III.1.2. *Let* (I.1.1) *be autonomous, i.e. of the form* (I.3.1). *Suppose that x_0 is an isolated fixed point of $P(\cdot,\varepsilon,\mu)$ with a nonzero Brouwer index (see A.1.5). Then $P_\eta(\cdot,\varepsilon,\mu)$ has a fixed point near x_0 for any $\eta > 0$ small, which implies the persistence of periodic orbits for* (I.3.1) *after smooth approximation.*

Proof. Since x_0 is an isolated fixed point of $P(\cdot,\varepsilon,\mu)$, there is its small neighborhood $O_{x_0} \subset \Sigma$ such that there is no other fixed point of $P(\cdot,\varepsilon,\mu)$ in O_{x_0}. Then the Brouwer index of x_0 is the Brouwer degree $\deg(\mathbb{I} - P(\cdot,\varepsilon,\mu), O_{x_0}) \neq 0$. By the homotopy invariance of the Brouwer degree, we have

$$\deg(\mathbb{I} - P_\eta(\cdot,\varepsilon,\mu), O_{x_0}) = \deg(\mathbb{I} - P(\cdot,\varepsilon,\mu), O_{x_0}) \neq 0,$$

for any $\eta > 0$ small. This means that $P_\eta(\cdot,\varepsilon,\mu)$ has a fixed point in O_{x_0}. The proof is finished. □

III.1.2. Approximating bifurcation functions

We need for (I.1.1) higher order approximations of $\widetilde{P}_\eta$ and P_η. To simplify our computations, we consider a simpler continuous approximation instead of (III.1.3). So we take a Lipschitz approximation instead of the smooth one, given by

$$\dot{x} = F_\pm(x,t,\alpha,\varepsilon,\mu), \quad x \in \overline{\Omega}_{\pm\eta},$$
$$\dot{x} = \frac{\eta + h(x)}{2\eta}F_+(x,t,\alpha,\varepsilon,\mu) + \frac{\eta - h(x)}{2\eta}F_-(x,t,\alpha,\varepsilon,\mu), \quad x \in \overline{\Omega}_\eta, \tag{III.1.6}$$

which is a singular system of ODEs on $\overline{\Omega}_\eta$ (see [7]). Note that now

$$\theta(r) = \begin{cases} -1 & \text{if } r \in (-\infty, -1], \\ r & \text{if } r \in [-1, 1], \\ 1 & \text{if } r \in [1, \infty). \end{cases} \tag{III.1.7}$$

Clearly, the results of Section III.1.1 remain valid also for (III.1.6). To study (III.1.6) on $\overline{\Omega}_\eta$, we change the time by

$$t = \tau + \eta s, \quad s \in [0, 2/c_1]$$

to consider t slowly varying. So for $y(s) = x(\tau + \eta s) - \xi$, we solve the problem

$$
\begin{aligned}
y' &= \frac{\eta + h(y + \xi)}{2} F_+(y + \xi, \tau + \eta s, \alpha, \varepsilon, \mu) \\
&+ \frac{\eta - h(y + \xi)}{2} F_-(y + \xi, \tau + \eta s, \alpha, \varepsilon, \mu),
\end{aligned}
\tag{III.1.8}
$$

$$
y(0) = 0
$$

for $s \in [0, 2/c_1]$ such that $y(s) + \xi \in \overline{\Omega}_\eta$. Now we consider $\xi \in B_\eta(x_1, r_1/2)$. Then

$$
\|y(s)\| \leq c_2 \eta s \leq r_1/2
$$

for any $s \in [0, 2/c_1]$ whenever $\eta < \frac{r_1 c_1}{4 c_2}$, and thus

$$
-c_2 c_3 \leq \frac{1}{\eta} \frac{d}{ds} h(y(s) + \xi) < -c_1, \quad \forall s \in [0, 2/c_1].
\tag{III.1.9}
$$

Hence

$$
1 - \frac{2 c_2 c_3}{c_1} \leq \frac{1}{\eta} h(y(2/c_1) + \xi) < -1.
$$

Then the IFT yields the existence of $\bar{\tau}_1, \bar{\varepsilon}_1 > 0$ and C^r-function

$$
s_{1\eta}(\cdot, \cdot, \cdot, \cdot, \cdot) : (t_1 - \bar{\tau}_1, t_1 + \bar{\tau}_1) \times B_\eta(x_1, r_1/2) \times (-\bar{\varepsilon}_1, \bar{\varepsilon}_1) \times \mathbb{R}^p \times \mathbb{R} \to [0, 2/c_1]
$$

such that

$$
\frac{1}{\eta} h(y(s) + \xi) = -1
$$

for $\tau \in (t_1 - \bar{\tau}_1, t_1 + \bar{\tau}_1), \xi \in B_\eta(x_1, r_1/2), \varepsilon \in (-\bar{\varepsilon}_1, \bar{\varepsilon}_1), \mu \in \mathbb{R}^p, \alpha \in \mathbb{R}$ and $s \in [0, 2/c_1]$ if and only if $s = s_{1\eta}(\tau, \xi, \varepsilon, \mu, \alpha)$. We recall

$$
\|y(s)\| \leq \frac{2 c_2 \eta}{c_1}
\tag{III.1.10}
$$

for any $s \in [0, 2/c_1]$. According to the theory of ODE [8], we also derive from (III.1.8) that

$$
y(s) = y(\tau, \xi, \varepsilon, \mu, \alpha, s)
$$

is C^r-smoothly depending on its parameters. Setting

$$
v(s) = y_\tau(\tau, \xi, \varepsilon, \mu, \alpha, s),
$$

we get

$$v' = \frac{Dh(y+\xi)v}{2}(F_+(y+\xi,\tau+\eta s,\alpha,\varepsilon,\mu) - F_-(y+\xi,\tau+\eta s,\alpha,\varepsilon,\mu))$$

$$+\frac{\eta + h(y+\xi)}{2}(F_{+y}(y+\xi,\tau+\eta s,\alpha,\varepsilon,\mu)v + F_{+\tau}(y+\xi,\tau+\eta s,\alpha,\varepsilon,\mu)) \qquad \text{(III.1.11)}$$

$$+\frac{\eta - h(y+\xi)}{2}(F_{-y}(y+\xi,\tau+\eta s,\alpha,\varepsilon,\mu)v + F_{-\tau}(y+\xi,\tau+\eta s,\alpha,\varepsilon,\mu)),$$

which implies

$$\|v(s)\| \le c_4(1+\eta)\int_0^s \|v(z)\|dz + c_4\eta, \quad s \in [0,2/c_1]$$

for a suitable constant $c_4 > 0$, and the Gronwall inequality [9] implies

$$\|v(s)\| \le c_4\eta e^{2c_4(1+\eta)/c_1}.$$

Similar arguments hold for other partial derivatives with respect to ε, μ, α. On the other hand, for the derivative with respect to ξ in the direction ζ, we derive

$$w' = \frac{Dh(y+\xi)(w+\zeta)}{2}(F_+(y+\xi,\tau+\eta s,\alpha,\varepsilon,\mu) - F_-(y+\xi,\tau+\eta s,\alpha,\varepsilon,\mu))$$

$$+\frac{\eta + h(y+\xi)}{2}F_{+y}(y+\xi,\tau+\eta s,\alpha,\varepsilon,\mu)(w+\zeta)$$

$$+\frac{\eta - h(y+\xi)}{2}F_{-y}(y+\xi,\tau+\eta s,\alpha,\varepsilon,\mu)(w+\zeta),$$

$$\text{(III.1.12)}$$

where

$$w(s) = y_\xi(\tau,\xi,\varepsilon,\mu,\alpha,s)\zeta, \quad Dh(\xi)\zeta = 0.$$

Hence (III.1.12) has the form

$$w' = \frac{Dh(y+\xi)w}{2}(F_+(y+\xi,\tau+\eta s,\alpha,\varepsilon,\mu) - F_-(y+\xi,\tau+\eta s,\alpha,\varepsilon,\mu))$$

$$+\frac{\eta + h(y+\xi)}{2}F_{+y}(y+\xi,\tau+\eta s,\alpha,\varepsilon,\mu)(w+\zeta)$$

$$+\frac{\eta - h(y+\xi)}{2}F_{-y}(y+\xi,\tau+\eta s,\alpha,\varepsilon,\mu)(w+\zeta)$$

$$+\frac{(Dh(y+\xi) - Dh(\xi))\zeta}{2}(F_+(y+\xi,\tau+\eta s,\alpha,\varepsilon,\mu) - F_-(y+\xi,\tau+\eta s,\alpha,\varepsilon,\mu)),$$

$$\text{(III.1.13)}$$

where, by (III.1.10), using

$$Dh(y+\xi) - Dh(\xi) = O(y) = O(\eta),$$

we also get $w = O(\eta)\|\zeta\|$. The smallness of the derivative $y_s(\tau, \xi, \varepsilon, \mu, \alpha, s)$ follows from (III.1.8). Summarizing, we see that $y(\tau, \xi, \varepsilon, \mu, \alpha, s)$ is $O(\eta)$-small in C^1-order. By induction, we can derive smallness for higher derivatives.

Furthermore, differentiating

$$\frac{1}{\eta}h(y(\tau, \xi, \varepsilon, \mu, \alpha, s_{1\eta}(\tau, \xi, \varepsilon, \mu, \alpha)) + \xi) = -1,$$

with respect to τ, we derive

$$Dh(y(s_{1\eta}) + \xi)\frac{1}{\eta}v(s_{1\eta}) + \frac{1}{\eta}Dh(y(s_{1\eta}) + \xi)y'(s_{1\eta})s_{1\eta,\tau} = 0$$

where we omitted the arguments for simplicity. From (III.1.9) and $v = O(\eta)$ we get

$$-c_2c_3 < \frac{1}{\eta}Dh(y(s_{1\eta}) + \xi)y'(s_{1\eta}) < -c_1, \quad Dh(y(s_{1\eta}) + \xi)\frac{1}{\eta}v(s_{1\eta}) = O(1).$$

Consequently,

$$s_{1\eta,\tau} = O(1).$$

Similar arguments hold for other partial derivatives with respect to ε, μ, α. On the other hand, for the derivative with respect to ξ in the direction ζ, we derive

$$\frac{1}{\eta}(Dh(y(s_{1\eta}) + \xi) - Dh(\xi))\zeta$$

$$+Dh(y(s_{1\eta}) + \xi)\frac{1}{\eta}w(s_{1\eta}) + \frac{1}{\eta}Dh(y(s_{1\eta}) + \xi)y'(s_{1\eta})s_{1\eta,\xi}\zeta = 0.$$

Since

$$Dh(y(s_{1\eta}) + \xi)\frac{1}{\eta}w(s_{1\eta}) = O(\|\zeta\|), \quad \frac{1}{\eta}(Dh(y(s_{1\eta}) + \xi) - Dh(\xi)) = O(1),$$

we get

$$s_{1\eta,\xi}\zeta = O(\|\zeta\|).$$

Summarizing, we see that $s_{1\eta}(\tau, \xi, \varepsilon, \mu, \alpha)$ is bounded in C^1-order. By induction, we can derive boundedness for higher derivatives.

Next, we note

$$t_{1\eta}(\tau, \xi, \varepsilon, \mu, \alpha) = \tau + \eta s_{1\eta}(\tau, \xi, \varepsilon, \mu, \alpha),$$
$$x_\eta(\tau, \xi)(t, \varepsilon, \mu, \alpha) = \xi + y(\tau, \xi, \varepsilon, \mu, \alpha, (t - \tau)/\eta).$$

Hence,

$$x_\eta(\tau, \xi)(t_{1\eta}(\tau, \xi, \varepsilon, \mu, \alpha), \varepsilon, \mu, \alpha) = \xi + y(\tau, \xi, \varepsilon, \mu, \alpha, s_{1\eta}(\tau, \xi, \varepsilon, \mu, \alpha)).$$

Summarizing these results, we see that the Poincaré mapping

$$(\tau, \xi, \varepsilon, \mu, \alpha) \mapsto x_\eta(\tau, \xi)(t_{1\eta}(\tau, \xi, \varepsilon, \mu, \alpha), \varepsilon, \mu, \alpha)$$

for $\xi \in \overline{\Omega}_\eta$ close to x_1, is C^r-close to the identity

$$(\tau, \xi, \varepsilon, \mu, \alpha) \mapsto \xi$$

as $\eta \to 0$. The same analysis can be done near x_2. This implies

$$\widetilde{P}_\eta(\xi, \varepsilon, \mu, \alpha) \to \widetilde{P}(\xi, \varepsilon, \mu, \alpha), \quad P_\eta(\xi, \varepsilon, \mu, \alpha) \to P(\xi, \varepsilon, \mu, \alpha)$$

for P and $\widetilde{P}$ given by (I.1.10) and (I.1.12), respectively, as $\eta \to 0$ uniformly in C^r-topology, since the restrictions of P and $\widetilde{P}$ on $\overline{\Omega}_\eta$ near x_1 and x_2 are also C^r-close to the identity

$$(\tau, \xi, \varepsilon, \mu, \alpha) \mapsto \xi.$$

By these restrictions we mean that we carry out the preceding analysis to (III.1.6), by considering (III.1.1) instead of (III.1.6), gradually on $\overline{\Omega}_\eta \cap \overline{\Omega}_+$ and $\overline{\Omega}_\eta \cap \overline{\Omega}_-$.

Consequently we arrive at

Theorem III.1.3. *The corresponding bifurcation equations of Chapters I.1, I.2 and I.3 can be approximated by* (III.1.6).

Remark III.1.4. Our analysis for (III.1.6) can be extended to (III.1.3). The problem analogous to (III.1.8), corresponding to (III.1.3), is to solve

$$y' = \frac{\eta + \eta\theta(h(y + \xi)/\eta)}{2} F_+(y + \xi, \tau + \eta s, \alpha, \varepsilon, \mu)$$
$$+ \frac{\eta - \eta\theta(h(y + \xi)/\eta)}{2} F_-(y + \xi, \tau + \eta s, \alpha, \varepsilon, \mu), \tag{III.1.14}$$
$$y(0) = 0$$

for $s \in [0, 2/c_1]$ such that $y(s) + \xi \in \overline{\Omega}_\eta$ with $h(\xi) = \eta$. We again derive (III.1.10). Hence we take $y(s) = \eta z(s)$ in (III.1.14) to get

$$z' = \frac{1 + \theta(h(\eta z + \xi)/\eta)}{2} F_+(\eta z + \xi, \tau + \eta s, \alpha, \varepsilon, \mu)$$
$$+ \frac{1 - \theta(h(\eta z + \xi)/\eta)}{2} F_-(\eta z + \xi, \tau + \eta s, \alpha, \varepsilon, \mu), \tag{III.1.15}$$
$$z(0) = 0$$

Using $h(\xi) = \eta$, we derive

$$\theta(h(\eta z + \xi)/\eta) = \theta\left(\frac{h(\eta z + \xi) - h(\xi)}{\eta} + 1\right)$$

$$= \theta\left(\int_0^1 Dh(\eta \varsigma z + \xi)d\varsigma z + 1\right) = \Theta(\eta, \xi, z)$$

when $\Theta(\eta, \xi, z)$ is a C^{r-1}-smooth function. Consequently, (III.1.15) has the form

$$z' = \frac{1 + \Theta(\eta, \xi, z)}{2} F_+(\eta z + \xi, \tau + \eta s, \alpha, \varepsilon, \mu)$$

$$+ \frac{1 - \Theta(\eta, \xi, z)}{2} F_-(\eta z + \xi, \tau + \eta s, \alpha, \varepsilon, \mu), \qquad \text{(III.1.16)}$$

$$z(0) = 0$$

with

$$|\Theta(\eta, \xi, z)| \le 1.$$

Summarizing, (III.1.3) is suitable for C^{r-1}-approximation of Poincaré mappings in all parameters including also η. On the other hand, using the Gronwall inequality as above, we derive that in fact it is a C^r-approximation in all parameters excluding η but uniformly with respect to it. The same consequence is true for other functions mentioned in Remark III.1.1.

Remark III.1.5. If (III.1.1) is symmetric with respect to a matrix $A: \mathbb{R}^n \to \mathbb{R}^n$, i.e. it holds

$$A(\overline{\Omega}_\pm) \subset \overline{\Omega}_\pm, \quad AF_\pm(x, t, \alpha, \varepsilon, \mu) = F_\pm(Ax, t, \alpha, \varepsilon, \mu),$$

$$h(Ax) = h(x) \quad x \in \overline{\Omega}_\pm,$$

then the approximation (III.1.3) is also symmetric.

III.1.3. Examples

Continuous approximations of discontinuous systems are often used for numerical computation. Let us first consider a simple discontinuous ODE

$$\dot{x} = 2 \pm 1, \quad \pm x > 0, \quad x(0) = \xi < 0$$

with the solution

$$x(t) = \begin{cases} t + \xi & \text{if } t \in [0, -\xi], \\ 3(t + \xi) & \text{if } t \in [-\xi, \infty). \end{cases} \qquad \text{(III.1.17)}$$

Its approximation (III.1.3) has the form

$$\dot{x}(t) = \begin{cases} 1 & \text{if } x \leq -\eta, \\ 2 + \tanh \frac{2\eta x}{\eta^2 - x^2} & \text{if } x \in (-\eta, \eta), \\ 3 & \text{if } x \geq \eta. \end{cases} \tag{III.1.18}$$

By following Remark III.1.4, the solution of (III.1.18) is as follows

$$x_\eta(t) = \begin{cases} t + \xi & \text{if } t \in [0, -\xi - \eta], \\ \eta y((t + \eta + \xi)/\eta) & \text{if } t \in [-\xi - \eta, \eta s_1 - \eta - \xi], \\ 3(t - \eta s_1 + \eta + \xi) + \eta & \text{if } t \in [\eta s_1 - \eta - \xi, \infty), \end{cases} \tag{III.1.19}$$

where $y(s)$ is the solution of

$$y'(s) = 2 + \tanh \frac{2y(s)}{1 - y(s)^2}, \quad y(0) = -1,$$

and $s_1 \doteq 1.19878$ is the unique solution of $y(s_1) = 1$ (see Figure III.1.3).

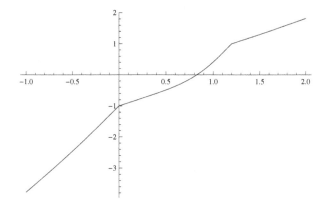

Figure III.1.3 The plot of $y(s)$ on interval $[-1, 2]$

The solution (III.1.17) and its approximation (III.1.19) are illustrated in Figure III.1.4.

Clearly $x_\eta(t) \rightrightarrows x(t)$ uniformly on compact intervals as $\eta \to 0^+$.

Now, we consider the reflection pendulum [10] (see also An introductory example and Section II.1.3) given by

$$\ddot{x} + x + a \operatorname{sgn} x = p(t) \tag{III.1.20}$$

for $a > 0$ and periodic $p(t)$ with phase portrait on Figure III.1.6 for $p(t) = 0$.

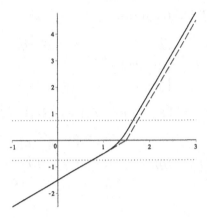

Figure III.1.4 Solution (III.1.17) (dashed) approximated by (III.1.19) (solid) with $\eta = 0.75$, $\xi = -1.5$, and depicted area of approximation

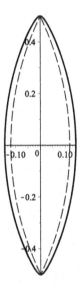

Figure III.1.5 Solution of (III.1.21) (solid), and (III.1.20) (dashed) with $p(t) = 0$, $a = 1$ and the same initial conditions

Considering a concrete case with

$$\ddot{x} + x + \frac{2}{\pi} \arctan[80x] = 0, \quad x(0) = 0, \ \dot{x}(0) = 0.5 \qquad \text{(III.1.21)}$$

we see numerically on Figure III.1.5 that approximation fits very well with the original solution: the higher multiple of x in arctan, the better approximation. We note that

(III.1.20) with $p(t) = 0$ is reversible and odd, and all solutions are periodic around $(0, 0)$ (see Figure III.1.6). This remains true also for (III.1.21).

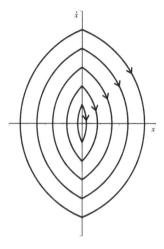

Figure III.1.6 Phase portrait of (III.1.20) with $p(t) = 0$

Finally, we refer the reader to [11] for more details and references in this direction.

CHAPTER III.2

Sliding periodic orbits

III.2.1. Setting of the problem

In this chapter, we present an approximation method for sliding orbits studied in Chapter I.4. We consider for simplicity that $\Omega_\pm = \{\pm z > 0\}$ when $x = (z, y)$ and $\mathbb{R}^n = \mathbb{R} \times \mathbb{R}^{n-1}$, so $\dim z = 1$, $\dim y = n - 1$ and $h(z, y) = z$. Then (I.4.1) has the form

$$\begin{aligned}\dot{z} &= H_\pm(z, y, t + \alpha, \varepsilon, \mu) \\ \dot{y} &= G_\pm(z, y, t + \alpha, \varepsilon, \mu)\end{aligned} \qquad \text{if } z \gtrless 0 \qquad \text{(III.2.1)}$$

where $F_\pm(z, y, t, \varepsilon, \mu) = (H_\pm(z, y, t, \varepsilon, \mu), G_\pm(z, y, t, \varepsilon, \mu))^*$, while (I.4.2) is given by

$$\dot{y} = W(y, t + \alpha, \varepsilon, \mu) :=$$
$$\frac{G_+(0, y, t + \alpha, \varepsilon, \mu) - G_-(0, y, t + \alpha, \varepsilon, \mu)}{2} V(y, t + \alpha, \varepsilon, \mu)$$
$$+ \frac{G_+(0, y, t + \alpha, \varepsilon, \mu) + G_-(0, y, t + \alpha, \varepsilon, \mu)}{2}, \quad z = 0 \qquad \text{(III.2.2)}$$

for any $y \in \mathbb{R}^{n-1}$, where

$$V(y, t + \alpha, \varepsilon, \mu) = \frac{H_+(0, y, t + \alpha, \varepsilon, \mu) + H_-(0, y, t + \alpha, \varepsilon, \mu)}{H_-(0, y, t + \alpha, \varepsilon, \mu) - H_+(0, y, t + \alpha, \varepsilon, \mu)}. \qquad \text{(III.2.3)}$$

We suppose assumptions H1) and H2) of Chapter I.4. By splitting $\gamma(t) = (p_1(t), p_2(t)) \in \mathbb{R} \times \mathbb{R}^{n-1}$, they obtain the forms (see (I.4.4))

$$p_1(t) > 0, \quad \forall t \in [0, t_1) \cup (t_2, T], \qquad p_1(t) = 0, \quad \forall t \in [t_1, t_2],$$
$$H_-^0(0, p_2(t))) > 0, \quad \forall t \in [t_1, t_2], \qquad H_+^0(0, p_2(t)) < 0, \quad \forall t \in [t_1, t_2), \qquad \text{(III.2.4)}$$
$$H_+^0(0, p_2(t_2)) = 0, \qquad D_y H_+^0(0, p_2(t_2)) \dot{p}_2(t_2) > 0,$$

where $H_\pm^0(z, y) = H_\pm(z, y, t, 0, \mu)$. These properties (III.2.4) coincide with assumptions (A1) and (A2) of [12].

Now we take $\kappa > 0$ small and consider a continuous approximation of (III.2.1) given by

$$\begin{aligned}\dot{z} &= H_\pm(z, y, t + \alpha, \varepsilon, \mu) \\ \dot{y} &= G_\pm(z, y, t + \alpha, \varepsilon, \mu)\end{aligned} \qquad \text{if } z \gtrless \kappa \qquad \text{(III.2.5)}$$

Poincaré-Andronov-Melnikov Analysis for Non-Smooth Systems.
http://dx.doi.org/10.1016/B978-0-12-804294-6.50015-1

and for $|z| \leq \kappa$:

$$\dot{z} = \frac{H_+(\kappa, y, t + \alpha, \varepsilon, \mu) - H_-(-\kappa, y, t + \alpha, \varepsilon, \mu)}{2\kappa} z$$

$$+ \frac{H_+(\kappa, y, t + \alpha, \varepsilon, \mu) + H_-(-\kappa, y, t + \alpha, \varepsilon, \mu)}{2},$$

$$\dot{y} = \frac{G_+(\kappa, y, t + \alpha, \varepsilon, \mu) - G_-(-\kappa, y, t + \alpha, \varepsilon, \mu)}{2\kappa} z \qquad\qquad (\text{III.2.6})$$

$$+ \frac{G_+(\kappa, y, t + \alpha, \varepsilon, \mu) + G_-(-\kappa, y, t + \alpha, \varepsilon, \mu)}{2}.$$

Our aim is to study the relationship between Poincaré maps of (III.2.1) and (III.2.5), (III.2.6). We show that they are smoothly close as $\kappa \to 0^+$ when the right-hand sides of (III.2.1) are C^{r+2}-smooth.

III.2.2. Planar illustrative examples

We start with an unperturbed problem to illustrate our approach by following [12]. Let us consider a one-degree of freedom mechanical system consisting of a mass moving on a belt and connected to a nonlinear oscillator. Such a model can be described by the equation of a dry friction oscillator

$$\ddot{y} - y + y^3 - \frac{0.6}{1 + |\dot{y} - 1|} \operatorname{sgn}(1 - \dot{y}) = 0. \qquad\qquad (\text{III.2.7})$$

By putting $\dot{y} = -z + 1$, we get the system

$$\dot{z} = y^3 - y - \frac{0.6}{1 + |z|} \operatorname{sgn} z$$

$$\dot{y} = -z + 1. \qquad\qquad (\text{III.2.8})$$

Now we approximate (III.2.8) by the system (see (III.2.5), (III.2.6))

$$\dot{z} = y^3 - y - f_\kappa(z)$$

$$\dot{y} = -z + 1 \qquad\qquad (\text{III.2.9})$$

for $\kappa > 0$ small and a function $f_\kappa : \mathbb{R} \to \mathbb{R}$ defined as follows

$$f_\kappa(z) := \begin{cases} \dfrac{0.6}{1 + |z|} \operatorname{sgn} z & \text{for} \quad |z| \geq \kappa, \\[3mm] \dfrac{0.6z}{(1 + \kappa)\kappa} & \text{for} \quad |z| \leq \kappa. \end{cases}$$

Then for $z \geq \kappa$, (III.2.9) has the form

$$\dot{z} = y^3 - y - \frac{0.6}{1 + z}$$

$$\dot{y} = -z + 1$$
(III.2.10)

which is (III.2.8) for $z > 0$. For $|z| \leq \kappa$, we take $z = \kappa w$, $|w| \leq 1$, and (III.2.9) has the form of a singular perturbation

$$\kappa \dot{w} = y^3 - y - \frac{0.6}{1 + \kappa} w$$

$$\dot{y} = -\kappa w + 1.$$
(III.2.11)

Checking the vector field of (III.2.8) near the line $z = 0$ for $z > 0$, we see that for $y < y_0$: the line $z = 0$ is attracting, and for $y > y_0$: the line $z = 0$ is repelling. Of course, the variable y is increasing. Here $y_0^3 - y_0 = 0.6$, $y_0 \doteq 1.2212$.

Now we find numerically (see Figure III.2.1) that the solution of (III.2.10) with the initial conditions $z(0) = 0$, $y(0) = y_0$ transversally crosses the line $z = 0$ at the time $t_0 \doteq 6.8006$ in $y(t_0) := \bar{y}_0 \doteq -0.1257 \in (-1, 2)$. Moreover, checking the vector field of (III.2.8) near the line $z = 0$ for $z < 0$, we see that for $y \in [-1, 2]$: the line $z = 0$ is attracting.

The sliding mode equation (III.2.2) is just $\dot{y} = 1$, so we easily see that

$$T = t_0 + y_0 - \bar{y}_0 \doteq 8.1475.$$

Of course, for the discontinuous system (III.2.8), we get a periodic solution $\gamma(t)$ starting from the point $(0, y_0)$, which is infinitely stable, i.e. all solutions starting near periodic solution $\gamma(t)$ collapse after a finite time to $\gamma(t)$. We expect that its approximation (III.2.9) will also possess a unique periodic solution near $\gamma(t)$ with a rapid attractivity. This phenomenon is numerically demonstrated in [13–15]. To show this property for our simple system (III.2.9) analytically, we consider the dynamics of a Poincaré map of (III.2.9) near periodic orbit $\gamma(t)$ of (III.2.8). For the construction of this Poincaré map, we take the interval

$$I := [\bar{y}_0 - \delta, \bar{y}_0 + \delta]$$

for a fixed small $\delta > 0$. Concerning (III.2.11), we put $\kappa = 0$ and we get

$$V(y) := \frac{5}{3}(y^3 - y)$$

as a solution of the equation

$$f(V, y) := y^3 - y - 0.6V = 0.$$

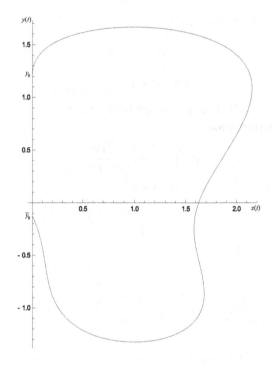

Figure III.2.1 Periodic solution of (III.2.10) with $z(0) = 0$, $y(0) = y_0$

Moreover, we have

$$\frac{\partial f}{\partial V}(V(y), y) = -0.6 < 0.$$

Let us consider the rectangle

$$Q := [-1, 1] \times [-1, 2].$$

The curve $(V(y), y)$ leaves Q only through the points $(0, -1)$ and $(1, y_0)$. We can apply the Tichonov theorem [7, 16] to the singularly perturbed system (III.2.11) (see also Theorem A.9). We have already verified all assumptions of this theorem except V. of Section A.3.2: We must show that for any $y \in I$, the solution $w(\tau)$ of the equation

$$\dot{w}(\tau) = y^3 - y - 0.6w(\tau), \quad w(0) = 1$$

satisfies

$$w(\tau) \in [-1, 1] \quad \text{for} \quad \tau \geq 0,$$
$$w(\tau) \to V(y) \quad \text{as} \quad \tau \to \infty.$$

This is elementary, since

$$w(\tau) = e^{-0.6\tau}\left(1 - \frac{5}{3}(y^3 - y)\right) + \frac{5}{3}(y^3 - y)$$

$$\approx e^{-0.6\tau}\left(1 - \frac{5}{3}(\bar{y}_0^3 - \bar{y}_0)\right) + \frac{5}{3}(\bar{y}_0^3 - \bar{y}_0) \doteq 0.7938e^{-0.6\tau} + 0.2062.$$

Summarizing, we can apply Tichonov theorem A.9 to (III.2.11). Consequently, we take $w(0) = 1$, $y(0) = y \in I$, and the corresponding solution $(w(t), y(t))$ of (III.2.11) transversally leaves Q near y_0 at the time $\tilde{t}$. It has the asymptotic expansion (see (A.12) and (A.14))

$$w(t) = \frac{5}{3}((y + t)^3 - y - t) + e^{-\frac{0.6t}{\kappa}}\left(1 - \frac{5}{3}(y^3 - y)\right) + O(\kappa),$$

$$y(t) = y + t + O(\kappa).$$

So $w(\tilde{t}) = 1$, and thus $\tilde{t} = y_0 - y + O(\kappa)$. We set

$$\Psi_\kappa(y) = y(\tilde{t})$$

and note $\Psi_\kappa(y) = y_0 + O(\kappa)$. Next, we consider the solution $(z(t), y(t))$ of (III.2.10) with the initial value conditions $y(0) = y \in \bar{I} = [y_0 - \delta_1, y_0 + \delta_1]$, $z(0) = \kappa$. Then for a small $\delta_1 > 0$, there is $\tilde{t} \sim t_0$ such that $z(\tilde{t}) = \kappa$. We put

$$\Phi_\kappa(y) := y(\tilde{t}).$$

We get a mapping $\Phi_\kappa : \bar{I} \to I$ for $\delta_1 > 0$ small. The smaller κ, the smaller $\delta_1 > 0$ can be taken. Finally, we put

$$P_\kappa(y) := \Phi_\kappa(\Psi_\kappa(y))$$

for $y \in I$. Clearly $P_\kappa : I \to I$ and this is the desired Poincaré map of (III.2.9) near periodic solution $\gamma(t)$ of (III.2.8). The map Φ_κ depends smoothly on κ small and $y \in \bar{I}$. Similarly, the map Ψ_κ depends smoothly on $\kappa > 0$ small and $y \in I$, and it holds

$$P_\kappa(y) = \Phi_\kappa(\Psi_\kappa(y)) = \Phi_0(y_0) + O(\kappa) = \bar{y}_0 + O(\kappa). \qquad \text{(III.2.12)}$$

We note that the limit map $P_0(y) = \bar{y}_0$ in (III.2.12) is just the Poincaré mapping along the periodic solution $\gamma(t)$ of (III.2.8). Furthermore, the identity (III.2.12) holds also in the C^1-topology, i.e. it holds

$$P_\kappa'(y) = O(\kappa). \qquad \text{(III.2.13)}$$

Hence the map $P_\kappa : I \to I$ has a unique fixed point $y_\kappa \in I$ of the form $y_\kappa = \bar{y}_0 + O(\kappa)$, which is according to (III.2.13) also rapidly attractive. Summarizing, we get the next theorem.

Theorem III.2.1. *The discontinuous system* (III.2.8) *has the periodic solution* $\gamma(t)$ *starting from the point* $(0, \bar{y}_0)$, *which is infinitely stable, i.e. all solutions starting near* $\gamma(t)$ *collapse after a finite time to* $\gamma(t)$. *Its approximation* (III.2.9) *has also a unique periodic solution* p_κ *starting from the point* (κ, y_κ) *which approximates* $\gamma(t)$ *and which is rapidly attracting. This coincides with the infinite asymptotic stability of* $\gamma(t)$.

Related results are presented in [12, 17]. The above constructed Poincaré mapping is different from the one in Chapter I.4. We present here an alternative approach which is suitable for autonomous discontinuous systems. Poincaré mapping (I.4.7) is derived for periodically forced discontinuous systems such as

$$\ddot{y} - y + y^3 - \frac{0.6}{1 + |\dot{y} - 1|}\,\mathrm{sgn}(1 - \dot{y}) = \varepsilon \cos \frac{2\pi t}{T}. \qquad (\text{III.2.14})$$

We do not go into detail for (III.2.14), since we study a more general case in the next section.

III.2.3. Higher dimensional systems

We derive the main results mentioned in Section III.2.1. Motivated by the approach of Section III.2.2, we put $z = \kappa w$, $|w| \leq 1$ in (III.2.6) to get the system

$$
\begin{aligned}
\kappa \dot{w} &= \frac{H_+(\kappa, y, t + \alpha, \varepsilon, \mu) - H_-(-\kappa, y, t + \alpha, \varepsilon, \mu)}{2}\,w \\
&\quad + \frac{H_+(\kappa, y, t + \alpha, \varepsilon, \mu) + H_-(\kappa, y, t + \alpha, \varepsilon, \mu)}{2} \\
\dot{y} &= \frac{G_+(\kappa, y, t + \alpha, \varepsilon, \mu) - G_-(-\kappa, y, t + \alpha, \varepsilon, \mu)}{2}\,w \\
&\quad + \frac{G_+(\kappa, y, t + \alpha, \varepsilon, \mu) + G_-(-\kappa, y, t + \alpha, \varepsilon, \mu)}{2}.
\end{aligned}
\qquad (\text{III.2.15})
$$

We follow Chapter I.4 to construct the Poincaré mapping of (III.2.5), (III.2.6). Since we start in Ω_+, we consider (III.2.5) just for $z \geq \kappa$:

$$
\begin{aligned}
\dot{z} &= H_+(z, y, t + \alpha, \varepsilon, \mu) \\
\dot{y} &= G_+(z, y, t + \alpha, \varepsilon, \mu)
\end{aligned}
\qquad \text{if } z \geq \kappa.
\qquad (\text{III.2.16})
$$

So first, we have a mapping

$$\Omega_+ \supset B(x_0, r_0) \to \Omega_\kappa := \{z = \kappa\}$$

given by (see (I.4.5))

$$\xi \mapsto x_+(0, \xi)(\tilde{t}_1(\kappa, \xi, \varepsilon, \mu, \alpha), \varepsilon, \mu, \alpha) \qquad (\text{III.2.17})$$

where $\tilde{t}_1(\kappa, \xi, \varepsilon, \mu, \alpha)$ is uniquely determined by the equation

$$x_+^z(0, \xi)(t, \varepsilon, \mu, \alpha) = \kappa$$

where

$$x_+(0, \xi)(t, \varepsilon, \mu, \alpha) = (x_+^z(0, \xi)(t, \varepsilon, \mu, \alpha), x_+^y(0, \xi)(t, \varepsilon, \mu, \alpha)).$$

Note $\tilde{t}_1(0, x_0, 0, \mu, \alpha) = t_1$. Next, we take the solution

$$w(t) = w(\kappa, \xi, \alpha, \varepsilon, \mu; t), \quad y(t) = y(\kappa, \xi, \alpha, \varepsilon, \mu; t)$$

of (III.2.15) with the initial value condition

$$w(\tilde{t}_1(\kappa, \xi, \varepsilon, \mu, \alpha)) = 1,$$
$$y(\tilde{t}_1(\kappa, \xi, \varepsilon, \mu, \alpha)) = x_+^y(0, \xi)(\tilde{t}_1(\kappa, \xi, \varepsilon, \mu, \alpha), \varepsilon, \mu, \alpha).$$

Note $x_+^y(0, x_0)(\tilde{t}_1(0, x_0, 0, \mu, \alpha), \varepsilon, \mu, \alpha) = p_2(t_1)$. We intend to apply the Tichonov theorem A.9. For this reason, we shift the time $t \to t + \tilde{t}_1$ in (III.2.15) to get

$$\begin{aligned}
\kappa \dot{w}_0 &= \frac{H_+(\kappa, y_0, t + \tilde{t}_1 + \alpha, \varepsilon, \mu) - H_-(-\kappa, y_0, t + \tilde{t}_1 + \alpha, \varepsilon, \mu)}{2} w_0 \\
&\quad + \frac{H_+(\kappa, y_0, t + \tilde{t}_1 + \alpha, \varepsilon, \mu) + H_-(\kappa, y_0, t + \tilde{t}_1 + \alpha, \varepsilon, \mu)}{2} \\
\dot{y}_0 &= \frac{G_+(\kappa, y_0, t + \tilde{t}_1 + \alpha, \varepsilon, \mu) - G_-(-\kappa, y_0, t + \tilde{t}_1 + \alpha, \varepsilon, \mu)}{2} w_0 \\
&\quad + \frac{G_+(\kappa, y_0, t + \tilde{t}_1 + \alpha, \varepsilon, \mu) + G_-(-\kappa, y_0, t + \tilde{t}_1 + \alpha, \varepsilon, \mu)}{2},
\end{aligned} \tag{III.2.18}$$

and we consider (III.2.18) on $[0, t_2 - t_1 + \delta]$ for $\delta > 0$ sufficiently small with the initial value condition

$$w_0(0) = 1, \quad y_0(0) = x_+^y(0, \xi)(\tilde{t}_1, \varepsilon, \mu, \alpha). \tag{III.2.19}$$

For simplicity, we write $\tilde{t}_1$ without its arguments. Solving the first equation of (III.2.18) by w_0 with $\kappa = 0$, we get

$$w_0 = V(y_0, t + \tilde{t}_1 + \alpha, \varepsilon, \mu)$$

for V given by (III.2.3), and plugging it into the second equation of (III.2.18), we get the reduced equation of (III.2.18), which has the form (III.2.2) shifted by $\tilde{t}_1$ in time,

$$\dot{y}_0 = W(y_0, t + \tilde{t}_1 + \alpha, \varepsilon, \mu). \tag{III.2.20}$$

So the reduced equation is precisely the sliding mode equation. We know that (III.2.20) for $\varepsilon = 0$ has a solution $p_2(t + t_1)$ on $[0, t_2 - t_1 + \delta]$, which is its unique solution with $y_0(0) = p_2(t_1)$. Considering

$$V_0(t) = \frac{H_+^0(0, p_2(t + t_1)) + H_-^0(0, p_2(t + t_1))}{H_-^0(0, p_2(t + t_1)) - H_+^0(0, p_2(t + t_1))},$$

by (III.2.4) we get $V_0(t) > -1$ for $t \in [0, t_2 - t_1 + \delta]$, $1 > V_0(t)$ for $t \in [0, t_2 - t_1)$, $V_0(t_2 - t_1) = 1$ and

$$\frac{dV_0}{dt}(t_2 - t_1) = 2\frac{D_y H_+^0(0, p_2(t_2))\dot{p}_2(t_2)}{H_-^0(0, p_2(t_2))} > 0.$$

This means that function $V_0(t)$ crosses the level $V_0^{-1}(1)$ on $[0, t_2 - t_1 + \delta]$ only at $t = t_2 - t_1$, and transversally. Moreover, (III.2.4) implies

$$H_+^0(0, p_2(t + t_1)) - H_-^0(0, p_2(t + t_1)) < 0$$

for $t \in [0, t_2 - t_1 + \delta]$. So, we can apply the Tichonov theorem A.9 to get the solution

$$w_0(t) = w_0(\kappa, \xi, \alpha, \varepsilon, \mu; t), \quad y_0(t) = y_0(\kappa, \xi, \alpha, \varepsilon, \mu; t)$$

of (III.2.18) with the initial value condition (III.2.19) over $[0, t_2 - t_1 + \delta]$ such that

$$
\begin{aligned}
w_0(\kappa, \xi, \alpha, \varepsilon, \mu; t) &\to w_0(\xi, \alpha, \varepsilon, \mu; t) &\quad \text{on } (0, t_2 - t_1 + \delta], \\
\dot{w}_0(\kappa, \xi, \alpha, \varepsilon, \mu; t) &\to \dot{w}_0(\xi, \alpha, \varepsilon, \mu; t) &\quad \text{on } (0, t_2 - t_1 + \delta], \\
y_0(\kappa, \xi, \alpha, \varepsilon, \mu; t) &\rightrightarrows y_0(\xi, \alpha, \varepsilon, \mu; t) &\quad \text{on } [0, t_2 - t_1 + \delta], \\
\dot{y}_0(\kappa, \xi, \alpha, \varepsilon, \mu; t) &\rightrightarrows \dot{y}_0(\xi, \alpha, \varepsilon, \mu; t) &\quad \text{on } [0, t_2 - t_1 + \delta],
\end{aligned}
\tag{III.2.21}
$$

as $\kappa \to 0^+$, where $w_0(\xi, \alpha, \varepsilon, \mu; t) = V(y_0(\xi, \alpha, \varepsilon, \mu; t), t + \tilde{t}_1 + \alpha, \varepsilon, \mu)$ and $y_0(\xi, \alpha, \varepsilon, \mu; t)$ solves (III.2.20) with the second initial value condition of (III.2.19).

Remark III.2.2. The singular part of (III.2.18) is a scalar linear differential equation in w of the form

$$\kappa \dot{w}_0 = a(t)w_0 + b(t), \quad w_0(0) = 1 \tag{III.2.22}$$

with $a, b \in C^2([0, t_2 - t_1 + \delta])$ and $a(t) < -a_0 < 0$ for a constant $a_0 > 0$. Then (III.2.22) has the solution (see [16, (3.12)])

$$
\begin{aligned}
w_0(t) = &-\frac{b(t)}{a(t)} - \kappa\frac{1}{a(t)}\left(\frac{b(t)}{a(t)}\right)' \\
&+ \left(1 + \frac{b(0)}{a(0)} + \kappa\frac{1}{a(0)}\left(\frac{b(t)}{a(t)}\right)'_{t=0}\right)\exp\left(\frac{1}{\kappa}\int_0^t a(s)ds\right) \\
&+ \kappa\int_0^t\left(\frac{1}{a(s)}\left(\frac{b(s)}{a(s)}\right)'\right)'\exp\left(\frac{1}{\kappa}\int_s^t a(q)dq\right)ds.
\end{aligned}
\tag{III.2.23}
$$

By (III.2.22) we derive

$$\dot{w}_0(t) = -\left(\frac{b(t)}{a(t)}\right)'$$

$$+a(t)\left(1 + \frac{b(0)}{a(0)} + \kappa\frac{1}{a(0)}\left(\frac{b(t)}{a(t)}\right)'_{t=0}\right)\frac{1}{\kappa}\exp\left(\frac{1}{\kappa}\int_0^t a(s)ds\right) \qquad \text{(III.2.24)}$$

$$+a(t)\int_0^t \left(\frac{1}{a(s)}\left(\frac{b(s)}{a(s)}\right)'\right)'\exp\left(\frac{1}{\kappa}\int_s^t a(q)dq\right)ds.$$

Using

$$\exp\left(\frac{1}{\kappa}\int_0^t a(s)ds\right) \le e^{-\frac{a_0}{\kappa}t}, \quad \int_0^t \exp\left(\frac{1}{\kappa}\int_s^t a(q)dq\right)ds \le \kappa\frac{1 - e^{-\frac{a_0}{\kappa}t}}{a_0},$$

and (III.2.23), (III.2.24), we derive the first two limits in (III.2.21).

Note $w_0(x_0, \alpha, 0, \mu; t) = V_0(t)$. Hence by the preceding arguments, there is a unique solution $\tilde{t}_2(\kappa, \xi, \varepsilon, \mu, \alpha)$ of

$$w(\kappa, \xi, \alpha, \varepsilon, \mu; \tilde{t}_2(\kappa, \xi, \varepsilon, \mu, \alpha)) = 1$$

for $\kappa > 0$, ε small and ξ close to x_0 with $\tilde{t}_2(0^+, x_0, 0, \mu, \alpha) = t_2$. Finally we consider (III.2.16) for $t \in [\tilde{t}_2(\kappa, \xi, \varepsilon, \mu, \alpha), T]$ with its solution

$$x_+(\tilde{t}_2(\kappa, \xi, \varepsilon, \mu, \alpha), (\kappa, y(\kappa, \xi, \alpha, \varepsilon, \mu; \tilde{t}_2(\kappa, \xi, \varepsilon, \mu, \alpha))))(t, \varepsilon, \mu, \alpha).$$

Summarizing, we get the stroboscopic Poincaré mapping of (III.2.5), (III.2.6) given by

$$\widetilde{P}_\kappa(\xi, \varepsilon, \mu, \alpha) := x_+(\tilde{t}_2(\kappa, \xi, \varepsilon, \mu, \alpha), (\kappa, y(\kappa, \xi, \alpha, \varepsilon, \mu; \tilde{t}_2(\kappa, \xi, \varepsilon, \mu, \alpha))))(T, \varepsilon, \mu, \alpha).$$
$$\text{(III.2.25)}$$

By our arguments we obtain

$$\widetilde{P}_\kappa(\xi, \varepsilon, \mu, \alpha) \to \widetilde{P}(\xi, \varepsilon, \mu, \alpha) \qquad \text{(III.2.26)}$$

C^r-smoothly with respect to $(\xi, \varepsilon, \mu, \alpha)$ as $\kappa \to 0^+$ for Poincaré mapping (I.4.7) of (III.2.1). Consequently, the corresponding bifurcation equations for (III.2.5), (III.2.6) tend to the bifurcation equation of (III.2.1) in Chapter I.4.

III.2.4. Examples

Certainly, the example of Section I.4.2 can be used to demonstrate our theory. Furthermore, dry friction between a mass and belt in (III.2.7) is given by

$$T(v) = \frac{0.6}{1 + |v|}\operatorname{sgn}(v)$$

where v is a relative velocity between the mass and belt. More sophisticated dry friction formulas are presented in [11] such as

$$T(v) = \begin{cases} \frac{1}{2}(1 - c_2 v), & v > 0, \\ -1 + \frac{c_2}{10}\left(e^{-c_2(11v+3)} + e^{-c_2 v - 1}\right), & v < 0 \end{cases}$$

with a positive constant c_2. We do not go into analytical details since the formulas are rather awkward.

Now we extend the approach of Section III.2.2 to autonomous (III.2.1), i.e. we consider

$$\begin{aligned} \dot{z} &= H_{\pm}^0(z, y) \\ \dot{y} &= G_{\pm}^0(z, y) \end{aligned} \qquad \text{if } \pm z > 0 \qquad (\text{III.2.27})$$

under assumptions (III.2.4) for setting

$$H_{\pm}^0(z, y) = H_{\pm}(z, y, 0, \mu), \quad G_{\pm}^0(z, y) = G_{\pm}(z, y, 0, \mu).$$

Then $K = \{y \in \mathbb{R}^{n-1} \mid H_+^0(0, y) = 0\}$ is a manifold near $p_2(t_2)$. Let $B(p_2(t_1), \delta)$ be a ball in $\mathbb{R}^{n-1}$ centered at $p_2(t_1)$ with the small radius $\delta > 0$. We consider

$$\begin{aligned} \dot{z} &= \frac{H_+^0(\kappa, y) - H_-^0(-\kappa, y)}{2\kappa} z + \frac{H_+^0(\kappa, y) + H_-^0(\kappa, y)}{2} \\ \dot{y} &= \frac{G_+(\kappa, y) - G_-(-\kappa, y)}{2\kappa} z + \frac{G_+^0(\kappa, y) + G_-^0(-\kappa, y)}{2} \end{aligned} \qquad (\text{III.2.28})$$

for $|z| \leq \kappa$, and its solution $(z(t), y(t))$ starting from the point (κ, y), $y \in B(p_2(t_1), \delta)$. By the arguments of Section III.2.3, it transversally crosses the surface $z = \kappa$ at a time $\tilde{t}$ near the point $(\kappa, p_2(t_2))$. We also consider

$$\dot{y} = W^0(y), \quad z = 0 \qquad (\text{III.2.29})$$

for $W^0(y) = W(y, t, \varepsilon, \mu)$ and its solution $\bar{y}(t)$ with the initial condition $\bar{y}(0) = y$ which transversally crosses K at the point $\Theta(y) \in K$. Thus we put

$$\Psi_\kappa(y) := y(\tilde{t}),$$

and by the arguments of Sections III.2.2, III.2.3, we note

$$\Psi_\kappa : B(p_2(t_1), \delta) \to \mathbb{R}^{n-1}, \qquad \Psi_\kappa(y) = \Theta(y) + O(\kappa).$$

Now we take

$$\begin{aligned} \dot{z} &= H_+^0(z, y) \\ \dot{y} &= G_+^0(z, y) \end{aligned} \qquad \text{if } z \geq \kappa \qquad (\text{III.2.30})$$

and its solution starting from the point $(z(0), y(0)) = (\kappa, y)$, $y \in B(p_2(t_2), \delta_1)$ with $\delta_1 > 0$ small. This solution transversally crosses the surface $z = \kappa$ near $\gamma(t_1)$ in the point

$(\kappa, y(\bar{t}))$. We consider the map

$$\Phi_\kappa(y) := y(\bar{t})$$

with $\Phi_\kappa \colon B(p_2(t_2), \delta_1) \to \mathbb{R}^{n-1}$. Since

$$\Psi_\kappa(p_2(t_1)) = \Theta(p_2(t_1)) + O(\kappa) = p_2(t_2) + O(\kappa)$$

and Θ are continuous, we have

$$\Theta, \Psi_\kappa \colon B(p_2(t_1), \delta) \to B(p_2(t_2), \delta_1)$$

for fixed $\delta > 0$ small and any $\kappa > 0$ small. So finally we may consider the Poincaré mapping

$$P_\kappa(y) := \Phi_\kappa(\Psi_\kappa(y)), \qquad y \in B(p_2(t_1), \delta)$$

of (III.2.28) and (III.2.30). By the arguments of Section III.2.3, we have

$$\begin{aligned} P_\kappa(y) &= \Phi_0(\Theta(y)) + O(\kappa), \\ DP_\kappa(y) &= D[\Phi_0(\Theta(y))] + O(\kappa). \end{aligned} \tag{III.2.31}$$

Setting $K_0 = \Phi_0(K \cap B(p_2(t_2), \delta_2)) \subset B(p_2(t_1), \delta)$ for $\delta_1 > \delta_2 > 0$ small, the mapping Φ_0 is a local diffeomorphism of K near $p_2(t_2)$ to K_0 near $p_2(t_1)$. So we may write

$$\Phi_0 \circ \Theta = \Phi_0 \circ \Omega \circ \Phi_0^{-1} \tag{III.2.32}$$

on K_0 for a map $\Omega \colon K \cap B(p_2(t_2), \delta_2) \to K$ defined by

$$\Omega(y) := \Theta(\Phi_0(y)).$$

The map Ω is a Poincaré map of $\gamma(t)$ for (III.2.27) and (III.2.29). Clearly, $\Omega(p_2(t_2)) = p_2(t_2)$.

Now we take a tubular neighborhood $K_0 \times W$ of K_0 in $\mathbb{R}^{n-1}$ near the point $p_2(t_1)$, where $W \subset \mathbb{R}$ is an open neighborhood of $0 \in \mathbb{R}$. The corresponding projections are $\Gamma_1 \colon K_0 \times W \to W$ and $\Gamma_2 \colon K_0 \times W \to K_0$. Then equations (III.2.31) are decomposed into the system

$$(\Gamma_1 P_\kappa(y), \Gamma_2 P_\kappa(y)) = (O(\kappa), \Phi_0(\Theta(y)) + O(\kappa)),$$

$$DP_\kappa(y) = \begin{pmatrix} 0 & 0 \\ D_{y_1} \Phi_0(\Theta(y)) & D_{y_2} \Phi_0(\Theta(y)) \end{pmatrix} + O(\kappa),$$

$$y = y_1 + y_2 \text{ for } (y_1, y_2) \in W \times K_0,$$

where we use $\Phi_0(\Theta(y)) \in K_0$. So fixed points of P_κ are given by

$$\begin{aligned} y_1 &= \Gamma_1 P_\kappa(y) = O(\kappa), \\ y_2 &= \Gamma_2 P_\kappa(y) = \Phi_0(\Theta(y)) + O(\kappa) \end{aligned} \tag{III.2.33}$$

with linearization

$$DP_\kappa(y) = \begin{pmatrix} 0 & 0 \\ D_{y_1}\Phi_0(\Theta(y_2)) & D_{y_2}\Phi_0(\Theta(y_2)) \end{pmatrix} + O(\kappa).$$

Hence, if the linearization $\mathbb{I} - D_{y_2}\Phi_0(\Theta(p_2(t_1)))$ is nonsingular, we can solve (III.2.33) near $p_2(t_1)$ by using the implicit function theorem. Moreover, if $D_{y_2}\Phi_0(\Theta(p_2(t_1)))$ is stable, i.e. all eigenvalues of $D_{y_2}\Phi_0(\Theta(p_2(t_1)))$ are inside the unit circle of $\mathbb{C}$, then also the corresponding approximate periodic solution $\gamma_\kappa(t)$ of (III.2.28) and (III.2.30) is stable. Of course, the stability of $D_{y_2}\Phi_0(\Theta(p_2(t_1)))$ gives the asymptotic stability of the periodic solution $\gamma(t)$ for (III.2.27) and (III.2.29).

Note

$$\Theta(p_2(t_1)) = p_2(t_2), \qquad \Phi_0(p_2(t_2)) = p_2(t_1),$$

which by (III.2.32) gives

$$D\left[\Phi_0(\Theta(p_2(t_1)))\right] = D\Phi_0(p_2(t_2))D\Omega(p_2(t_2))D\left[\Phi_0^{-1}(p_2(t_1))\right].$$

Next, using

$$D\Phi_0(p_2(t_2)): T_{p_2(t_2)}K \to T_{p_2(t_1)}K_0,$$
$$[D\Phi_0(p_2(t_2))]^{-1} = D\left[\Phi_0^{-1}(p_2(t_1))\right]: T_{p_2(t_1)}K_0 \to T_{p_2(t_2)}K,$$
$$D\Omega(p_2(t_2)): T_{p_2(t_2)}K \to T_{p_2(t_2)}K$$

and

$$D\left[\Phi_0(\Theta(p_2(t_1)))\right]\varsigma = D_{y_2}\Phi_0(\Theta(p_2(t_1)))\varsigma$$

for any $\varsigma \in T_{p_2(t_1)}K_0$, we derive that the spectrum of

$$D_{y_2}\Phi_0(\Theta(p_2(t_1))): T_{p_2(t_1)}K_0 \to T_{p_2(t_1)}K_0$$

is equal to the spectrum of $D\Omega(\gamma(t_2))$. Summarizing, we arrive at the following extension of Theorem III.2.1.

Theorem III.2.3. *If the linearization* $\mathbb{I} - D\Omega(p_2(t_2)): T_{p_2(t_2)}K \to T_{p_2(t_2)}K$ *is nonsingular, then the approximate system* (III.2.28), (III.2.30) *possesses a periodic solution* $\gamma_\kappa(t)$ *near* $\gamma(t)$. *If* $D\Omega(p_2(t_2))$ *is stable, then all these periodic solutions are stable.*

Remark III.2.4. Theorem III.2.3 contains some transversal/generic assumptions (see (III.2.4)), namely that K is a manifold, $\gamma(t)$ transversally crosses K and that the linearization $\mathbb{I} - D\Omega(p_2(t_2))$ is nonsingular. If one of them fails, then the construction of the Poincaré map P_κ as well as the solvability of equation (III.2.33) become problematic. Some bifurcations of periodic solutions are expected as in Chapter I.4.

Remark III.2.5. Theorem III.2.3 makes a statement about the persistence of a generic periodic solution of discontinuous systems hitting a discontinuity level under a continuous approximation. This persistence could be proved by using the Leray-Schauder degree theory [18, 19] (see also A.1.5), but since we use the implicit function theorem, we get uniqueness and asymptotic stability of periodic solutions as well. Furthermore, this approach is constructive. Finally, we assume that the discontinuity level has a codimension 1. Higher codimension problems could also be interesting to study.

For application of Theorem III.2.3, we need to derive

$$D\Omega(p_2(t_2)) \colon T_{p_2(t_2)}K \to T_{p_2(t_2)}K,$$

where $T_{p_2(t_2)}K$ is the tangent space of manifold K at point $p_2(t_2)$ given by

$$T_{p_2(t_2)}K = \left\{ \eta \in \mathbb{R}^{n-1} \,\middle|\, \eta \perp \nabla_y H_+^0(0, p_2(t_2)) \right\}.$$

It is not difficult to derive, using the constructions of maps Θ and Φ_0, that

$$D\Omega(p_2(t_2))\eta = w(t_2) - \frac{\langle \nabla_y H_+^0(\gamma(t_2)), w(t_2)\rangle}{\langle \nabla_y H_+^0(\gamma(t_2)), \dot{p}_2(t_2)\rangle} \dot{p}_2(t_2), \tag{III.2.34}$$

where $\langle \cdot, \cdot \rangle$ is the scalar product and the function $w(t)$ (depending on η) is the solution of the initial value problem

$$\dot{w}(t) = DW^0(p_2(t))w(t),$$
$$w(t_1) = -\frac{\dot{p}_2(t_1)}{\dot{p}_1(t_1)} z(T + t_1) + y(T + t_1), \tag{III.2.35}$$

where the functions $z(t), y(t)$ (also depending on η) are the solutions of the initial value problem

$$\dot{z}(t) = H_{+z}^0(\gamma(t))z(t) + H_{+y}^0(\gamma(t))y(t)$$
$$\dot{y}(t) = G_{+z}^0(\gamma(t))z(t) + G_{+y}^0(\gamma(t))y(t) \tag{III.2.36}$$
$$z(t_2) = 0, \qquad y(t_2) = \eta.$$

Here $W^0(y) = W(y, t, 0, \mu)$.

We apply the above formulas to an extension of system (III.2.7) given by a dry friction oscillator

$$\ddot{v} - v + v^3 - \frac{0.6}{1 + |\dot{v} - 1|} \operatorname{sgn}(1 - \dot{v}) + uf(u, v) = 0$$
$$\ddot{u} + \delta\dot{u} + ug(u, v) = 0, \tag{III.2.37}$$

where $u \in \mathbb{R}$, $\delta > 0$ and f, g are smooth functions. Equation (III.2.37) has the form

$$\dot{z} = y_1^3 - y_1 - \frac{0.6}{1 + |z|} \operatorname{sgn} z + y_2 f(y_2, y_1)$$

$$\dot{y}_1 = 1 - z \qquad\qquad\qquad\qquad\qquad (\text{III.2.38})$$

$$\dot{y}_2 = y_3$$

$$\dot{y}_3 = -\delta y_3 - y_2 g(y_2, y_1).$$

Like in Section III.2.2, we consider $\Sigma = \{z = 0\}$, so $t_2 = T$. Since now

$$H_\pm^0(z, y) = y_1^3 - y_1 \mp \frac{0.6}{1 \pm z} + y_2 f(y_2, y_1)$$

for $y = (y_1, y_2, y_3)$, we take $\gamma(t) = (w(t), y(t), 0, 0)$, $t \in [0, t_0]$ where $w(t)$, $y(t)$ solve (III.2.10) with the initial conditions $y(0) = y_0$ and $w(0) = 0$. So $t_1 = t_0$. Here we use the notation of Section III.2.2. We note that $\gamma(t_0) = (0, \bar{y}_0, 0, 0)$. Furthermore, the sliding system (III.2.29) now has the form

$$\dot{y}_1 = 1$$

$$\dot{y}_2 = y_3$$

$$\dot{y}_3 = -\delta y_3 - y_2 g(y_2, y_1).$$

Hence $p_2(t) = (\bar{y}_0 + t - t_0, 0, 0)$ with $t_2 = T = t_0 + y_0 - \bar{y}_0$. Since

$$D_y H_+^0(0, p_2(t_2)) \dot{p}_2(t_2) = 3y_0^2 - 1 \doteq 3.4740 > 0,$$

assumption (III.2.4) holds. So, in order to apply Theorem III.2.3 for system (III.2.38), we need to find the spectrum of $D\Omega(p_2(t_2))$. We note that now $p_2(t_2) = (y_0, 0, 0)$, and then

$$T_{p_2(t_2)} K := \left\{ \eta \in \mathbb{R}^3 \,\middle|\, \eta \perp (3y_0^2 - 1, f(0, y_0), 0) \right\}$$

$$= \left\{ \left(-\frac{f(0, y_0)}{3y_0^2 - 1} \eta_2, \eta_2, \eta_3 \right) \,\middle|\, \eta_2, \eta_3 \in \mathbb{R} \right\}.$$

According to (III.2.34), (III.2.35) and (III.2.36), we now derive

$$D\Omega(p_2(t_2))\eta = \left(-\frac{f(0, y_0)}{3y_0^2 - 1} w_2(T), w_2(T), w_3(T) \right), \qquad (\text{III.2.39})$$

where $(w_1(t), w_2(t), w_3(t))$ solves the system

$$\dot{w}_1 = 0$$

$$\dot{w}_2 = w_3 \qquad\qquad\qquad\qquad\qquad (\text{III.2.40})$$

$$\dot{w}_3 = -\delta w_3 - g(0, \bar{y}_0 + t - t_0) w_2$$

with the initial value condition

$$w_1(t_0) = y_1(t_0) - \frac{\dot{y}(t_0)}{\dot{w}(t_0)} z_1(t_0), \quad w_2(t_0) = y_2(t_0), \quad w_3(t_0) = y_3(t_0),$$

where $(z_1(t), y_1(t), y_2(t), y_3(t))$ solves the following system of ODEs

$$\dot{z}_1 = (3y(t)^2 - y(t))y_1 + \frac{0.6}{(1 + w(t))^2} z_1 + f(0, y(t))y_2$$
$$\dot{y}_1 = -z_1 \qquad\qquad\qquad (\text{III.2.41})$$
$$\dot{y}_2 = y_3$$
$$\dot{y}_3 = -\delta y_3 - g(0, y(t))y_2$$

with the initial value condition

$$z_1(0) = 0, \quad y_1(0) = -\frac{f(0, y_0)}{3y_0^2 - 1} \eta_2, \quad y_2(0) = \eta_2, \quad y_3(0) = \eta_3$$

for $\eta_2, \eta_3 \in \mathbb{R}$. We can easily see from (III.2.39) and from the initial value problems (III.2.40), (III.2.41) that the spectrum satisfies

$$\sigma(D\Omega(p_2(t_2))) = \{\lambda_1, \lambda_2\}$$

where $\{\lambda_1, \lambda_2\}$ is the spectrum of the fundamental matrix solution of the system

$$\dot{y}_2 = y_3$$
$$\dot{y}_3 = -\delta y_3 - q(t)y_2, \qquad\qquad (\text{III.2.42})$$

where

$$q(t) = \begin{cases} g(0, y(t)) & \text{for } t \in [0, t_0], \\ g(0, \bar{y}_0 + t - t_0) & \text{for } t \in [t_0, T]. \end{cases}$$

We note that $q(t)$ is T-periodic. We can use several known criteria for the asymptotic stability of (III.2.42) (see [20–22]). Indeed, (III.2.42) has the form

$$\ddot{y}_2 + \delta \dot{y}_2 + q(t)y_2 = 0,$$

which can be transformed by $z_2 = e^{\frac{\delta}{2}t} y_2$ to

$$\ddot{z}_2 + P(t)z_2 = 0, \qquad P(t) = q(t) - \frac{\delta^2}{4}. \qquad\qquad (\text{III.2.43})$$

Using the Lyapunov result [21, Theorem 2.5.4, p. 76] we obtain that if

$$P(t) > 0, \qquad T \int_0^T P(t)dt \le 4,$$

then all solutions of (III.2.43) and their derivatives are bounded on $\mathbb{R}_+$. Hence if

$$q(t) > \frac{\delta^2}{4}, \quad \int_0^T q(t)dt \le \frac{\delta^2 T}{4} + \frac{4}{T}, \tag{III.2.44}$$

then the spectrum $\sigma(D\Omega(p_2(t_2)))$ is inside the unit disc. Summarizing, condition (III.2.44) implies the asymptotic stability of the periodic solution $\gamma(t)$ as well as the existence and asymptotic stability of periodic solutions $\gamma_\kappa(t)$ from Theorem III.2.3 applied to system (III.2.38). Clearly, condition (III.2.44) trivially holds for a smooth function $g(z, v)$ when $\frac{\delta^2}{4} < g(0, v) \le \frac{\delta^2}{4} + \frac{4}{T^2}$ for any (see Figure III.2.2)

$$v \in \left[\min_{[0,t_0]} y(t), \max_{[0,t_0]} y(t) \right] \doteq [-1.3227, 1.6667].$$

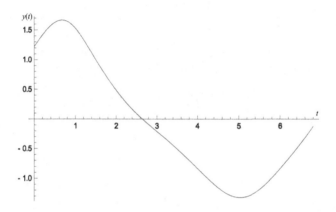

Figure III.2.2 Solution $y(t)$ of (III.2.10) with $z(0) = 0$, $y(0) = y_0$ over $[0, t_0]$

Now we take g with $g(0, v) = v$. Since $y(t)$ changes the sign over the interval $[0, t_0]$, $q(t)$ defined previously also changes the sign over the interval $[0, T]$. Hence, we cannot use criterion (III.2.44). Instead, we use a criterion by R. Einandi from [20, p. 63] of the form

$$\int_0^T q(t)dt \tan \frac{\delta T}{2} \le 2\delta,$$

since now $\int_0^T q(t)dt = \int_0^T p_2(t)dt \doteq 0.2040 > 0$. Consequently, we get the following inequality

$$\tan(4.0737\delta) \le 9.8053\delta$$

for the parameter δ, where our theory is applied to obtain the existence and asymptotic stability of periodic solutions $\gamma_\kappa(t)$ from Theorem III.2.3 for the concrete system

(III.2.38). This holds when $\delta < 0.3068$.

Furthermore, using criterion [20, (4.3.vii), p. 63], we see that if

$$P(t) < 0, \quad \forall t \in \mathbb{R}, \qquad -\frac{1}{T} \int_0^T P(t)dt < \frac{\delta^2}{4},$$

then (III.2.42) is asymptotically stable, i.e. if

$$\delta^2 > 4 \max_{t \in \mathbb{R}} q(t), \qquad \int_0^T q(t)dt > 0.$$

For $g(0, v) = v$, the above conditions state

$$\delta^2 > 4 \max_{t \in \mathbb{R}} p_2(t) \doteq 6.6668, \qquad \int_0^T p_2(t)dt \doteq 0.20397 > 0,$$

i.e. the concrete system (III.2.38) has asymptotically stable $\gamma_\kappa(t)$ and $\gamma(t)$, if $\delta > 2.58202$. Summarizing, system (III.2.38) with $g(0, v) = v$ has asymptotically stable $\gamma_\kappa(t)$ and $\gamma(t)$, if either $0 < \delta < 0.3068$ or $\delta > 2.58202$.

On the other hand, (III.2.42) is asymptotically unstable for any $\delta > 0$, whenever $\max_{t \in [0,T]} q(t) < 0$. Indeed, setting $\bar{q} := -\max_{t \in [0,T]} q(t)$, we derive $\max_{t \in [0,T]} P(t) = -\left(\bar{q} + \frac{\delta^2}{4}\right) < 0$. So by Lemma A.10 we see that the largest characteristic multiplier of (III.2.43) is greater than or equal to $e^{\sqrt{\bar{q} + \frac{\delta^2}{4}}T}$. Hence the largest characteristic multiplier of (III.2.42) is greater than or equal to

$$e^{\sqrt{\bar{q} + \frac{\delta^2}{4}}T - \frac{\delta}{2}T} > 1,$$

which gives its asymptotic instability. So the system (III.2.38) with $g(0, v) = v - a$, $a > \max_{t \in \mathbb{R}} p_2(t) \doteq 1.6667$, has asymptotically unstable $\gamma_\kappa(t)$ and $\gamma(t)$ for any $\delta > 0$.

CHAPTER III.3

Impact periodic orbits

III.3.1. Setting of the problem

Let $\Omega \subset \mathbb{R}^n$ be an open subset and $G: \Omega \to \mathbb{R}$ a C^2-function, such that $DG(x) \neq 0$ for any $x \in S := \{x \in \Omega \mid G(x) = 0\} \subset \Omega$. Then the impact manifold S is a smooth hypersurface in Ω. We set $\Omega_{\pm} = \{x \in \Omega \mid \pm G(x) > 0\}$ and consider the following regular-singular perturbed system:

$$\begin{aligned} \varepsilon \dot{x} &= f_+(x) + \varepsilon g_+(t, x, \varepsilon) && \text{for } x \in \Omega_+ \\ \dot{x} &= f_-(x) + \varepsilon g_-(t, x, \varepsilon) && \text{for } x \in \Omega_- \end{aligned} \tag{III.3.1}$$

for $\varepsilon > 0$ small. We assume that the system

$$\begin{aligned} \dot{x} &= f_+(x) && \text{for } x \in \Omega_+ \\ \dot{x} &= f_-(x) && \text{for } x \in \Omega_- \end{aligned} \tag{III.3.2}$$

has a continuous periodic solution $q(t)$ crossing transversally the impact manifold S, given by:

$$q(t) = \begin{cases} q_-(t) \in \Omega_- & \text{for } -T_-^0 < t < 0, \\ q_+(t) \in \Omega_+ & \text{for } 0 < t < T_+^0 \end{cases}$$

and $q_-(0) = q_+(0) \in S$, $q_-(-T_-^0) = q_+(T_+^0) \in S$. By transversal crossing we mean that

$$DG(q(\pm T_{\pm}^0))\dot{q}_{\pm}(\pm T_{\pm}^0) < 0 < DG(q(0))\dot{q}_{\pm}(0).$$

We set $T_\varepsilon := T_-^0 + \varepsilon T_+^0$ and assume that $g_{\pm}(t, x, \varepsilon)$ are T_ε-periodic in t.

Transversal crossing implies [23, 24] that (III.3.2) has a family of continuous solutions $q(t, \alpha)$ where $\alpha \in I_0$ and I_0 is an open neighborhood of $0 \in \mathbb{R}^{n-1}$, crossing transversally the impact manifold S, and given by

$$q(t, \alpha) = \begin{cases} q_-(t, \alpha) \in \Omega_- & \text{for } -T_-(\alpha) < t < 0, \\ q_+(t, \alpha) \in \Omega_+ & \text{for } 0 < t < T_+(\alpha) \end{cases}$$

where we have $q_-(0, \alpha) = q_+(0, \alpha) \in S$, $q_-(-T_-(\alpha), \alpha), q_+(T_+(\alpha), \alpha) \in S$, $q_{\pm}(t, 0) = q_{\pm}(t)$ and $T_{\pm}(0) = T_{\pm}^0$. Moreover, $T_{\pm}(\alpha)$ is C^2 in α, and the maps $\alpha \mapsto q(0, \alpha)$ and $\alpha \mapsto q_{\pm}(\pm T_{\pm}(\alpha), \alpha)$ give smooth (C^2) parametrizations of the manifold S in small neighborhoods U_0 of $q(0)$ and $U_\pm$ of $q(T^0_+) = q(-T^0_-)$. Then the map

$$R: U_0 \cap S \to U_+ \cap S, \quad q(0, \alpha) \mapsto q_+(T_+(\alpha), \alpha)$$

is C^2-smooth. In this chapter we study the problem of existence of a T_ε-periodic solution of the singular problem (III.3.1) in a neighborhood of the set

$$\left\{ q_-(t) \,\middle|\, t \in [-T^0_-, 0] \right\} \cup \left\{ q_+(t) \,\middle|\, t \in [0, T^0_+] \right\}.$$

As a matter of fact, in the time interval $[0, \varepsilon T^0_+]$, or $[-T^0_-, 0]$, the periodic solutions will stay close to $q_+(\varepsilon^{-1} t)$, or $q_-(t)$. Thus it will pass from the point of S near $q(0)$ to the point of S near $q_+(T^0_+)$ in a very short time (of the size of εT^0_+). So, we may say that the behavior of the periodic solutions of (III.3.1) in the interval $[-T^0_-, \varepsilon T^0_+]$ is quite well simulated by the solution of the perturbed impact system

$$\begin{aligned} \dot{x} &= f_-(x) \\ R(q_-(0, \alpha)) &= q_+(T_+(\alpha), \alpha). \end{aligned} \tag{III.3.3}$$

Our results, Theorems III.3.2 and III.3.6, state that if a certain Poincaré-Andronov-Melnikov-like function has a simple zero then the above problem has an affirmative answer. This chapter is related to Part II.

III.3.2. Bifurcation equation

To start with obtaining our main results, we set $u_+(t, \alpha) = q_+(\varepsilon^{-1} t, \alpha)$, $u_-(t, \alpha) = q_-(t, \alpha)$ and

$$u(t, \alpha) = \begin{cases} u_-(t, \alpha) & \text{for } -T_-(\alpha) \le t < 0, \\ u_+(t, \alpha) & \text{for } 0 \le t < \varepsilon T_+(\alpha). \end{cases}$$

Note that

$$\varepsilon \dot{u}_+(t, \alpha) = f_+(u_+(t, \alpha))$$
$$\dot{u}_-(t, \alpha) = f_-(u_-(t, \alpha))$$
$$u_+(0, \alpha) = u_-(0, \alpha)$$
$$u_+(\varepsilon T_+(\alpha), \alpha), \ u_-(-T_-(\alpha), \alpha), 0) \in S,$$

and that $u(t, 0)$ is a continuous periodic solution, of period $T^0_- + \varepsilon T^0_+$, of the piecewise-continuous singular system

$$\begin{aligned} \varepsilon \dot{x} &= f_+(x) && \text{for } x \in \Omega_+ \\ \dot{x} &= f_-(x) && \text{for } x \in \Omega_-. \end{aligned}$$

Obviously, $u_-(t, \alpha)$ extends to a solution of the impact system

$$\dot{x} = f_-(x) \qquad \text{for } x \in \Omega_-$$
$$x(t^+) = q_+(T_+(\alpha), \alpha) \qquad \text{when } x(t^-) = q_-(0, \alpha)$$

that can be written as

$$\dot{x} = f_-(x) \qquad \text{for } x \in \Omega_-$$
$$x(t^+) = R(x(t^-)) \qquad \text{when } x(t^-) \in U_0 \cap S.$$

Our purpose is to find a T_ε-periodic solution $x(t, \varepsilon)$ of system (III.3.1) which is orbitally close to $u(t, \alpha)$ for some $\alpha = \alpha(\varepsilon) \to 0$ as $\varepsilon \to 0^+$, that is such that

$$\sup_{-T_-^0 \leq t \leq \varepsilon T_+^0} |x(t + \tau(\varepsilon), \varepsilon) - u(t, \alpha(\varepsilon))| \to 0 \quad \text{as} \quad \varepsilon \to 0^+ \qquad \text{(III.3.4)}$$

for some $(\tau(\varepsilon), \alpha(\varepsilon)) \to (\tau_0, 0)$ as $\varepsilon \to 0$. Thus, we may say that, in some sense, the impact periodic solution $u_-(t, 0)$ approximates the periodic solution $x(t, \varepsilon)$ of the singular perturbed equation (III.3.1).

To achieve this goal, we first set $x(t + \tau) = x_+(t) + u_+(t, \alpha)$ in the equation $\varepsilon \dot{x} = f_+(x) + \varepsilon g_+(t, x, \varepsilon)$. Then $x_+(t)$ satisfies

$$\varepsilon \dot{x} - Df_+(u_+(t, \alpha))x = h_+(t, \tau, x, \alpha, \varepsilon) \qquad \text{(III.3.5)}$$

where:

$$h_+(t, \tau, x, \alpha, \varepsilon) = f_+(x + u_+(t, \alpha)) - f_+(u_+(t, \alpha)) - Df_+(u_+(t, \alpha))x$$
$$+ \varepsilon g_+(t + \tau, x + u_+(t, \alpha), \varepsilon).$$

Since $u_+(0, \alpha)$ describes $U_0 \cap S$ we consider (III.3.5) with the initial condition $x_0 = 0$, i.e. we have $x(\tau) = u_+(0, \alpha) \in U_0 \cap S$. Let $X_+(t, \alpha)$ be the fundamental solution of $\dot{x} = Df_+(q_+(t, \alpha))x$ such that $X_+(0, \alpha) = \mathbb{I}$. Then $X_+(\varepsilon^{-1}t, \alpha)$ is the fundamental solution of $\varepsilon \dot{x} = Df_+(u_+(t, \alpha))x$ with $X_+(0, \alpha) = \mathbb{I}$. Let T_+ be near T_+^0. By the variation of constants formula [9] the solution of (III.3.5) with the initial condition $x_0 = 0$ satisfies

$$x_+(t) = \varepsilon^{-1} \int_0^t X_+(\varepsilon^{-1}t, \alpha) X_+^{-1}(\varepsilon^{-1}s, \alpha) h(s, \tau, x_+(s), \alpha, \varepsilon) ds.$$

Thus we conclude that for $\rho > 0$ and T_+ near T_+^0 equation $\varepsilon \dot{x} = f_+(x) + \varepsilon g(t, x, \varepsilon)$ has a solution $x(t)$ such that $\sup_{0 \leq t \leq \varepsilon T_+} |x(t + \tau) - u_+(t, \alpha)| < \rho$ if and only if the map $x(t) \mapsto \hat{x}(t)$ given by

$$\hat{x}(t) = \varepsilon^{-1} \int_0^t X_+(\varepsilon^{-1}t, \alpha) X_+^{-1}(\varepsilon^{-1}s, \alpha) h(s, \tau, x(s), \alpha, \varepsilon) ds \qquad \text{(III.3.6)}$$

has a fixed point whose supremum norm on $[0, \varepsilon T_+]$ is smaller than ρ. To show that (III.3.6) has a fixed point of norm less than ρ, we set $y(t) := x(\varepsilon T_+ t)$, $t \in [0, 1]$ and note that $x(t)$ is a fixed point of (III.3.6) of norm less than ρ, with $0 \leq t \leq \varepsilon T_+$, if and only if $y(t)$ is a fixed point of norm less than ρ of the map

$$\hat{y}(t) = T_+ \int_0^t X_+(T_+t, \alpha)X_+^{-1}(T_+\sigma, \alpha)h(\varepsilon T_+\sigma, \tau, y(\sigma), \alpha, \varepsilon)d\sigma \qquad \text{(III.3.7)}$$

for $0 \leq t \leq 1$. Note that

$$h_+(\varepsilon T_+t, \tau, x, \alpha, \varepsilon) = f_+(x + q_+(tT_+, \alpha)) - f_+(q_+(tT_+, \alpha))$$
$$-Df_+(q_+(tT_+, \alpha))x + \varepsilon g_+(\varepsilon tT_+ + \tau, x + q_+(tT_+, \alpha), \varepsilon),$$

and hence in the fixed point equation (III.3.7), we may also take $\varepsilon \leq 0$. Then since $(x, T_+, \alpha, \varepsilon) \mapsto h_+(\varepsilon T_+\tau, \tau, x, \alpha, \varepsilon)$ with $0 \leq \tau \leq 1$ is a C^2-map and

$$|h_+(t, \tau, x, \alpha, \varepsilon)| \leq \Delta(|x|)|x| + N_g|\varepsilon|$$

where

$$N_g = \sup\left\{|g_+(t, \tilde{x}, \varepsilon)| \,\middle|\, t \in \mathbb{R}, \, |\tilde{x}| \leq \rho + \sup_{t \in [0, T_+(\alpha)], \alpha \in I_0} |q_+(t, \alpha)|, \, |\varepsilon| \leq \varepsilon_0\right\},$$

$$\Delta(\rho) = \sup\left\{|Df(x + q_+(t, \alpha)) - Df(q_+(t, \alpha))| \,\middle|\, t \in [0, T_+(\alpha)], \, |x| \leq \rho, \, \alpha \in I_0\right\},$$

the map $y(t) \mapsto \hat{y}(t)$ is a C^2-contraction on the Banach space of bounded continuous functions on $[0, 1]$ whose supremum norm is less than or equal to ρ provided that ρ is sufficiently small, T_+ is near T_+^0, $|\varepsilon|$ is small, $\alpha \in I_0$ and $\tau \in \mathbb{R}$. Let $y_+(t, \tau, \alpha, T_+, \varepsilon)$ be the C^2-solution of the fixed point problem (III.3.7) given by the Banach fixed point theorem or the IFT (see Theorem A.4 or [25, 26]). We emphasize the fact that ε may also be nonpositive. Then $x_+(t, \tau, \alpha, \varepsilon) := y_+(\varepsilon^{-1}T_+^{-1}t, \tau, \alpha, T_+, \varepsilon)$ is a fixed point of (III.3.6) and

$$x_+(\varepsilon t, \tau, \alpha, \varepsilon) := y_+(T_+^{-1}t, \tau, \alpha, T_+, \varepsilon) \qquad \text{(III.3.8)}$$

is C^2 in all parameters including t.

Writing $T_+^{-1}t$ in place of t in (III.3.7) and using (III.3.8) we see that

$$x_+(\varepsilon t, \tau, \alpha, \varepsilon) = \int_0^t X_+(t, \alpha)X_+^{-1}(s, \alpha)h_+(\varepsilon s, \tau, x_+(\varepsilon s, \tau, \alpha, \varepsilon), \alpha, \varepsilon)ds \qquad \text{(III.3.9)}$$

for $0 \leq t \leq T_+$. We have, by definition, $x_+(0, \tau, \alpha, \varepsilon) + u_+(0, \alpha) = u_+(0, \alpha) \in S$ and

$$x_+(\varepsilon T_+, \tau, \alpha, \varepsilon) + u_+(\varepsilon T_+, \alpha) \in S$$

if and only if (recall that $u_+(\varepsilon T_+, \alpha) = q_+(T_+, \alpha)$)

$$G\left(q_+(T_+, \alpha) + \int_0^{T_+} X_+(T_+, \alpha)X_+^{-1}(s, \alpha)h_+(\varepsilon s, \tau, x_+(\varepsilon s, \tau, \alpha, \varepsilon), \alpha, \varepsilon)ds\right) = 0.$$
$$\text{(III.3.10)}$$

We remark that equation (III.3.10) has meaning also when $\varepsilon < 0$, but it is relevant for our problem only when $\varepsilon > 0$.

As a second step we consider the solution of the differential equation

$$\dot{x} = f_-(x) + \varepsilon g_-(t, x, \varepsilon), \qquad x(\tau) = q(0, \alpha)$$

on Ω_-, which is close to $u_-(t - \tau, \alpha)$ for $-T_- + \tau \le t \le \tau$, $T_- \sim T_-^0$. Let $X_-(t, \alpha)$ be the fundamental solution of the linear system $\dot{x} = Df_-(u_-(t, \alpha))x$ such that $X_-(0, \alpha) = \mathbb{I}$. Setting $x(t + \tau) = x_-(t) + u_-(t, \alpha)$ we see that (for $t \in [-T_-, 0]$) $x_-(t)$ satisfies the initial value problem

$$\dot{x} - Df_-(u_-(t, \alpha))x = h_-(t, \tau, x, \alpha, \varepsilon)$$
$$x(0) = 0 \tag{III.3.11}$$

where

$$h_-(t, \tau, x, \alpha, \varepsilon) = f_-(x + u_-(t, \alpha)) - f_-(u_-(t, \alpha))$$
$$-Df_-(u_-(t, \alpha))x + \varepsilon g_-(t + \tau, x + u_-(t, \alpha), \varepsilon).$$

Again, by the variation of constants formula we get the integral formula

$$x_-(t) = \int_0^t X_-(t, \alpha)X_-(s, \alpha)^{-1}h_-(s, \tau, x_-(s), \alpha, \varepsilon)ds$$

which, as before, has a unique solution of norm less than a given small ρ and $x_-(t, \tau, \alpha, \varepsilon)$ with $-T_- \le t \le 0$. At $t = -T_-$ the solution of (III.3.11) takes the value

$$-\int_{-T_-}^0 X_-(-T_-, \alpha)X_-(s, \alpha)^{-1}h_-(s, \tau, x_-(s, \alpha, \varepsilon), \alpha, \varepsilon)ds.$$

Now, we want to solve the equation

$$x_-(-T_-, \tau, \alpha, \varepsilon) + u_-(-T_-, \alpha) = x_+(\varepsilon T_+, \tau, \alpha, \varepsilon) + u_+(\varepsilon T_+, \alpha),$$

that is (again using $u_+(\varepsilon T_+, \alpha) = q_+(T_+, \alpha)$ and $u_-(-T_-, \alpha) = q_-(-T_-, \alpha)$)

$$q_+(T_+, \alpha) + \int_0^{T_+} X_+(T_+, \alpha)X_+^{-1}(s, \alpha)h_+(\varepsilon s, \tau, x_+(\varepsilon s, \tau, \alpha, \varepsilon), \alpha, \varepsilon)ds$$

$$= q_-(-T_-, \alpha) - \int_{-T_-}^0 X_-(-T_-, \alpha)X_-(s, \alpha)^{-1}h_-(s, \tau, x_-(s, \tau, \alpha, \varepsilon), \alpha, \varepsilon)ds. \tag{III.3.12}$$

Of course, when (III.3.12) holds then (III.3.10) is equivalent to

$$G\left(q_-(-T_-, \alpha) - \int_{-T_-}^0 X_-(-T_-, \alpha)X_-(s, \alpha)^{-1}h_-(s, \tau, x_-(s, \tau, \alpha, \varepsilon), \alpha, \varepsilon)ds\right) = 0. \tag{III.3.13}$$

So our task reduces to solving the system formed by equations (III.3.12), (III.3.13) together with the period equation

$$T_- + \varepsilon T_+ = T_-^0 + \varepsilon T_+^0,$$

i.e. the equation $\mathcal{F}(T_+, T_-, \tau, \alpha, \varepsilon) = 0$ where

$$\mathcal{F}(T_+, T_-, \tau, \alpha, \varepsilon) :=$$
$$\begin{pmatrix} x_-(-T_-, \tau, \alpha, \varepsilon) + q_-(-T_-, \alpha) - x_+(\varepsilon T_+, \tau, \alpha, \varepsilon) - q_+(T_+, \alpha) \\ G\left(q_-(-T_-, \alpha) - \int_{-T_-}^0 X_-(-T_-, \alpha)X_-^{-1}(s, \alpha)h_-(s, \tau, x_-(s, \tau, \alpha, \varepsilon), \alpha, \varepsilon), \varepsilon)ds\right) \\ T_- - T_-^0 + \varepsilon(T_+ - T_+^0) \end{pmatrix}.$$

According to the smoothness properties of $x_-(t, \tau, \alpha, \varepsilon)$ and $x_+(\varepsilon t, \tau, \alpha, \varepsilon)$, it results that $\mathcal{F}(T_+, T_-, \tau, \alpha, \varepsilon)$ is C^2.

III.3.3. Bifurcation from a single periodic solution

In this section, we study the case, which we call non-degenerate, when

$$D_\alpha \left[q_+(T_+(\alpha), \alpha) - q_-(-T_-(\alpha), \alpha) \right]_{\alpha=0} w \neq 0, \quad \forall w \in \mathbb{R}^{n-1} : DT_-(0)w = 0. \tag{III.3.14}$$

Remark III.3.1. Condition (III.3.14) has a simple geometrical meaning. The impact system (III.3.3) has a T_-^0-periodic solution if and only if the following condition holds

$$q_+(T_+(\alpha), \alpha) = q_-(-T_-(\alpha), \alpha), \qquad T_-(\alpha) = T_-^0. \tag{III.3.15}$$

Now, suppose there is a sequence $0 \neq \alpha_n \to 0$ as $n \to \infty$ such that (III.3.15) holds. Possibly passing to a subsequence, we can suppose that $\lim_{n\to\infty} \frac{\alpha_n}{|\alpha_n|} = w$, $|w| = 1$. Then, taking the limit in the equalities

$$\frac{q_+(T_+(\alpha_n), \alpha_n) - q_-(-T_-(\alpha_n), \alpha_n)}{|\alpha_n|} = 0, \qquad \frac{T_-(\alpha_n) - T_-^0}{|\alpha_n|} = 0$$

we see that condition (III.3.14) does not hold. Thus (III.3.14) implies that, in a neighborhood of $\alpha = 0$, there are no other T_-^0-periodic solutions of (III.3.3) apart from $q_-(t)$.

Now we are ready to prove our first main result of this chapter.

Theorem III.3.2. *Assume condition* (III.3.14) *holds and let* $(\psi, \psi_1, \psi_2) \in \mathbb{R}^n \times \mathbb{R} \times \mathbb{R}$

be the unique (up to a multiplicative constant) solution of the linear system

$$\psi^* \dot{q}_+(T_+^0, 0) = 0$$

$$\psi_2 = \left[\psi^* + \psi_1 DG(q(-T_-^0, 0)) \right] \dot{q}_-(-T_-^0, 0)$$

$$\psi^* \left[D_\alpha q_-(-T_-^0, 0) - D_\alpha q_+(T_+^0, 0) \right] + \psi_1 DG(q(-T_-^0, 0)) \dot{q}_-(-T_-^0, 0) DT_-(0) = 0.$$

$$(\text{III.3.16})$$

If the Poincaré-Andronov-Melnikov function

$$M(\tau) := \psi^* \int_0^{T_+^0} X_+(T_+^0, 0) X_+(s, 0)^{-1} g_+(\tau, u(0,0), 0) ds$$

$$+ \psi^* \int_{-T_-^0}^0 X_-(-T_-^0, 0) X_-(s, 0)^{-1} g_-(s + \tau, u_-(s, 0), 0) ds \qquad (\text{III.3.17})$$

$$+ \psi_1 DG(q(-T_-^0, 0)) \int_{-T_-^0}^0 X_-(-T_-^0, 0) X_-^{-1}(s, 0) g_-(s + \tau, q_-(s, 0), 0) ds$$

has a simple zero at $\tau = \tau_0$, then system (III.3.1) has a T_ε-periodic solution $x(t, \varepsilon)$ satisfying (III.3.4).

Proof. To start with, we make few remarks on the functions $x_\pm(t, \tau, \alpha, \varepsilon)$. First we note that when $\varepsilon = 0$, equation (III.3.11) reads

$$\dot{x} = f_-(x + u_-(t, \alpha)) - f_-(u_-(t, \alpha))$$

$$x(0) = 0$$

which has the (unique) solution $x(t) = 0$. Thus

$$x_-(t, \tau, \alpha, 0) = 0.$$

Next, differentiating equation (III.3.11) with respect to ε we see that $D_\varepsilon x_-(t, \tau, \alpha, 0)$ satisfies the equation:

$$\dot{x} - Df_-(u_-(t, \alpha))x = g_-(t + \tau, u_-(t, \alpha), 0)$$

$$x(0) = 0.$$

Hence,

$$x_{-,\varepsilon}(t, \tau, \alpha, 0) := D_\varepsilon x_-(t, \tau, \alpha, 0) = \int_0^t X_-(t, \alpha) X_-(s, \alpha)^{-1} g_-(s + \tau, u_-(s, \alpha), 0) ds.$$

Next, $x_+(0, \tau, \alpha, \varepsilon) = 0$ by definition. Differentiating equation (III.3.9) with respect to ε at $\varepsilon = 0$ and using the equalities

$$x_+(0, \tau, \alpha, \varepsilon) = 0, \qquad h_{-,t}(0, \tau, 0, \alpha, 0) = 0, \qquad h_{-,x}(0, \tau, 0, \alpha, 0) = 0,$$

we obtain

$$t\dot{x}_+(0,\tau,\alpha,0) = \int_0^t X_+(t,\alpha)X_+^{-1}(s,\alpha)g_+(\tau,u_+(0,\alpha),0)ds.$$

So equation (III.3.12) at $\varepsilon = 0$ and $T_\pm = T_\pm(\alpha)$ becomes

$$q_-(-T_-(\alpha),\alpha) = q_+(T_+(\alpha),\alpha)$$

which is satisfied for $\alpha = 0$. Now we look at equation (III.3.13). Since $h_-(t,\tau,0,\alpha,0) = 0$, we see that when $\varepsilon = 0$ and $T_- = T_-(\alpha)$, the equality is satisfied. As a consequence we get

$$\mathcal{F}(T_+(\alpha),T_-(\alpha),\tau,\alpha,0) = \begin{pmatrix} q_-(-T_-(\alpha),\alpha) - q_+(T_+(\alpha),\alpha) \\ 0 \\ T_-(\alpha) - T_-^0 \end{pmatrix} \qquad \text{(III.3.18)}$$

and $\mathcal{F}(T_+^0,T_-^0,\tau,0,0) = 0$. Next we look at derivatives of $\mathcal{F}$ with respect to T_+, T_-, α and ε at the point $(T_+^0, T_-^0, \tau, 0, 0)$. We have

$$D_{T_-}\left[x_-(-T_-,\tau,\alpha,\varepsilon) + q_-(-T_-,\alpha) - x_+(\varepsilon T_+,\tau,\alpha,\varepsilon) - q_+(T_+,\alpha)\right]$$
$$= -\dot{x}_-(-T_-,\tau,\alpha,\varepsilon) - \dot{q}_-(-T_-,\alpha) \to -\dot{q}_-(-T_-,\alpha) \quad \text{as } \varepsilon \to 0,$$

and similarly, using

$$\varepsilon\dot{x}_+(\varepsilon T_+,\tau,\alpha,\varepsilon) = f(x_+(\varepsilon T_+,\tau,\alpha,\varepsilon) + q_+(T_+,\alpha)) - f(q_+(T_+,\alpha))$$
$$+\varepsilon g(t+\tau,x_+(\varepsilon T_+,\tau,\alpha,\varepsilon) + q_+(T_+,\alpha),\varepsilon),$$

we get

$$D_{T_+}\left[x_-(-T_-,\tau,\alpha,\varepsilon) + q_-(-T_-,\alpha) - x_+(\varepsilon T_+,\tau,\alpha,\varepsilon) - q_+(T_+,\alpha)\right]$$
$$= -\varepsilon\dot{x}_+(\varepsilon T_+,\tau,\alpha,\varepsilon) - \dot{q}_+(T_+,\alpha) \to -\dot{q}_+(T_+,\alpha) \quad \text{as } \varepsilon \to 0.$$

Next,

$$D_\alpha\left[x_-(-T_-,\tau,\alpha,\varepsilon) + q_-(-T_-,\alpha) - x_+(\varepsilon T_+,\tau,\alpha,\varepsilon) - q_+(T_+,\alpha)\right]$$
$$\to D_\alpha q_-(-T_-,\alpha) - D_\alpha q_+(T_+,\alpha) \quad \text{as } \varepsilon \to 0,$$

and

$$D_\tau\left[x_-(-T_-,\tau,\alpha,\varepsilon) + q_-(-T_-,\alpha) - x_+(\varepsilon T_+,\tau,\alpha,\varepsilon) - q_+(T_+,\alpha)\right] \to 0$$

as $\varepsilon \to 0$. So the Jacobian matrix L of $\mathcal{F}$ at the point $(T_+^0, T_-^0, \tau, 0, 0)$ is

$$L := \frac{\partial\mathcal{F}}{\partial(T_+,T_-,\tau,\alpha)}(T_+^0,T_-^0,\tau,0,0) =$$
$$\begin{pmatrix} -\dot{q}_+(T_+^0,0) & -\dot{q}_-(-T_-^0,0) & 0 & D_\alpha q_-(-T_-^0,0) - D_\alpha q_+(T_+^0,0) \\ 0 & -DG(q(-T_-^0,0))\dot{q}_-(-T_-^0,0) & 0 & DG(q(-T_-^0,0))D_\alpha q_-(-T_-^0,0) \\ 0 & 1 & 0 & 0 \end{pmatrix},$$

and $(\mu_+, \mu_-, \tau, w) \in \mathbb{R} \times \mathbb{R} \times \mathbb{R} \times \mathbb{R}^{n-1}$ belongs to the kernel $\mathcal{N}L$ of L if and only if

$$\mu_- = 0, \qquad \left[D_\alpha q_-(-T_-^0, 0) - D_\alpha q_+(T_+^0, 0) \right] w = \dot{q}_+(T_+^0, 0)\mu_+,$$
$$DG(q_-(-T_-^0, 0))D_\alpha q_-(-T_-^0, 0)w = 0. \tag{III.3.19}$$

From $G(q_-(-T_-(\alpha), \alpha) = 0$ we get

$$DG(q(-T_-^0, 0)) \left[-\dot{q}_-(-T_-^0, 0)DT_-(0) + D_\alpha q_-(-T_-^0, 0) \right] = 0. \tag{III.3.20}$$

Thus, on account of the transversality condition $DG(q(T_-^0, 0))\dot{q}_-(-T_-^0, 0) \neq 0$, (III.3.19) is equivalent to

$$\left[D_\alpha q_-(-T_-^0, 0) - D_\alpha q_+(T_+^0, 0) \right] w = \dot{q}_+(T_+^0, 0)\mu_+,$$
$$DT_-(0)w = 0, \qquad \mu_- = 0. \tag{III.3.21}$$

Next, from $G(q_+(T_+(\alpha), \alpha) = 0$ we get

$$DG(q(T_+^0, 0)) \left[\dot{q}_+(T_+^0, 0)DT_+(0) + D_\alpha q_+(T_+^0, 0) \right] = 0. \tag{III.3.22}$$

Consequently, subtracting (III.3.20) from (III.3.22) and using $q(T_+^0, 0) = q(-T_-^0, 0)$ we obtain

$$DG(q(T_+^0, 0)) \left[\dot{q}_+(T_+^0, 0)DT_+(0) + \dot{q}_-(-T_-^0, 0)DT_-(0) \right]$$
$$= DG(q(T_+^0, 0)) \left[D_\alpha q_-(-T_-^0, 0) - D_\alpha q_+(T_+^0, 0) \right].$$

So, if $w \in \mathbb{R}^{n-1}$ satisfies (III.3.21), we see that

$$DG(q(T_+^0, 0))\dot{q}_+(T_+^0, 0)DT_+(0)w = DG(q(T_+^0, 0))\dot{q}_+(T_+^0, 0)\mu_+$$

and then, on account of transversality, $DT_+(0)w = \mu_+$. Summarizing, we have seen that, if $(\mu_+, \mu_-, \tau, w) \in \mathcal{N}L$ then $\mu_+ = DT_+(0)w$, $\mu_- = 0$, and $w \in \mathbb{R}^{n-1}$ satisfies

$$\left[D_\alpha q_-(-T_-^0, 0) - D_\alpha q_+(T_+^0, 0) \right] w = \dot{q}_+(T_+^0, 0)DT_+(0)w$$
$$DT_-(0)w = 0. \tag{III.3.23}$$

On the other hand, if $w \in \mathbb{R}^{n-1}$ satisfies (III.3.23) then $(DT_+(0)w, 0, \tau, w)$ belongs to $\mathcal{N}L$. So $\mathcal{N}L = [(0, 0, 1, 0)]$ if and only if system (III.3.23) has the trivial solution $w = 0$ only. But (III.3.23) is equivalent to

$$D_\alpha \left[q_-(-T_-(\alpha), \alpha) - q_+(T_+(\alpha), \alpha) \right]_{\alpha=0} w = 0$$
$$DT_-(0)w = 0,$$

and hence (III.3.23) has the trivial solution if and only if the non-degeneracy condition (III.3.14) holds. We emphasize the fact that, assuming condition (III.3.14), equation $\mathcal{F}(T_+, T_-, \tau, \alpha, 0) = 0$ has the manifold of fixed points $(T_+, T_-, \tau, \alpha) = (T_+^0, T_-^0, \tau, 0)$, and the linearization of $\mathcal{F}$ at these points is Fredholm with index zero with the one-dimensional kernel $[(0, 0, 1, 0)]$. Hence there is a unique vector, up to a multiplicative constant, $\widetilde{\psi} \in \mathbb{R}^{n+2}$ such that $\widetilde{\psi}^* L = 0$. Writing $\widetilde{\psi}^* = (\psi^*, \psi_1, \psi_2)$, $\psi \in \mathbb{R}^n$, $\psi_1, \psi_2 \in \mathbb{R}$ we see that (ψ, ψ_1, ψ_2) satisfies (III.3.16).

We apply Theorem A.5 to the map $\mathcal{F}(T_+, T_-, \tau, \alpha, \varepsilon)$ with $\mu = \tau$. Then $L(\tau) = L$ is independent of τ, and hence so is $\Pi(\tau) = \Pi$. Next $(\mathbb{I} - \Pi)z = \frac{\widetilde{\psi}^* z}{|\widetilde{\psi}|^2} \widetilde{\psi}$ where $\mathcal{R}L = \{\widetilde{\psi}\}^\perp$ and $\widetilde{\psi}^* = (\psi^*, \psi_1, \psi_2) \in \mathbb{R}^{n+2}$, $\psi \in \mathbb{R}^n$, $\psi_1, \psi_2 \in \mathbb{R}$, is any vector satisfying (III.3.16). To apply Theorem A.5, we look at the derivative of $\mathcal{F}(T_+^0, T_-^0, \tau, 0, \varepsilon)$ with respect to ε at $\varepsilon = 0$. First we have

$$D_\varepsilon(x_+(\varepsilon T_+, \alpha, \varepsilon) - x_-(-T_-, \alpha, \varepsilon))_{\varepsilon=0}$$

$$= \int_0^{T_+} X_+(T_+, \alpha) X_+(s, \alpha)^{-1} g_+(\tau, u(0, \alpha), 0) ds$$

$$+ \int_{-T_-}^0 X_-(-T_-, \alpha) X_-(s, \alpha)^{-1} g_-(s + \tau, u_-(s, \alpha), 0) ds,$$

whereas differentiating (III.3.13) with respect to ε at $\varepsilon = 0$, we get

$$-DG(q_-(-T_-, \alpha)) \int_{-T_-}^0 X_-(-T_-, \alpha) X_-^{-1}(s, \alpha) g_-(s + \tau, q_-(s, \alpha), 0) ds.$$

As a consequence, we obtain

$$D_\varepsilon \mathcal{F}(T_+^0, T_-^0, \tau, 0, 0) =$$

$$\begin{pmatrix} - \int_0^{T_+^0} X_+(T_+^0, 0) X_+(s, 0)^{-1} g_+(\tau, q_+(0, 0), 0) ds \\ - \int_{-T_-^0}^0 X_-(-T_-^0, 0) X_-(s, 0)^{-1} g_-(s + \tau, q_-(s, 0), 0) ds \\ -DG(q(-T_-^0, 0)) \int_{-T_-^0}^0 X_-(-T_-^0, 0) X_-^{-1}(s, 0) g_-(s + \tau, q_-(s, 0), 0) ds \\ 0 \end{pmatrix},$$

and then the Poincaré-Andronov-Melnikov function is

$$M(\tau) := \psi^* \int_0^{T_+^0} X_+(T_+^0, 0) X_+(s, 0)^{-1} g_+(\tau, u(0, 0), 0) ds$$

$$+ \psi^* \int_{-T_-^0}^0 X_-(-T_-^0, 0) X_-(s, 0)^{-1} g_-(s + \tau, u_-(s, 0), 0) ds \qquad \text{(III.3.24)}$$

$$+ \psi_1 DG(q(-T_-^0, 0)) \int_{-T_-^0}^0 X_-(-T_-^0, 0) X_-^{-1}(s, 0) g_-(s + \tau, q_-(s, 0), 0) ds.$$

The conclusion of Theorem III.3.2 now easily follows from (III.3.24) and Theorem A.5. $\qquad \square$

III.3.4. Poincaré-Andronov-Melnikov function and adjoint system

In this section we want to give a suitable definition of the adjoint system of the
linearization of

$$\dot{x} = f_-(x)$$
$$x(0) = q_-(0, \alpha) \in S \cap U_0$$
$$x(-T(\alpha)) = R(x(0)) \qquad \text{(III.3.25)}$$
$$G(x(-T(\alpha))) = 0$$
$$-T(\alpha) \le t \le 0$$

along $q_-(t)$ in such a way that the Poincaré-Andronov-Melnikov function (III.3.17)
can be put in a relation with solutions of such an adjoint system.

For $\alpha = 0$, (III.3.25) has the solution $x(t) = q_-(t, 0)$, $-T_-^0 \le t \le 0$. We denote $x(t, \alpha)$
the solution of the impact system (III.3.25) on $[-T(\alpha), 0]$. Then its derivative with
respect to α at $\alpha = 0$ satisfies the linearized equation

$$\dot{u} = Df_-(q_-(t, 0))u$$
$$u(0) = D_\alpha q_-(0, 0)$$
$$DR(q(0, 0))u(0) = u(-T_-^0) - \dot{q}_-(-T_-^0, 0)T_1 \qquad \text{(III.3.26)}$$
$$DG(q_-(-T_-^0, 0))[u(-T_-^0) - \dot{q}_-(-T_-^0, 0)T_1] = 0$$
$$DT(0) = T_1 : \mathbb{R}^{n-1} \to \mathbb{R}.$$

Next, recalling (III.3.1), we consider a perturbed impact system of (III.3.25) (see also
(III.3.11)) of the form

$$\dot{x} = f_-(x) + \varepsilon g_-(t + \tau, x, \varepsilon)$$
$$x(0) = q_-(0, \alpha) \in S \cap U_0$$
$$x(-T(\alpha, \varepsilon)) = R(\tau; x(0), \varepsilon) \qquad \text{(III.3.27)}$$
$$G(x(-T(\alpha, \varepsilon))) = 0$$
$$-T(\alpha, \varepsilon) \le t \le 0$$

where $R: \mathbb{R} \times U_0 \cap S \times (-\delta, \delta) \to U_+ \cap S$ is defined as $R(\tau; \xi, \varepsilon) = x_+(\varepsilon T_+(\xi, \tau, \varepsilon), \tau, \varepsilon)$
and $x_+(t, \tau, \varepsilon)$ is the solution of

$$\varepsilon \dot{x} = f_+(x) + \varepsilon g_+(t + \tau, x, \varepsilon)$$
$$x(0) = \xi.$$

Note that R is a C^2-smooth map on $\mathbb{R} \times U_0 \cap S \times \mathbb{R}$ taking values on $U_+ \cap S$, and
$R(\tau; q(0, \alpha), 0) = q_+(T_+(\alpha), \alpha)$. Moreover, when g_+ is autonomous, then R is independent of τ, so we may take $\tau = 0$ in its definition. We recall that for simplicity we write
$R(\xi)$ instead of $R(\tau; \xi, 0)$, $\xi \in S$.

To study the problem of existence of solutions of system (III.3.27), we are then led to find conditions on $h(t)$, d and T_1 so that the nonhomogeneous linear equation

$$\dot{u} - Df_-(q_-(t, 0))u = h(t)$$
$$u(0) = D_\alpha q_-(0, 0)\theta, \ \theta \in \mathbb{R}^{n-1}$$
$$u(-T^0_-) - \dot{q}_-(-T^0_-, 0)T - DR(q(0, 0))u(0) = d \in \mathbb{R}^n \qquad \text{(III.3.28)}$$
$$DG(q_-(-T^0_-, 0))[u(-T^0_-) - \dot{q}_-(-T^0_-, 0)T] = 0$$
$$T = T_1$$

has a solution $(u(t), \theta, T)$. Let us comment on equation (III.3.28) (and similarly on (III.3.26)) that condition $u(-T^0_-) - \dot{q}_-(-T^0_-, 0)T - DR(q(0, 0))u(0) = d$ involves only the derivative of $R(\xi)$ on the tangent space $T_\xi S$ since $u(0) = D_\alpha q_-(0, 0) \in T_\xi S$, $\xi = q_-(0, 0)$. So, it is independent of any extension we take of $R(\xi)$ to a neighborhood of $q_-(0, 0)$. We also note that for simplicity we denote again by T_1 the value of the linear functional T_1 in (III.3.28).

Since $G(R(q_-(0, \alpha))) = 0$, we get

$$DG(R(q_-(0, 0)))DR(q_-(0, 0))D_\alpha q_-(0, 0)\theta = 0$$

for any $\theta \in \mathbb{R}^{n-1}$, and then

$$DG(R(q_-(0, 0)))d$$
$$= DG(R(q_-(0, 0)))\left[u(-T^0_-) - \dot{q}_-(-T^0_-, 0)T - DR(q(0, 0))D_\alpha q_-(0, 0)\theta\right] = 0.$$

So, if equation (III.3.28) has a solution, we must necessarily have

$$DG(R(q_-(0, 0)))d = 0 \qquad \left(\Leftrightarrow DG(q_+(T^0_+, 0))d = 0\right).$$

Next, we define two Hilbert spaces

$$X := \left\{(u, \theta, T) \in W^{1,2}([-T^0_-, 0], \mathbb{R}^n) \times \mathbb{R}^{n-1} \times \mathbb{R} \, \middle| \, u(0) = D_\alpha q_-(0, 0)\theta\right\},$$
$$Y := \left\{(h, d, T) \in L^2([-T^0_-, 0], \mathbb{R}^n) \times \mathbb{R}^{n-1} \times \mathbb{R} \times \mathbb{R} \, \middle| \, DG(R(q_-(0, 0)))d = 0\right\}.$$

Note Y is a Hilbert space and X is a closed subspace of a Hilbert space

$$W^{1,2}([-T^0_-, 0], \mathbb{R}^n) \times \mathbb{R}^{n-1} \times \mathbb{R}.$$

Then (III.3.28) can be written as

$$A(u, \theta, T) = (h, d, 0, T_1)$$

with

$$A(u,\theta,T) := \begin{pmatrix} \dot{u} - \mathrm{D}f_-(q_-(t,0))u \\ u(-T_-^0) - \dot{q}_-(-T_-^0,0)T - \mathrm{D}R(q(0,0))u(0) \\ \mathrm{D}G(q_-(-T_-^0,0))[u(-T_-^0) - \dot{q}_-(-T_-^0,0)T] \\ T \end{pmatrix}$$

and $A\colon X \to Y$.

Lemma III.3.3. *The range $\mathcal{R}A$ is closed.*

Proof. Let $A(u_n,\theta_n,T_n) = (h_n,d_n,0,T_1^n) \to (\bar{h},\bar{d},0,\overline{T}_1)$ as $n \to \infty$. Then

$$u_n(t) = \mathrm{D}_\alpha q_-(t,0)\theta_n - \int_t^0 X_-(t)X_-^{-1}(t,s)h_n(s)ds$$

and

$$\mathrm{D}R(q(0,0))\mathrm{D}_\alpha q_-(0,0)\theta_n - \mathrm{D}_\alpha q_-(-T_-^0,0)\theta_n$$
$$= -d_n - \int_{-T_-^0}^0 X_-(-T_-^0,0)X_-(s,0)^{-1}h_n(s)ds - \dot{q}_-(-T_-^0,0)T_1^n,$$
$$\mathrm{D}G(q_-(-T_-^0,0))d_n = 0.$$

Since

$$-d_n - \int_{-T_-^0}^0 X_-(-T_-^0,0)X_-(s,0)^{-1}h_n(s)ds - \dot{q}_-(-T_-^0,0)T_1^n \to$$
$$-\bar{d} - \int_{-T_-^0}^0 X_-(-T_-^0,0)X_-(s,0)^{-1}\bar{h}(s)ds - \dot{q}_-(-T_-^0,0)\overline{T}_1$$

and $\mathcal{R}\left[\mathrm{D}R(q(0,0))\mathrm{D}_\alpha q_-(0,0) \cdot -\mathrm{D}_\alpha q_-(-T_-^0,0)\cdot\right]$ is closed, then $\mathrm{D}G(q_-(-T_-^0,0))\bar{d} = 0$ and there exists $\bar{\theta} \in \mathbb{R}^{n-1}$ so that

$$\mathrm{D}R(q(0,0))\mathrm{D}_\alpha q_-(0,0)\bar{\theta} - \mathrm{D}_\alpha q_-(-T_-^0,0)\bar{\theta}$$
$$= -\bar{d} - \int_{-T_-^0}^0 X_-(-T_-^0,0)X_-(s,0)^{-1}\bar{h}(s)ds - \dot{q}_-(-T_-^0,0)\overline{T}_1.$$

Denoting

$$\bar{u}(t) := \mathrm{D}_\alpha q_-(t,0)\bar{\theta} - \int_t^0 X_-(t)X_-^{-1}(t,s)\bar{h}(s)ds, \qquad \overline{T} = \overline{T}_1,$$

we derive $(\bar{h},\bar{d},0,\overline{T}_1) = A(\bar{u},\bar{\theta},\overline{T}) \in \mathcal{R}A$. The proof is finished. □

Next, we prove the following result.

Proposition III.3.4. *Let $(h, d, T) \in Y$. Then the nonhomogeneous system (III.3.28) has a solution $(u(t), \theta, T) \in X$ if and only if equation*

$$\int_{-T_-^0}^{0} v(t)^* h(t) dt + \psi^* d + \psi_2 T_1 = 0 \qquad (III.3.29)$$

holds for any solution $v(t)$ of the adjoint system

$$\dot{v}(t) + Df_-(q_-(t, 0))^* v(t) = 0$$

$$D_\alpha q_-(0, 0)^* [v(0) - DR(q_-(0, 0))^* \psi] = 0$$

$$v(-T_-^0) = \psi + \psi_1 DG(q_-(-T_-^0, 0))^* \qquad (III.3.30)$$

$$\psi^* \dot{q}_+(T_+^0, 0) = 0$$

and $\psi_2 = \psi^* \dot{q}_-(-T_-^0, 0) + \psi_1 DG(q_-(-T_-^0, 0)) \dot{q}_-(-T_-^0, 0)$.

Proof. Before starting with the proof we observe that, because of $DG((q_+(T_+^0, 0))d = 0$, ψ is not uniquely determined by equation (III.3.29), since changing it with $\psi + \lambda DG(q_-(-T_-^0, 0))^*$, $\lambda \in \mathbb{R}$, the equation remains the same. So, in equation (III.3.29) we look for ψ in a subspace of $\mathbb{R}^n$ which is transverse to $DG(q_-(-T_-^0, 0))^*$. It turns out that the best choice, from a computational point of view, is to take ψ so that $\psi^* \dot{q}_+(T_+^0, 0) = 0$ (see equation (III.3.16)).

First we prove necessity. Assume that (III.3.28) can be solved for $(u, \theta, T) \in X$ and let $(v(t), \psi, \psi_1)$, $v \in W^{1,2}([-T_-^0, 0], \mathbb{R}^n)$, be a solution of equation (III.3.30). Then

$$h(t) = \dot{u}(t) - Df(q_-(t, 0))u(t)$$

$$d = u(-T_-^0) - \dot{q}_-(-T_-^0, 0)T - DR(q(0, 0))D_\alpha q_-(0, 0)\theta$$

$$0 = DG(q_-(-T_-^0, 0))[u(-T_-^0) - \dot{q}_-(-T_-^0, 0)T]$$

$$T_1 = T.$$

Plugging these equalities in the left-hand side of (III.3.29) and integrating by parts, (III.3.29) reads

$$v(0)^* D_\alpha q_-(0, 0)\theta - v(-T_-^0)^* u(-T_-^0) - \int_{-T_-^0}^{0} [\dot{v}(t) + Df_-(q_-(t, 0))^* v(t)]^* u(t) dt$$

$$+ \psi^* [u(-T_-^0) - \dot{q}_-(-T_-^0, 0)T - DR(q(0, 0))D_\alpha q_-(0, 0)\theta]$$

$$+ \psi_1 DG(q_-(-T_-^0, 0))[u(-T_-^0) - \dot{q}_-(-T_-^0, 0)T] + \psi_2 T = 0,$$

or

$$\left(\mathrm{D}_\alpha q_-(0,0)^* \left[v(0) - \mathrm{D}R(q_-(0,0))^*\psi\right]\right)^* \theta$$

$$+\left(\psi - v(-T_-^0) + \psi_1 \mathrm{D}G(q_-(-T_-^0,0))^*\right)^* u(-T_-^0)$$

$$- \int_{-T_-^0}^0 \left(\dot{v}(t) + \mathrm{D}f_-(q_-(t,0))^* v(t)\right)^* u(t) dt \qquad (\mathrm{III.3.31})$$

$$+[\psi_2 - \psi^* \dot{q}_-(-T_-^0,0) - \psi_1 \mathrm{D}G(q_-(-T_-^0,0))\dot{q}_-(-T_-^0,0)]T = 0$$

because of the definition of ψ_2 and the fact that $(v(t),\psi,\psi_1)$ satisfies (III.3.30).

To prove the sufficiency we show that if $(h,d,T) \in Y$ does not belong to $\mathcal{R}A$, then there exists a solution of the variational equation (III.3.30) such that (III.3.31) does not hold. So, assume that $(h,d,0,T_1) \notin \mathcal{R}A$. By Lemma III.3.3 and Theorem A.3, there is $(\bar{v},\bar{\psi},\bar{\psi}_1,\bar{\psi}_2) \in Y$ such that

$$\langle(\bar{v},\bar{\psi},\bar{\psi}_1,\bar{\psi}_2), A(u,\theta,T)\rangle = 0, \quad \forall(u,\theta,T) \in X, \qquad (\mathrm{III.3.32})$$

and

$$\langle(\bar{v},\bar{\psi},\bar{\psi}_1,\bar{\psi}_2),(h,d,0,T_1)\rangle = 1, \qquad (\mathrm{III.3.33})$$

where $\langle\cdot,\cdot\rangle$ is the usual scalar product on Y. We already noted that we can assume that $\bar{\psi}^* \dot{q}_+(T_+^0,0) = 0$, and (III.3.32), (III.3.33) remain valid. Repeating our previous arguments we see that $v(t) \in W^{1,2}([-T_-^0,0],\mathbb{R}^n)$ and that (III.3.32) implies that $(\bar{v},\bar{\psi},\bar{\psi}_1,\bar{\psi}_2)$ solves the adjoint system (III.3.30). Summarizing, if $(h,d,0,T_1) \notin \mathcal{R}A$, there exists a solution of the adjoint system for which (III.3.30) does not hold. This completes the proof. $\qquad\square$

Again we note that equation (III.3.30) depends only on the derivative $\mathrm{D}R(q_-(0,0))$ on $T_{q_-(0,0)}S$ since

$$\mathrm{D}_\alpha q_-(0,0)^* \mathrm{D}R(q_-(0,0))^*\psi$$

$$= \left[\dot{q}_+(T_+^0,0)\mathrm{D}T_+(0) + \mathrm{D}_\alpha q_+(T_+^0,0)\right]^* \psi = \mathrm{D}_\alpha q_+(T_+^0,0)^*\psi,$$

where we use $\psi^* \dot{q}_+(T_+^0,0) = 0$. In other words, it is independent of any C^1-extension we take of $R(\xi)$ to the whole U_0.

Now, we prove the following statement.

Proposition III.3.5. *The adjoint system* (III.3.30) *has a solution if and only if* (ψ,ψ_1) *satisfies the first and the third equation in* (III.3.16) *(and we take the second equation in* (III.3.16) *as a definition of* ψ_2*).*

Proof. Let $v(t)$ be a solution of (III.3.30). Then

$$v(t) = Y(t)Y(-T_-^0)^{-1}v(-T_-^0)$$

for $Y(t) = X_-^{-1}(t)^*$ being the fundamental matrix of the linear equation

$$\dot{v}(t) + Df_-(q_-(t,0))^*v(t) = 0.$$

Then, taking $v(-T_-^0) = \psi + \psi_1 DG(q_-(-T_-^0,0))^*$, the two remaining conditions in equation (III.3.30) read

$$D_\alpha q_-(0,0)^*\left[Y(-T_-^0)^{-1}[\psi + \psi_1 DG(q_-(-T_-^0,0))^*] - DR(q_-(0,0)^*\psi\right] = 0$$
$$\psi^*\dot{q}_+(T_+^0,0) = 0.$$

These can be rewritten as

$$D_\alpha q_-(-T_-^0,0)^*[\psi + \psi_1 DG(q_-(-T_-^0,0))^*] - [\psi^*DR(q_-(0,0)D_\alpha q_-(0,0)]^* = 0$$
$$\psi^*\dot{q}_+(T_+^0,0) = 0$$

or, on account of $R(q_-(0,\alpha)) = q_+(T_+(\alpha),\alpha)$,

$$\psi^*\left[D_\alpha q_-(-T_-^0,0) - D_\alpha q_+(T_+^0,0)\right] + \psi_1 DG(q_-(-T_-^0,0))D_\alpha q_-(-T_-^0,0) = 0$$
$$\psi^*\dot{q}_+(T_+^0,0) = 0.$$

The proof is finished.   □

We conclude this section giving another expression of the Poincaré-Andronov-Melnikov function (III.3.17) in terms of the solution of the adjoint system (III.3.30). Denote $v(t)$ a solution of the adjoint system (III.3.30). Since a fundamental matrix of the linear equation

$$\dot{v} + Df_-(q_-(t,0))v = 0$$

is $X_-^{-1}(t)^*$, we see that

$$v(t) = X_-^{-1}(t)^*X_-(-T_-^0)^*v(-T_-^0) = X_-^{-1}(t)^*X_-(-T_-^0)^*\left[\psi + \psi_1 DG(q_-(-T_-^0,0))^*\right],$$

i.e.

$$v(t)^* = \left[\psi^* + \psi_1 DG(q_-(-T_-^0,0))\right]X_-(-T_-^0)X_-^{-1}(t).$$

Then

$$M(\tau) = \psi^* \int_0^{T_+^0} X_+(T_+^0)X_+(t)^{-1}g_+(\tau, q_+(0,0), 0)dt$$
$$+ \int_{-T_-^0}^0 v(t)^*g_-(t+\tau, q_-(t,0), 0)dt.$$

For the first term in the previous equality we can show that it is related to the impact $R(\tau; \xi, \varepsilon)$. Indeed, we know that the solution of the singular equation

$$\varepsilon \dot{x} = f_+(x) + \varepsilon g(t, x, \varepsilon)$$

can be written as:

$$x(t + \tau) = x_+(t) + q_+(\varepsilon^{-1} t, \alpha)$$

with $x_+(\varepsilon t)$ as in equation (III.3.9). Thus $\xi = x(\tau) = q_+(0, \alpha) \in S$ and

$$R(\tau; \xi, \varepsilon) = x_+(\varepsilon T_+) + q_+(T_+, \alpha)$$

$$= \int_0^{T_+} X_+(T_+, \alpha) X_+^{-1}(s, \alpha) h_+(\varepsilon s, \tau, x_+(\varepsilon s), \alpha, \varepsilon) ds + q_+(T_+, \alpha)$$

for some $T_+ = T_+(\tau; \alpha, \varepsilon)$. Then

$$D_\varepsilon R(\tau; q_-(0, 0), 0) = \dot{q}_+(T_+^0, 0) D_\varepsilon T_+ + \int_0^{T_+^0} X_+(T_+^0) X_+^{-1}(s) g_+(\tau, q_+(0, 0), 0) ds$$

and therefore, using again $\psi^* \dot{q}_+(T_+^0, 0) = 0$, we see that

$$\psi^* \int_0^{T_+^0} X_+(T_+^0) X_+^{-1}(s) g_+(\tau, q_+(0, 0), 0) ds = \psi^* D_\varepsilon R(\tau; q_-(0, 0), 0),$$

i.e.

$$M(\tau) = \psi^* D_\varepsilon R(\tau; q_-(0, 0), 0) + \int_{-T_-^0}^0 v(t)^* g_-(t + \tau, q_-(t, 0), 0) dt. \qquad \text{(III.3.34)}$$

The expression (III.3.34) of the Poincaré-Andronov-Melnikov function should be compared with the one given in Theorem II.1.4.

III.3.5. Bifurcation from a manifold of periodic solutions

In this section we assume that $q_-(-T_-(\alpha), \alpha) = q_+(T_+(\alpha), \alpha)$ for any α in an open neighborhood of 0 in $\mathbb{R}^{n-1}$. Hence, from (III.3.18) we see that

$$\mathcal{F}(T_+(\alpha), T_-(\alpha), \tau, \alpha, 0) = \begin{pmatrix} 0 \\ 0 \\ T_-(\alpha) - T_-^0 \end{pmatrix}.$$

We distinguish the two cases:
a) $DT_-(0) \neq 0$,
b) $T_-(\alpha) = T_-^0$ for all α in an open neighborhood of 0 in $\mathbb{R}^{n-1}$.
First we assume a). Then there exists a C^2-, $(n-2)$-dimensional submanifold S of an open neighborhood of $\alpha = 0$ in $\mathbb{R}^{n-1}$, such that $T_-(\alpha) = T_-^0$ for any $\alpha \in S$. So, for

$\varepsilon = 0$, $\mathcal{F}(T_+, T_-, \tau, \alpha, 0) = 0$ has the $(n-1)$-dimensional manifold of solutions

$$(T_+, T_-, \tau, \alpha) = \xi(\alpha, \tau) := (T_+(\alpha), T_-^0, \tau, \alpha), \quad (\alpha, \tau) \in \mathcal{S} \times \mathbb{R}.$$

Hence, we are in a position to apply Theorem A.5. First we have to verify that the kernel $\mathcal{N}D_1\mathcal{F}(\xi(\alpha, \tau), 0)$ equals to the tangent space $T_{\xi(\alpha, \tau)}X$, $X = \{\xi(\alpha, \tau) \mid (\alpha, \tau) \in \mathcal{S} \times \mathbb{R}\}$, and then that the Poincaré-Andronov-Melnikov function (vector)

$$[\mathbb{I} - \Pi(\alpha, \tau)] D_2\mathcal{F}(\xi(\alpha, \tau), 0)$$

has a simple zero at $(\alpha, \tau) = (0, \tau_0)$. Note that

$$T_{\xi(\alpha, \tau)}X = [(DT_+(\alpha)v, 0, 0, v)^*, (0, 0, 1, 0)^* \mid v \in T_\alpha \mathcal{S}].$$

From (III.3.18) we get:

$$D_1\mathcal{F}(\xi(\alpha, \tau), 0) =$$
$$\begin{pmatrix} -\dot{q}_+(T_+(\alpha), \alpha) & -\dot{q}_-(-T_-^0, \alpha) & 0 & D_\alpha q_-(-T_-^0, \alpha) - D_\alpha q_+(T_+(\alpha), \alpha) \\ 0 & -DG(q_-(-T_-^0, \alpha))\dot{q}_-(-T_-^0, \alpha) & 0 & DG(q_-(-T_-^0, \alpha))D_\alpha q_-(-T_-^0, \alpha) \\ 0 & 1 & 0 & 0 \end{pmatrix}.$$

Note that $D_1\mathcal{F}(\xi(\alpha, \tau), 0)$ does not depend on τ. Using $G(q_-(-T_-^0, \alpha)) = 0$ and $q_-(-T_-^0, \alpha) = q_+(T_+(\alpha), \alpha)$ for any $\alpha \in \mathcal{S}$ we easily see that:

$$D_1\mathcal{F}(\xi(\alpha, \tau), 0)|_{T_{\xi(\alpha, \tau)}X} = 0$$

for any $v \in T_\alpha \mathcal{S}$. On the other hand assume that

$$\begin{pmatrix} \mu_+ \\ \mu_- \\ w \end{pmatrix} \in \mathcal{N}D_1\mathcal{F}(\xi(\alpha, \tau), 0)$$

for some $\mu_+, \mu_- \in \mathbb{R}$ and $w \in \mathbb{R}^{n-1}$. Then $\mu_- = 0$ and (μ_+, w) satisfies

$$-\dot{q}_+(T_+(\alpha), \alpha)\mu_+ + \left[D_\alpha q_-(-T_-^0, \alpha) - D_\alpha q_+(T_+(\alpha), \alpha)\right]w = 0$$
$$DG(q_-(-T_-^0, \alpha))D_\alpha q_-(-T_-^0, \alpha)w = 0$$

that, on account of $q_-(-T_-^0, \alpha) = q_+(T_+(\alpha), \alpha)$, is equivalent to

$$\dot{q}_+(T_+(\alpha), \alpha)[DT_+(\alpha)w - \mu_+] = 0$$
$$DG(q_-(-T_-^0, \alpha))D_\alpha q_-(-T_-^0, \alpha)w = 0.$$

Now, from $G(q_-(-T_-(\alpha), \alpha)) = 0$ we get, for any $w \in \mathbb{R}^{n-1}$,

$$DG(q_-(-T_-(\alpha), \alpha))D_\alpha q_-(-T_-(\alpha), \alpha)w = DG(q_-(-T_-(\alpha), \alpha))\dot{q}_-(-T_-(\alpha), \alpha)DT_-(\alpha)w,$$

and hence

$$DG(q_-(-T_-^0, \alpha))D_\alpha q_-(-T_-, \alpha)w = 0 \Leftrightarrow DG(q_-(-T_-^0, \alpha))\dot{q}_-(-T_-^0, \alpha)DT_-(\alpha)w = 0$$

which, in turn, is equivalent to $w \in T_\alpha S$ because of transversality and the fact that $T_\alpha S = \mathcal{N}DT_-(\alpha)$. Hence we conclude that $\mathcal{N}D_1 \mathcal{F}(\xi(\alpha,\tau),0) = T_{\xi(\alpha,\tau)}\mathcal{X}$.

Now we consider the case b). The Poincaré-Andronov-Melnikov function (vector) $[\mathbb{I} - \Pi(\alpha,\tau)]D_2 \mathcal{F}(\xi(\alpha,\tau),0)$, $\alpha \in \mathcal{S}$ can be written as

$$\psi(\alpha,\tau)^* D_2 \mathcal{F}(\xi(\alpha,\tau),0) \tag{III.3.35}$$

where $\psi(\alpha,\tau)^*$ is a matrix whose rows are left eigenvectors of the zero eigenvalue of the matrix $D_1 \mathcal{F}(\xi(\alpha,\tau),0)$, that is

$$\psi(\alpha,\tau)^* D_1 \mathcal{F}(\xi(\alpha,\tau),0) = 0. \tag{III.3.36}$$

Note that $\psi(\alpha,\tau) = \psi(\alpha)$ is independent of τ, since the same holds for $D_1 \mathcal{F}(\xi(\alpha,\tau),0)$. Then (III.3.35) reads

$$M(\alpha,\tau) := \psi(\alpha)^* \int_0^{T_+^0} X_+(T_+^0,\alpha)X_+(s,\alpha)^{-1}g_+(\tau,q(0,\alpha),0)ds$$

$$+ \psi(\alpha)^* \int_{-T_-^0}^0 X_-(-T_-^0,\alpha)X_-(s,\alpha)^{-1}g_-(s+\tau,q_-(s,\alpha),0)ds$$

$$+ \psi_1(\alpha)DG(q(-T_-^0,\alpha)) \int_{-T_-^0}^0 X_-(-T_-^0,\alpha)X_-^{-1}(s,\alpha)g_-(s+\tau,q_-(s,\alpha),0)ds.$$

Arguing as in Section III.3.3, equation (III.3.36) is equivalent to

$$\psi(\alpha)^* \dot{q}_+(T_+(\alpha),\alpha) = 0$$

$$\psi_2(\alpha) = \left[\psi(\alpha)^* + \psi_1(\alpha)DG(q(-T_-^0,\alpha))\right]\dot{q}_-(-T_-^0,\alpha)$$

$$\psi(\alpha)^* \left[D_\alpha q_-(-T_-^0,\alpha) - D_\alpha q_+(T_+(\alpha),\alpha)\right] + \psi_1(\alpha)DG(q(-T_-^0,\alpha))D_\alpha q_-(-T_-^0,\alpha) = 0. \tag{III.3.37}$$

Moreover, the adjoint variational system along $q_-(t,\alpha)$ is defined as

$$\dot{v}(t) + Df_-(q_-(t,\alpha))^* v(t) = 0$$

$$D_\alpha q_-(0,\alpha)^* \left[v(0) - DR(q_-(0,\alpha))^* \psi(\alpha)\right] = 0$$

$$v(-T_-^0) = \psi(\alpha) + \psi_1(\alpha)DG(q_-(-T_-^0,\alpha))^* \tag{III.3.38}$$

$$\psi(\alpha)^* \dot{q}_+(T_+(\alpha),\alpha) = 0$$

where $(\psi(\alpha)^*, \psi_1(\alpha), \psi_2(\alpha))$ satisfies equation (III.3.36). Then the Poincaré-Andronov-Melnikov vector function can be written as

$$M(\alpha,\tau) = \psi(\alpha)^* \int_0^{T_+(\alpha)} X_+(T_+(\alpha),\alpha)X_+(t,\alpha)g_+(\tau,q_+(0,\alpha),0)dt$$

$$+ \int_{-T_-^0}^0 v(t,\alpha)^* g_-(t+\tau,q_-(t,\alpha),0)dt, \tag{III.3.39}$$

or

$$M(\alpha, \tau) = \psi(\alpha)^* D_\varepsilon R(\tau; q_-(0, \alpha), 0) + \int_{-T_-^0}^0 v(t, \alpha)^* g_-(t + \tau, q_-(t, \alpha), 0) dt \quad \text{(III.3.40)}$$

for $v(t, \alpha)$ being the solution of (III.3.38) and $X_+(t, \alpha)$ the fundamental matrix of the linear equation

$$\dot{x} = Df_+(q_+(t, \alpha))x.$$

Of course, the only difference between the cases a) and b) is that in the first case the Poincaré-Andronov-Melnikov function is defined for $(\alpha, \tau) \in S \times \mathbb{R}$, while in the second it is defined for $(\alpha, \tau) \in O \times \mathbb{R}$ for an open neighborhood O of $0 \in \mathbb{R}^{n-1}$. Summarizing, we proved the following result.

Theorem III.3.6. *Assume that $q_-(-T_-(\alpha), \alpha) = q_+(T_+(\alpha), \alpha)$ for any α in a neighborhood of $\alpha = 0$, and that* a) *(or* b)) *holds. Then system (III.3.37) has a d-dimensional space of solutions where $d = n$ (or $d = n + 1$). Moreover, if the Poincaré-Andronov-Melnikov function (III.3.39) (or (III.3.40)) has a simple zero at $(0, \tau_0)$, then system (III.3.1) has a T_ε-periodic solution $x(t, \varepsilon)$ satisfying (III.3.4).*

Finally we note that when we can show that a Brouwer degree of a Poincaré-Andronov-Melnikov function from either Theorem III.3.2 or III.3.6 is nonzero, then by following [18, 27, 28] or the proof of Theorem III.1.2, we can prove existence results.

III.3.6. Examples

We start with a general second-order equation

$$\varepsilon^2 \ddot{x} = f_+(x, \dot{x}) + \varepsilon g_+(t, x, \dot{x}, \varepsilon) \qquad \text{for } x > 0$$
$$\ddot{x} = f_-(x) + \varepsilon g_-(t, x, \dot{x}, \varepsilon) \qquad \text{for } x < 0$$

with $G(x, \dot{x}) = x$. We write $q_\pm(t, \alpha) = \begin{pmatrix} q_1^\pm(t,\alpha) \\ \dot{q}_1^\pm(t,\alpha) \end{pmatrix}$ with $q_-(0, \alpha) = \begin{pmatrix} 0 \\ \alpha+\alpha_0 \end{pmatrix}$, i.e. $q_1^\pm(0, \alpha) = 0$ and $\dot{q}_1^-(0, \alpha) = \alpha + \alpha_0$. We also write $q_+(T_+(\alpha), \alpha) = \begin{pmatrix} 0 \\ \varphi(\alpha) \end{pmatrix}$ so that

$$R: \begin{pmatrix} 0 \\ \alpha + \alpha_0 \end{pmatrix} \mapsto \begin{pmatrix} 0 \\ \varphi(\alpha) \end{pmatrix},$$

i.e. we take

$$R(x_1, x_2) = \begin{pmatrix} 0 \\ \varphi(x_2 - \alpha_0) \end{pmatrix}$$

in the plane coordinates (x_1, x_2). According to equation (III.3.38) with $\psi(\alpha) = \begin{pmatrix} \psi'(\alpha) \\ \psi''(\alpha) \end{pmatrix}$, the adjoint variational system reads

$$\dot{v}_1 = -Df_-(q_1^-(t, \alpha))v_2$$

$$\dot{v}_2 = -v_1$$

$$v_2(0) - \varphi'(\alpha)\psi''(\alpha) = 0$$

$$v_1(-T_-^0) = \psi'(\alpha) + \psi_1(\alpha)$$

$$v_2(-T_-^0) = \psi''(\alpha)$$

$$\psi'(\alpha)\varphi(\alpha) + \psi''(\alpha)f_+(0, \varphi(\alpha)) = 0$$

which can be written as (with $v_2 = w$ and $v_1 = -\dot{w}$)

$$\ddot{w} = Df_-(q_1^-(t, \alpha))w$$

$$w(0) - \varphi'(\alpha)w(-T_-^0) = 0$$

$$\psi'' = w(-T_-^0) \qquad\qquad \text{(III.3.41)}$$

$$\psi'\varphi(\alpha) + \psi'' f_+(0, \varphi(\alpha)) = 0$$

$$\psi_1 = -\dot{w}(-T_-^0) - \psi'.$$

Note that (when $\varphi(\alpha) \neq 0$) the last three equations are actually the definitions of $\psi(\alpha) = \begin{pmatrix} \psi'(\alpha) \\ \psi''(\alpha) \end{pmatrix}$ and $\psi_1(\alpha)$ in terms of the unique (up to a multiplicative constant) bounded solution of the boundary value problem

$$\ddot{w} = Df_-(q_1^-(t, \alpha))w$$

$$w(0) = \varphi'(\alpha)w(-T_-^0),$$

and the Poincaré-Andronov-Melnikov function (III.3.40) reads

$$M(\alpha, \tau) = w(-T_-^0, \alpha)\left(-\frac{f_+(0, \varphi(\alpha))}{\varphi(\alpha)} \quad 1\right)D_\varepsilon R(\tau; q_-(0, \alpha), 0)$$

$$+ \int_{-T_-^0}^0 w(t, \alpha)g_-(t + \tau, q_-(t, \alpha), 0)dt,$$

whereas (III.3.34) reads

$$M(\tau) = w(-T_-^0)\left(-\frac{f_+(0, \varphi(0))}{\varphi(0)} \quad 1\right)D_\varepsilon R(\tau; q_-(0, 0), 0)$$

$$+ \int_{-T_-^0}^0 w(t)g_-(t + \tau, q_-(t, 0), 0)dt.$$

To be more concrete, we consider the equation

$$\ddot{x} + x = \varepsilon g_-(t, x, \dot{x}, \varepsilon).$$

The unperturbed equation $\ddot{x} + x = 0$ with the condition $\dot{x}(0) = 0$ has the solutions

$$q_-(t, \alpha) = (\alpha + \alpha_0) \begin{pmatrix} \sin t \\ \cos t \end{pmatrix}, \quad -\pi \le t \le 0,$$

and $T_-(\alpha) = \pi$. Note that to have $q_-(t, \alpha) \in \{(x_1, x_2) \mid x_1 < 0\}$ for $-\pi < t < 0$ we need $\alpha + \alpha_0 > 0$.

First, we consider the non-degenerate case (III.3.14), which now has the form

$$R(q_-(0, 0)) = q_-(-\pi, 0)$$
$$D_\alpha [R(q_-(0, \alpha)) - q_-(-\pi, \alpha)]_{\alpha=0} \ne 0. \tag{III.3.42}$$

Note $DT_-(0) = 0$ for this case. Since

$$R(q_-(0, \alpha)) - q_-(-\pi, \alpha) = \begin{pmatrix} 0 \\ \varphi(\alpha) + \alpha + \alpha_0 \end{pmatrix},$$

(III.3.42) is equivalent to

$$\varphi(0) = -\alpha_0, \qquad \varphi'(0) + 1 \ne 0. \tag{III.3.43}$$

Then it is easy to see that system (III.3.41) with $\alpha = 0$ reads

$$\ddot{w} + w = 0$$
$$w(0) - \varphi'(0)w(-\pi) = 0$$
$$\psi'' = w(-\pi)$$
$$-\psi' \alpha_0 + \psi'' f_+(0, -\alpha_0) = 0$$
$$\psi_1 = -\dot{w}(-\pi) - \psi'.$$

Solving $\ddot{w} + w = 0$ we get $w(t) = a \cos(t + t_0)$, and the boundary condition is $a(1 + \varphi'(0)) \cos t_0 = 0$. So, we can take $w(t) = \cos(t - \frac{\pi}{2}) = \sin t$. Since $\varphi(0) = -\alpha_0 \ne 0$ then

$$\psi'' = 0, \qquad \psi' = 0, \qquad \psi_1 = 1,$$

and the Poincaré-Andronov-Melnikov function reads

$$M(\tau) = \int_{-\pi}^0 g_-(t + \tau, \alpha_0 \sin t, \alpha_0 \cos t, 0) \sin t \, dt.$$

For example, taking $g_-(t, x, \dot{x}, \varepsilon) = -\dot{x} \cos^2 \left(\frac{\pi}{\pi + \varepsilon T_+^0} \right) t$, where T_+^0 is the time the solution of the problem

$$\ddot{x} = f_+(x, \dot{x}), \qquad x(0) = 0, \qquad \dot{x}(0) = \alpha_0$$

needs to reach the discontinuity manifold $x = 0$, leads to

$$M(\tau) = \frac{\pi}{8} \alpha_0 \sin(2\tau)$$

which has a simple zero at $\tau = 0$.

To conclude the example we need to find a second order equation $\ddot{x} = f_+(x, \dot{x})$ such that (III.3.43) holds. We consider

$$\ddot{x} + x = f_+(x, \dot{x}) := f(x^2 + \dot{x}^2 - 1)g(x, \dot{x})$$

with $f(0) = 0$ and $Df(0) \neq 0$. It has the solution $x = \sin t$. So we take $q_+(t) = (\sin t, \cos t)^*$, and then $T_+^0 = \pi$. Note $q_+(T_+(\alpha), \alpha) = \begin{pmatrix} 0 \\ \varphi(\alpha) \end{pmatrix}$ is equivalent to

$$q_1^+(T_+(\alpha), \alpha) = 0, \qquad \dot{q}_1^+(T_+(\alpha), \alpha) = \varphi(\alpha).$$

Then $\varphi(0) = -1$, so we take $\alpha_0 = 1$. Furthermore,

$$\varphi'(0) = \ddot{q}_1^+(\pi, 0)DT_+(0) + D_\alpha \dot{q}_1^+(\pi, 0) = D_\alpha \dot{q}_1^+(\pi, 0).$$

Setting $\zeta(t) := D_\alpha q_1^+(t, 0)$ we have

$$\ddot{\zeta} + \zeta = 2f'(0)g(\sin t, \cos t)(\zeta \sin t + \dot{\zeta} \cos t). \tag{III.3.44}$$

Since $q_1^+(0, \alpha) = 0$ and $\dot{q}_1^+(0, \alpha) = \alpha + 1$, we obtain $\zeta(0) = 0$ and $\dot{\zeta}(0) = 1$. Clearly, (III.3.44) has a solution $\zeta_1(t) = \cos t$. Then the second solution is

$$\zeta_2(t) = \cos t \int_0^t \frac{e^{2f'(0) \int_0^s \cos \sigma g(\sin \sigma, \cos \sigma)d\sigma}}{\cos^2 s} ds = \sin t\, e^{2f'(0) \int_0^t \cos \sigma g(\sin \sigma, \cos \sigma)d\sigma}$$

$$+ 2f'(0) \cos t \int_0^t g(\sin s, \cos s)e^{2f'(0) \int_0^s \cos \sigma g(\sin \sigma, \cos \sigma)d\sigma} \sin s\, ds.$$

Hence

$$\dot{\zeta}_2(t) = \cos t\, e^{2f'(0) \int_0^t g(\sin \sigma, \cos \sigma) \cos \sigma d\sigma}$$

$$+ f'(0) \sin 2t\, g(\sin t, \cos t)e^{2f'(0) \int_0^t g(\sin \sigma, \cos \sigma) \cos \sigma d\sigma}$$

$$- 2f'(0) \sin t \int_0^t \sin s\, g(\sin s, \cos s)e^{2f'(0) \int_0^s g(\sin \sigma, \cos \sigma) \cos \sigma d\sigma} ds$$

$$+ f'(0) \sin 2t\, g(\sin t, \cos t)e^{2f'(0) \int_0^t g(\sin \sigma, \cos \sigma) \cos \sigma d\sigma}.$$

This implies

$$\varphi'(0) = \dot{\zeta}_2(\pi) = -e^{2f'(0) \int_0^\pi g(\sin \sigma, \cos \sigma) \cos \sigma d\sigma}.$$

Consequently if

$$\int_0^\pi g(\sin \sigma, \cos \sigma) \cos \sigma\, d\sigma \neq 0$$

then $\varphi'(0) \neq -1$. So we conclude with the following.

Corollary III.3.7. *Let $f(r)$ and $g(x, \dot{x})$, $g_-(t, x, \dot{x}, \varepsilon)$ be C^2-functions such that $f(0) =$*

$0 \neq f'(0)$, $g_-(t, x, \dot{x}, \varepsilon) = g_-(t + (1 + \varepsilon)\pi, x, \dot{x}, \varepsilon)$ and

$$\int_0^\pi g(\sin t, \cos t) \cos t \, dt \neq 0.$$

Suppose, also, that the function

$$M(\tau) := \int_{-\pi}^0 g_-(t + \tau, \sin t, \cos t, 0) \sin t \, dt$$

has a simple zero at $\tau = 0$. Then, for $\varepsilon > 0$ sufficiently small the singularly perturbed system

$$\varepsilon^2 \ddot{x} + x = f(x^2 + \dot{x}^2 - 1)g(x, \dot{x}) \qquad \text{if } x > 0$$
$$\ddot{x} + x = \varepsilon g_-(t, x, \dot{x}, \varepsilon) \qquad \text{if } x < 0$$

has a $(1 + \varepsilon)\pi$-periodic solution orbitally close to the set $\{(\sin t, \cos t) \mid -\pi \leq t \leq \pi\}$.

To get a second example we change the preceding as follows. We take

$$\Omega_+ = \{(x, \dot{x}) \mid x < 0, \dot{x} > 0\}, \qquad \Omega_- = \mathbb{R}^2 \backslash \Omega_+$$

with equations

$$\ddot{x} + x = \varepsilon g_-(t, x, \dot{x}, \varepsilon) \qquad \text{for } (x, \dot{x}) \in \Omega_-$$
$$\varepsilon^2 \ddot{x} = -2x - \frac{3}{2}x^2 \qquad \text{for } (x, \dot{x}) \in \Omega_+.$$

It should be noted that the discontinuity line is the union of the two half lines $\{x = 0, \ \dot{x} > 0\}$ and $\{x < 0, \ \dot{x} = 0\}$ which is not C^1. However, all results hold true as long as we remain outside a (small) neighborhood of $(0, 0)$.

The unperturbed equation on Ω_- has the solutions

$$q_-(t, \alpha) = (\alpha + 1)\begin{pmatrix} -\cos t \\ \sin t \end{pmatrix}, \qquad -\frac{3}{2}\pi \leq t \leq 0$$

with $q_-(0, \alpha) = -(\alpha + 1)\begin{pmatrix} 1 \\ 0 \end{pmatrix}$ and $q_-(-\frac{3}{2}\pi, \alpha) = (\alpha + 1)\begin{pmatrix} 0 \\ 1 \end{pmatrix}$. Consequently, $q_+(T_+(\alpha), \alpha)$ $= R(q_-(0, \alpha))$ is the value of the solution $\begin{pmatrix} z_+(t,\alpha) \\ \dot{z}_+(t,\alpha) \end{pmatrix}$ of

$$\ddot{x} + 2x + \frac{3}{2}x^2 = 0, \qquad x(0) = -(1 + \alpha), \qquad \dot{x}(0) = 0$$

at the time $T_+(\alpha)$ such that $z_+(T_+(\alpha), \alpha) = 0$. Since the equation has the Hamiltonian $H_+(x, \dot{x}) = \dot{x}^2 + (x + 2)x^2$ we see that $z_+(t, \alpha)$ satisfies

$$\dot{z}^2(t) + (z(t) + 2)z^2(t) = (1 - \alpha)(1 + \alpha)^2, \qquad z(0) = -1 - \alpha, \qquad \text{(III.3.45)}$$

and hence

$$R(q_-(0,\alpha)) = q_+(T_+(\alpha),\alpha) = \begin{pmatrix} 0 \\ (1+\alpha)\sqrt{1-\alpha} \end{pmatrix}. \qquad \text{(III.3.46)}$$

We observe that T_+^0 is the first positive time such that $x(T_+^0) = 0$, where $x(t) = z_+(t,1)$ is the solution of

$$\dot{x}^2 + (x+2)x^2 = 1, \qquad x(0) = -1.$$

Thus,

$$T_+^0 = \int_{-1}^0 \frac{dx}{\sqrt{1 - x^2(x+2)}} \doteq 1.88292. \qquad \text{(III.3.47)}$$

More related results are derived at the end of this section.

Then equations (III.3.42) have to be changed to

$$R(q_-(0,0)) = q_-\left(-\frac{3}{2}\pi, 0\right)$$
$$D_\alpha \left[R(q_-(0,\alpha)) - q_-\left(-\frac{3}{2}\pi, \alpha\right) \right]_{\alpha=0} \neq 0. \qquad \text{(III.3.48)}$$

However,

$$R(q_-(0,\alpha)) - q_-\left(-\frac{3}{2}\pi, \alpha\right) = (1+\alpha)\begin{pmatrix} 0 \\ \sqrt{1-\alpha} - 1 \end{pmatrix}$$

and (III.3.48) easily follows. Now we compute the variational equation and the Poincaré-Andronov-Melnikov function. From (III.3.46) and $q_-(0,\alpha) = \begin{pmatrix} -(\alpha+1) \\ 0 \end{pmatrix}$ it follows that we can take

$$R(x_1, x_2) = \begin{pmatrix} 0 \\ -x_1\sqrt{x_1 + 2} \end{pmatrix}$$

from which we get

$$DR(q_-(0,0)) = \begin{pmatrix} 0 & 0 \\ -\frac{1}{2} & 0 \end{pmatrix}.$$

Note that equation $\ddot{x} + 2x + \frac{3}{2}x^2 = 0$ has a homoclinic solution

$$-\frac{2}{3}\left(3\tanh\left[\frac{t}{\sqrt{2}}\right]^2 - 1 \right),$$

so the solution $q_+(t,\alpha)$ is a part of a periodic solution inside of $\overline{\Omega}_+$ bounded by the homoclinic one (see Figure III.3.1).

Then, since in a neighborhood of $q_-(-\frac{3}{2}\pi, 0) = \begin{pmatrix} 0 \\ 1 \end{pmatrix}$ we have $G(x_1, x_2) = -x_1$, we

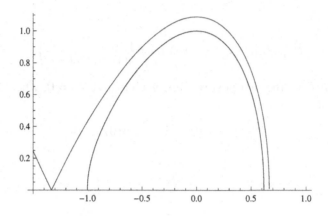

Figure III.3.1 The upper parts of homoclinic and periodic orbits of $\ddot{x} + 2x + \frac{3}{2}x^2 = 0$

get

$$DG\left(q_-\left(-\frac{3}{2}\pi, 0\right)\right) = \begin{pmatrix} -1 \\ 0 \end{pmatrix}.$$

Finally, since the equations on Ω_+ can be written as

$$\dot{x}_1 = x_2$$

$$\dot{x}_2 = -2x_1 - \frac{3}{2}x_1^2,$$

we obtain

$$f_+(q_+(T_+(\alpha), \alpha)) = f_+\left(0, (1+\alpha)\sqrt{1-\alpha}\right) = \begin{pmatrix} (1+\alpha)\sqrt{1-\alpha} \\ 0 \end{pmatrix}.$$

Putting all together we see that the adjoint variational system reads

$$\ddot{w} + w = 0 \qquad\qquad \ddot{w} + w = 0$$
$$\begin{pmatrix} -1 \\ 0 \end{pmatrix}^* \left[\begin{pmatrix} -\dot{w}(0) \\ w(0) \end{pmatrix} - \begin{pmatrix} -\frac{1}{2}\psi'' \\ 0 \end{pmatrix} \right] = 0 \qquad \dot{w}(0) = \frac{1}{2}\psi''$$
$$\begin{pmatrix} -\dot{w}(-\frac{3}{2}\pi) \\ w(-\frac{3}{2}\pi) \end{pmatrix} = \begin{pmatrix} \psi' \\ \psi'' \end{pmatrix} + \psi_1 \begin{pmatrix} -1 \\ 0 \end{pmatrix} \quad \Leftrightarrow \quad w\left(-\frac{3}{2}\pi\right) = \psi''$$
$$\left\langle \begin{pmatrix} \psi' \\ \psi'' \end{pmatrix}, \begin{pmatrix} 1 \\ 0 \end{pmatrix} \right\rangle = 0 \qquad\qquad \dot{w}\left(-\frac{3}{2}\pi\right) = -\psi' + \psi_1$$
$$\psi' = 0.$$

The first three equations give the boundary value problem

$$\ddot{w} + w = 0, \qquad 2\dot{w}(0) - w\left(-\frac{3}{2}\pi\right) = 0$$

possessing the unique solution (up to a multiplicative constant) $w(t) = \cos t$ which gives

$$\psi' = 0, \qquad \psi'' = 0, \qquad \psi_1 = -1.$$

Therefore, since $g_+(t, x, \varepsilon) = 0$, the Poincaré-Andronov-Melnikov function is

$$M(\tau) = \int_{-\frac{3}{2}\pi}^{0} g_-(t + \tau, -\cos t, \sin t, 0) \cos t \, dt. \tag{III.3.49}$$

We conclude with the following.

Corollary III.3.8. *Let T_+^0 be as in equation (III.3.47), $g_-(t, x, \dot{x}, \varepsilon)$ be a $(\pi + \varepsilon T_+^0)$-periodic C^2-function and suppose that the function (III.3.49) has a simple zero at $\tau = 0$. Then, for $\varepsilon > 0$ sufficiently small the singularly perturbed system*

$$\begin{aligned}
\varepsilon^2 \ddot{x} + 2x + \tfrac{3}{2}x^2 &= 0 && \text{if } x < 0 \text{ and } \dot{x} > 0 \\
\ddot{x} + x &= \varepsilon g_-(t, x, \dot{x}, \varepsilon) && \text{elsewhere}
\end{aligned}$$

has a $(\pi + \varepsilon T_+^0)$-periodic solution orbitally near the set $\{(-\cos t, \sin t) \mid -\frac{3}{2}\pi \le t \le 0\} \cup \{(z_+(t, 0), \dot{z}_+(t, 0) \mid 0 \le t \le T_+^0\}$.

Finally, we present an example of the degenerate case in Section III.3.5. We consider the case where $f_+(x) = f_-(x) = -x$, $\Omega_- = \{(x, \dot{x}) \mid x < 0\}$, $\Omega_+ = \{(x, \dot{x}) \mid x > 0\}$, i.e. we take

$$\begin{aligned}
\varepsilon^2 \ddot{x} + x &= 0 && \text{if } x > 0 \\
\ddot{x} + x &= \varepsilon g_-(t, x, \dot{x}, \varepsilon) && \text{if } x < 0,
\end{aligned} \tag{III.3.50}$$

where $g_-(t, x, \dot{x}, \varepsilon)$ is a $(1 + \varepsilon)\pi$-periodic C^2-function. Since

$$q_+(t, \alpha) = (\alpha + \alpha_0)\begin{pmatrix} \sin t \\ \cos t \end{pmatrix}, \qquad 0 \le t \le \pi,$$

we get $\varphi(\alpha) = -\alpha - \alpha_0$ for any α in a neighborhood of $\alpha = 0$, and $\alpha_0 > 0$. Hence by (III.3.43), we are in the degenerate case considered in Section III.3.5. The adjoint variational equation along $q_-(t, \alpha)$ reads now

$$\begin{aligned}
\ddot{w} + w &= 0 \\
w(0) + w(-\pi) &= 0 \\
\psi'' &= w(-\pi) \\
-(\alpha + \alpha_0)\psi' &= 0 \\
\psi_1 &= -\dot{w}(-\pi) - \psi'.
\end{aligned}$$

The first two equations have the two-dimensional family of solutions $w(t) = c \cos(t + t_0)$. We take the two independent solutions $w_1(t) = \cos t$ and $w_2(t) = \sin t$ with the corresponding vectors

$$\psi_1'' = -1, \quad \psi_1' = 0, \quad \psi_1^{(1)} = 0,$$
$$\psi_2'' = 0, \quad \psi_2' = 0, \quad \psi_1^{(2)} = 1.$$

With $g_+(t, x, \dot{x}, \varepsilon) = 0$ (which implies $R(\tau; \xi, \varepsilon)$ is independent of ε) the Poincaré-Andronov-Melnikov vector function is

$$M(\alpha, \tau) = \begin{pmatrix} \int_{-\pi}^{0} g_-(t + \tau, \alpha \sin t, \alpha \cos t, 0) \cos t \, dt \\ \int_{-\pi}^{0} g_-(t + \tau, \alpha \sin t, \alpha \cos t, 0) \sin t \, dt \end{pmatrix}.$$

Then we obtain:

Corollary III.3.9. *Let $g_-(t, x, \dot{x}, \varepsilon)$ be a $(1 + \varepsilon)\pi$-periodic, C^2-function and suppose that $M(\alpha, \tau)$ has a simple zero at $\alpha = \alpha_0$, $\tau = 0$. Then the singularly perturbed system* (III.3.50) *has a $(1 + \varepsilon)\pi$-periodic solution orbitally near the set $\{(\sin t, \cos t) \mid -\pi \leq t \leq \pi\}$.*

We close this section presenting some interesting properties of the solution of (III.3.45). From the identity

$$\dot{z}_+^2(t, \alpha) + (z_+(t, \alpha) + 2)z_+^2(t, \alpha) = (1 - \alpha)(1 + \alpha)^2$$

we derive

$$\frac{\dot{z}_+^2(t, \alpha)}{(1 - \alpha)(1 + \alpha)^2 - (z_+(t, \alpha) + 2)z_+^2(t, \alpha)} = 1. \qquad \text{(III.3.51)}$$

Note that

$$(1 - \alpha)(1 + \alpha)^2 - (x + 2)x^2 = -(1 + \alpha + x)(-1 + \alpha^2 + x - \alpha x + x^2)$$
$$= (A_\alpha - x)(x - B_\alpha)(x - C_\alpha),$$

where

$$A_\alpha = \frac{-1 + \alpha + \sqrt{5 - 2\alpha - 3\alpha^2}}{2}, \qquad B_\alpha = -1 - \alpha,$$

$$C_\alpha = \frac{-1 + \alpha - \sqrt{5 - 2\alpha - 3\alpha^2}}{2}$$

and, for α sufficiently small (in fact for $\alpha \in (-1, 1/3)$),

$$A_\alpha > 0 > B_\alpha > C_\alpha.$$

Using formula 3.131.5 in [29, page 254] we know that, for any $A \geq u > B > C$,

$$\int_B^u \frac{dx}{\sqrt{(A-x)(x-B)(x-C)}} = \frac{2F(\kappa, p)}{\sqrt{A-C}}$$

where

$$\kappa := \arcsin \sqrt{\frac{(A-C)(u-B)}{(A-B)(u-C)}}, \qquad p := \sqrt{\frac{A-B}{A-C}}$$

and F is the elliptic integral of the first kind [30].

Next note that $-1 - \alpha \leq z_+(t, \alpha) \leq 0$, $\dot{z}_+(t, \alpha) \geq 0$ for $t \in [0, T_+(\alpha)]$ and $z_+(0, \alpha) = -1 - \alpha$, $z_+(T_+(\alpha), \alpha) = 0$. Hence (III.3.51) gives

$$T_+(\alpha) = \int_0^{T_+(\alpha)} dt = \int_0^{T_+(\alpha)} \frac{\dot{z}_+(t, \alpha)\, dt}{\sqrt{(1-\alpha)(1+\alpha)^2 - (z_+(t, \alpha) + 2)z_+^2(t, \alpha)}}$$

$$= \int_{B_\alpha}^0 \frac{dx}{\sqrt{(A_\alpha - x)(x - B_\alpha)(x - C_\alpha)}} = \frac{2F(\kappa_\alpha, p_\alpha)}{\sqrt{A_\alpha - C_\alpha}}$$

where

$$\kappa_\alpha = \arcsin\left[\sqrt{\frac{10 + 6\alpha}{5 + 3\alpha + 3\sqrt{5 - 2\alpha - 3\alpha^2}}}\right], \qquad p_\alpha = \sqrt{\frac{1 + 3\alpha + \sqrt{5 - 2\alpha - 3\alpha^2}}{2\sqrt{(1-\alpha)(5+3\alpha)}}}.$$

Thus,

$$T_+(\alpha) = \frac{2F(\kappa_\alpha, p_\alpha)}{\sqrt[4]{5 - 2\alpha - 3\alpha^2}}.$$

We are interested in $\alpha = 0$. Then

$$\kappa_0 = \arcsin\left[\sqrt{\frac{10}{5 + 3\sqrt{5}}}\right], \qquad p_0 = \frac{\sqrt{1 + \sqrt{5}}}{\sqrt{2}\sqrt[4]{5}},$$

and hence

$$T_+(0) = \frac{2F\left(\arcsin\left[\sqrt{\frac{10}{5+3\sqrt{5}}}\right], \frac{\sqrt{1+\sqrt{5}}}{\sqrt{2}\sqrt[4]{5}}\right)}{\sqrt[4]{5}} \doteq 1.88292. \tag{III.3.52}$$

On the other hand, by (III.3.47), we directly verify that $T_+(0) \doteq 1.88292$ by a numerical integration. But we derived (III.3.52) to get an explicit formula for $T_+(0)$ and, in general, for $T_+(\alpha)$.

Furthermore, the previous computations also give

$$t = \int_{B_\alpha}^{z_+(t,\alpha)} \frac{dx}{\sqrt{(A_\alpha - x)(x - B_\alpha)(x - C_\alpha)}} = \frac{2F(\kappa_\alpha(t), p_\alpha)}{\sqrt[4]{5 - 2\alpha - 3\alpha^2}} \tag{III.3.53}$$

for any $0 \le t \le T_+(\alpha)$ and

$$\kappa_\alpha(t) = \arcsin\left[2\sqrt{\frac{\sqrt{(1-\alpha)(5+3\alpha)}(1 + \alpha + z_+(t,\alpha))}{(1 + 3\alpha + \sqrt{5 - 2\alpha - 3\alpha^2})(1 - \alpha + \sqrt{5 - 2\alpha - 3\alpha^2} + 2z_+(t,\alpha))}}\right]. \tag{III.3.54}$$

Solving (III.3.53) we obtain

$$\kappa_\alpha(t) = \text{am}\left(\sqrt[4]{5 - 2\alpha - 3\alpha^2}\frac{t}{2}, p_\alpha\right)$$

where am is the Jacobi amplitude function. Solving (III.3.54) we obtain

$$z_+(t, \alpha) =$$

$$\frac{(1+\alpha)\left(-3 + \sqrt{5 - 2\alpha - 3\alpha^2} - 3\alpha(-1 + H_\alpha(t)) + \left(3 + \sqrt{5 - 2\alpha - 3\alpha^2}\right)H_\alpha(t)\right)}{1 - \sqrt{5 - 2\alpha - 3\alpha^2} - 3\alpha(-1 + H_\alpha(t)) - H_\alpha(t) - \sqrt{5 - 2\alpha - 3\alpha^2}H_\alpha(t)} \tag{III.3.55}$$

for

$$H_\alpha(t) := \cos^2(\kappa_\alpha(t)) = \text{cn}^2\left(\sqrt[4]{5 - 2\alpha - 3\alpha^2}\frac{t}{2}, p_\alpha\right) \tag{III.3.56}$$

where cn is the Jacobi elliptic function. Formulas (III.3.55) and (III.3.56) give explicit solution $z_+(t, \alpha)$. For $\alpha = 0$ we derive

$$z_+(t, 0) = \frac{-3 + \sqrt{5} + \left(3 + \sqrt{5}\right)\text{cn}\left(\sqrt[4]{5}\frac{t}{2}, \frac{\sqrt{1+\sqrt{5}}}{\sqrt{2}\sqrt[4]{5}}\right)^2}{1 - \sqrt{5} - \left(1 + \sqrt{5}\right)\text{cn}\left(\sqrt[4]{5}\frac{t}{2}, \frac{\sqrt{1+\sqrt{5}}}{\sqrt{2}\sqrt[4]{5}}\right)^2}. \tag{III.3.57}$$

We can also compute the Taylor series of (III.3.57) integrating by series the equation $\ddot{x} + 2x + \frac{3}{2}x^2 = 0$ with $x(0) = -1$, $\dot{x}(0) = 0$. Setting

$$x(t) = -1 + \sum_{n=2}^{\infty} \frac{a_n}{n!}t^n$$

we see that the following recurrence condition holds

$$\frac{a_{n+2}}{n!} + 2\frac{a_n}{n!} + \frac{3}{2}\sum_{h=0}^{n} \frac{a_{n-h}}{(n-h)!}\frac{a_h}{h!} = 0 \quad \Leftrightarrow \quad a_{n+2} + 2a_n + \frac{3}{2}\sum_{h=0}^{n}\binom{n}{h}a_{n-h}a_h = 0,$$

where we set $\binom{0}{0} = 1$ and $a_0 = -1$, $a_1 = 0$. Since $a_1 = 0$, we see by induction that

$a_{2k+1} = 0$ for any $k \in \mathbb{N}$ (note that in the product $a_h a_{2k+1-h}$ one of the two indices is odd). So,

$$x(t) = -1 + \sum_{n=1}^{\infty} a_{2n} \frac{t^{2n}}{(2n)!}$$

and

$$a_{2(n+1)} + 2a_{2n} + \frac{3}{2} \sum_{h=0}^{n} \binom{2n}{2h} a_{2(n-h)} a_{2h} = 0.$$

For the first few indices we get

$$a_2 = a_4 = \frac{1}{2}, \qquad a_6 = a_4 - 9a_2^2 = -\frac{7}{4}, \qquad a_8 = a_6 - 45a_2 a_4 = -13,$$

$$a_{10} = a_8 - 84a_2 a_6 - 105a_4^2 = \frac{137}{4}, \qquad a_{12} = a_{10} - 135a_2 a_8 - 630a_4 a_6 = 1463.$$

Therefore,

$$z_+(t,0) = -1 + \frac{1}{2 \cdot 2!} t^2 + \frac{1}{2 \cdot 4!} t^4 - \frac{7}{4 \cdot 6!} t^6 - \frac{13}{8!} t^8 + \frac{137}{4 \cdot 10!} t^{10} + \frac{1463}{12!} t^{12} + \dots.$$

On the other hand, using Mathematica, we can expand (III.3.57) to get

$$z_+(t,0) = -1 + \frac{t^2}{4} + \frac{t^4}{48} - \frac{7t^6}{2880} - \frac{13t^8}{40320} + \frac{137t^{10}}{14515200} + \frac{19t^{12}}{6220800} + \frac{2531t^{14}}{63402393600}$$

$$- \frac{82291t^{16}}{3804143616000} - \frac{179107t^{18}}{166295420928000} + \frac{1972291t^{20}}{17013300756480000} + \dots,$$

which coincides with our previous analytical expansion.

CHAPTER III.4

Approximation and dynamics

III.4.1. Asymptotic properties under approximation

We conclude this part with a discussion on continuous approximations from a global aspect of dynamics. We already know from Remark III.1.5 that a symmetry is preserved under a suitable continuous approximation like (III.1.3). Now we study asymptotic properties by using the Lyapunov function method. We consider (III.1.1) with $\varepsilon = 0$, i.e.

$$\dot{x} = f_{\pm}(x), \quad x \in \overline{\Omega}_{\pm}. \tag{III.4.1}$$

Let $V \in C^1(\overline{\Omega}, \mathbb{R})$ be such that

$$\dot{V}_{\pm}(x) = DV(x)f_{\pm}(x) < 0, \quad x \in \overline{\Omega}_{\pm} \backslash \overline{\Omega}_0. \tag{III.4.2}$$

Let $x(t) \in \Omega, t \geq 0$ be a bounded and locally absolutely continuous solution of (III.4.1) with

$$\dot{x}(t) \in \mathrm{con}[f_-(x(t)), f_+(x(t))] \tag{III.4.3}$$

when $x(t) \in \Omega_0$ and $\dot{x}(t)$ exists. So we mean:

1. If $x(\bar{t}) \in \Omega_{\pm}$ then there is a $\delta_{\bar{t}} > 0$ so that $x(t) \in \Omega_{\pm}$ for any $t \in (\bar{t} - \delta_{\bar{t}}, \bar{t} + \delta_{\bar{t}})$, and $x \in C^1((\bar{t} - \delta_{\bar{t}}, \bar{t} + \delta_{\bar{t}}), \Omega_{\pm})$.
2. (III.4.3) holds for almost each (a.e.) $t \geq 0$ with $x(t) \in \Omega_0$.

Then its omega limit set $\omega(x)$ is compact, connected and nonempty [8, 9, 24]. We have the following result.

Lemma III.4.1. $\omega(x) \subset \overline{\Omega}_0$.

Proof. Note $\omega(x) \subset \overline{\Omega}$. Let $\{t_n\}_{n \in \mathbb{N}}$ be an increasing sequence $t_n > 0, t_n \to \infty$ as $n \to \infty$, and $x(t_n) \to \bar{x} \in \overline{\Omega}_+ \backslash \overline{\Omega}_0$. Then there is a neighborhood $O_{\bar{x}}$ of $\bar{x}$ and a constant $\bar{c} > 0$ such that

$$\dot{V}_+(x) \leq -\bar{c}, \quad \forall x \in O_{\bar{x}}.$$

We take a solution $\bar{x}(t)$ of $\dot{x} = f_+(x)$, $x(0) = \bar{x}$. If $\bar{x} \in \partial\Omega_+$, we consider a C^r-smooth extension of f_+ on $O_{\bar{x}}$, since $f_+ \in C_b^r(\overline{\Omega})$. Then there is a $\delta_{\bar{x}} > 0$ such that $\bar{x}(t) \in O_{\bar{x}}$ for any $t \in [-\delta_{\bar{x}}, \delta_{\bar{x}}]$. Hence $x(t) \in O_{\bar{x}}$ for any $t \in [t_n - \delta_{\bar{x}}, t_n + \delta_{\bar{x}}]$ and any n large, which implies

$$V(x(t_n + \delta_{\bar{x}})) - V(x(t_n - \delta_{\bar{x}})) \leq -2\bar{c}\delta_{\bar{x}}. \tag{III.4.4}$$

Furthermore, $\frac{d}{dt}V(x(t)) = V'(x(t))\dot{x}(t) = \dot{V}_{\pm}(x(t)) < 0$ when $x(t) \in \Omega_{\pm}$ by (III.4.2), and $\frac{d}{dt}V(x(t)) \leq 0$ when $x(t) \in \Omega_0$ and $\dot{x}(t)$ exists, since $\dot{V}_{\pm}(x(t)) \leq 0$ by (III.4.2) and $\dot{x}(t) \in \mathrm{con}[f_-(x(t)), f_+(x(t))]$. Thus $\frac{d}{dt}V(x(t)) \leq 0$ for a.e. $t \geq 0$. On the other hand, since $V(x)$ is locally Lipschitz and $x(t)$ locally absolutely continuous in its arguments, then $V(x(t))$ is also locally absolutely continuous on $[0, \infty)$ implying

$$V(x(t_1)) - V(x(t_2)) = \int_{t_1}^{t_2} \frac{d}{ds}V(x(s))ds \leq 0$$

for any $t_2 > t_1 \geq 0$. Thus $V(x(t))$ is nonincreasing. Since $x(t)$ is bounded and V is continuous on $\bar{\Omega}$, we get $\inf_{t \geq 0} V(x(t)) > -\infty$. So there is $t_0 > 0$ such that

$$V(x(t_2)) - V(x(t_1)) > -2\bar{c}\delta_{\bar{x}}$$

for any $t_2 > t_1 > t_0$, which contradicts (III.4.4). This means that $\omega(x) \cap \bar{\Omega}_+ \backslash \bar{\Omega}_0 = \emptyset$. Similarly we derive $\omega(x) \cap \bar{\Omega}_- \backslash \bar{\Omega}_0 = \emptyset$, so the proof is finished.   □

Remark III.4.2. We consider (III.4.2), since if (III.4.1) has a stable invariant set in $\Omega_{\pm}$ determined by a Lyapunov function, then this invariant set is preserved after approximation (III.1.3). So the interesting case occurs when the invariant set is not in $\Omega_{\pm}$, which is expressed by (III.4.2).

Now we consider approximation (III.1.3) having the form

$$\dot{x} = \frac{1 + \theta(h(x)/\eta)}{2}f_+(x) + \frac{1 - \theta(h(x)/\eta)}{2}f_-(x), \quad x \in \Omega. \tag{III.4.5}$$

Assumption (III.4.2) implies $DV(x)f_{\pm}(x) \leq 0$ for any $x \in \bar{\Omega}_0$. Since it could be $DV(x)f_{\pm}(x) = 0$ for some $x \in \bar{\Omega}_0$, we suppose in addition to (III.4.2), the existence of $\eta_0 > 0$ such that

$$\text{either } DV(x)(f_+(x) + f_-(x)) \leq 0$$
$$\text{and } h(x)DV(x)(f_+(x) - f_-(x)) < 0, \quad \forall x \in \bar{\Omega}_{\eta_0} \backslash \bar{\Omega}_0,$$
$$\text{or } DV(x)(f_+(x) + f_-(x)) < 0 \tag{III.4.6}$$
$$\text{and } h(x)DV(x)(f_+(x) - f_-(x)) \leq 0, \quad \forall x \in \bar{\Omega}_{\eta_0} \backslash \bar{\Omega}_0.$$

Set

$$\dot{V}_{\theta}(x) = DV(x)\left(\frac{1 + \theta(h(x)/\eta)}{2}f_+(x) + \frac{1 - \theta(h(x)/\eta)}{2}f_-(x)\right).$$

Then

$$\dot{V}_\theta(x) = \frac{1}{2}DV(x)(f_+(x) + f_-(x)) + \frac{\theta(h(x)/\eta)}{2}DV(x)(f_+(x) - f_-(x)). \qquad \text{(III.4.7)}$$

Let $\eta < \eta_0$. If $x \in \overline{\Omega}_\pm \backslash \overline{\Omega}_0$, then either $x \in \overline{\Omega}_+ \backslash \overline{\Omega}_0$ or $x \in \overline{\Omega}_- \backslash \overline{\Omega}_0$. If $x \in \overline{\Omega}_+ \backslash \overline{\Omega}_0$ then either $x \in \overline{\Omega}_+ \backslash \overline{\Omega}_{\eta_0}$ and $\dot{V}_\theta(x) = DV(x)f_+(x) < 0$ by (III.4.2), or $x \in \overline{\Omega}_{\eta_0} \backslash \overline{\Omega}_0$ and $\dot{V}_\theta(x) < 0$ by (III.4.6) and (III.4.7), since $\theta(r)r > 0$ for any $r \neq 0$. Similar arguments are applied for $x \in \overline{\Omega}_- \backslash \overline{\Omega}_0$. Summarizing, we obtain an analogy of (III.4.2) of the form

$$\dot{V}_\theta(x) < 0, \quad x \in \overline{\Omega}_\pm \backslash \overline{\Omega}_0. \qquad \text{(III.4.8)}$$

Hence we can repeat the proof of Lemma III.4.1 to arrive at the following

Lemma III.4.3. *Suppose* (III.4.2) *and* (III.4.6). *Let* $x(t) \in \Omega$, $t \geq 0$ *be a bounded* C^1-*solution of* (III.4.5) *with* $0 < \eta < \eta_0$. *Then* $\omega(x) \subset \overline{\Omega}_0$.

The restriction of (III.4.5) on $\overline{\Omega}_0$ is

$$\dot{x} = \frac{f_+(x) + f_-(x)}{2}, \qquad \text{(III.4.9)}$$

so $\omega(x)$ in Lemma III.4.3 must be invariant by (III.4.9), i.e.

$$\omega(x) \subset \{z \in \overline{\Omega}_0 \mid Dh(z)(f_+(z) + f_-(z)) = 0\}. \qquad \text{(III.4.10)}$$

Note $\omega(x)$ is connected.

Remark III.4.4. Lemma III.4.3 remains valid by taking in (III.4.5) any smooth, odd θ with $\theta(r)r > 0$ for any $r \neq 0$.

III.4.2. Application to pendulum with dry friction

Consider the known example of a pendulum with dry friction [10] given by

$$\ddot{x} + \operatorname{sgn} \dot{x} + x = 0$$

in a form of the system

$$\dot{x} = y, \qquad \dot{y} = -\operatorname{sgn} y - x. \qquad \text{(III.4.11)}$$

Now $\Omega = \mathbb{R}^2$, $h(x, y) = y$ and we take $V(x, y) = x^2 + y^2$. Then

$$\dot{V}_\pm(x) = \mp 2y,$$

and condition (III.4.2) holds. So by Lemma III.4.1, $\omega(x)$ of any bounded solution x of (III.4.11) is a subset of the line $y = 0$. Analyzing the vector field of (III.4.5) on $y = 0$ we see that the half-lines $\{|x| > 1, y = 0\}$ are crossed transversally, while the segment

$I = \{|x| \le 1, y = 0\}$ is a sticking region and it is reached in a finite time. We note that (I.4.2) for (III.4.11) on I is just $\dot{x} = 0$, so it is really a sticking part of (III.4.11) (see Figure III.4.1). We also see from the form of $V(x, y)$ that all solutions of (III.4.5) are bounded. So by Lemma III.4.1, $\omega(x)$ is a single point of I, i.e. any solution stops after finite time on the segment I.

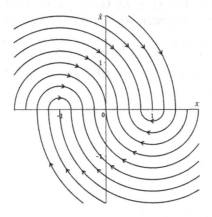

Figure III.4.1 The phase portrait of (III.4.11)

Next, considering an approximation (see (III.1.3))

$$\dot{x} = y, \qquad \dot{y} = -\theta(y/\eta) - x \qquad \qquad \text{(III.4.12)}$$

of (III.4.11) for a smooth odd function $\theta \colon \mathbb{R} \to \mathbb{R}$ with $\theta(y)y > 0$ for $y \ne 0$, we get $\dot{V}(x, y) = -y\theta(y/\eta) \le 0$. By the LaSalle invariance principle [31, Theorem 9.22], we see that $(0, 0)$ is globally asymptotically stable for (III.4.12). This simple example shows that a continuously approximated ODE can have different dynamics than its original discontinuous ODE. To better understand the approximate dynamics, for simplicity we consider (III.1.6), which now has the form

$$\dot{x} = y, \quad \dot{y} = \mp 1 - x, \quad \pm y > \eta,$$
$$\dot{x} = y, \quad \dot{y} = -y/\eta - x, \quad |y| \le \eta. \qquad \text{(III.4.13)}$$

The linear system of (III.4.13) has eigenvalues

$$\frac{-1 - \sqrt{1 - 4\eta^2}}{2\eta} \sim -\infty, \qquad \frac{-1 + \sqrt{1 - 4\eta^2}}{2\eta} \sim 0$$

with corresponding eigenvectors

$$\left(-\frac{1-\sqrt{1-4\eta^2}}{2\eta},1\right)^* \sim (0,1)^*, \qquad \left(1,-\frac{2\eta}{1+\sqrt{1-4\eta^2}}\right)^* \sim (1,0)^*$$

as $\eta \to 0$. Hence all solutions are attracted to $(0,0)$ approximately along the x-axis. This can also be seen in Figure III.4.2 when (III.1.2) is considered with $\eta = 0.01$.

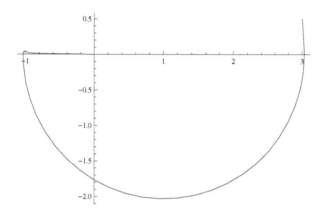

Figure III.4.2 The solution of (III.4.13) with $x(0) = 3$, $y(0) = 0.5$, (III.1.2) and $\eta = 0.01$

On the other hand, we have

$$f_\pm(y,x) = (y,\mp 1 - x), \quad f_+(x,y) + f_-(x,y) = 2(y,-x),$$
$$f_+(x,y) - f_-(x,y) = -2(0,1), \quad DV(x,y)(f_+(x,y) + f_-(x,y)) = 0,$$
$$h(x,y)DV(x,y)(f_+(x,y) - f_-(x,y)) = -4y^2,$$
$$Dh(x,y)(f_+(x,y) + f_-(x,y)) = -2x.$$

So the first alternative of assumption (III.4.6) is satisfied and Remark III.4.4 can be applied to (III.4.12). We know that all orbits of (III.4.12) are bounded, so $\omega(x,y)$ is nonempty. Next, (III.4.10) gives

$$\omega(x,y) \subset \{(x,0) \mid x = 0\} = \{(0,0)\},$$

and we arrive at the same result as previously.

When we take a smooth odd θ such that $\theta(r) = 0$ if and only if $|r| \le r_0$ for some $r_0 > 0$, then (III.4.12) has a more interesting dynamics, again determined by the LaSalle invariance principle: the circle S_η centered at $(0,0)$ with the radius ηr_0 is attracting all solutions from outside of the bounded disc D_η bordered by S_η, while solutions inside D_η are circles centered at $(0,0)$. So $(0,0)$ is stable but not asymptotically.

Next, let us compare (III.4.13) with (III.2.9). For $|y| \leq \eta$, we take $y = \eta w$, $|w| \leq 1$ and (III.4.13) has the form

$$\dot{x} = \eta w, \qquad \eta \dot{w} = -w - x. \qquad (\text{III.4.14})$$

Taking $\eta = 0$, we get the reduced ODE of (III.4.14) (see (A.8))

$$\dot{x} = 0, \qquad w = -x, \qquad (\text{III.4.15})$$

while the reduced ODE of (III.2.11) is

$$\dot{y} = 1, \qquad w = \frac{5}{3}(y^3 - y). \qquad (\text{III.4.16})$$

We see that the solution of (III.4.15) is sticking while the solution of (III.4.16) is sliding on the corresponding graphs $w = -x$ and $w = \frac{5}{3}(y^3 - y)$, respectively. This approves different kinds of studied problems: sticking and sliding. Nevertheless, taking the first-order approximations of the reduced equations, we get

$$\dot{x} = -\eta x \qquad (\text{III.4.17})$$

and

$$\dot{y} = 1 - \kappa \frac{5}{3}(y^3 - y), \qquad (\text{III.4.18})$$

respectively. Note the Tichonov theorem A.9 can be applied to (III.4.14). So (III.4.18) keeps the sliding property for $\kappa > 0$ small, while (III.4.17) loses the sticking property for $\eta > 0$ small, since $x = 0$ becomes a slowly attracting equilibrium, which gives another proof of $\omega(x, y) = \{(0, 0)\}$ for (III.4.13). This is a difference between these two concrete approximations: a sliding solution is continuously approximated by preserving the sliding property, while continuous approximation of sticking solutions changes the dynamics near them.

Of course, the persistence of other aspects of dynamical systems like hyperbolic equilibria or invariant manifolds can be studied. These questions are widely investigated in numerical analysis [32–36].

REFERENCE

[1] A. M. Teixeira, P. R. Silva, Regularization and singular perturbation techniques for non-smooth systems, *Phys. D* **241** (2012) 1948–1955.

[2] N. Fenichel, Geometric singular perturbation theory for ordinary differential equations, *J. Differential Equations* **31** (1979) 53–98.

[3] C. Jones, *Geometric Singular Perturbation Theory*, C.I.M.E. Lectures, Montecatini Terme 1994, in: Lec. Notes Math., vol. 1609, Springer-Verlag 1995.

[4] A. Ivanov, Bifurcations in impact systems, *Chaos Solitons Fractals* **7** (1996) 1615–1634.

[5] O. Makarenkov, F. Verhulst, Bifurcation of asymptotically stable periodic solutions in nearly impact oscillators (unpublished manuscript), preprint [arXiv:0909.4354v1].

[6] D. D. Novaes, M. R. Jeffrey, Regularization of hidden dynamics in piecewise smooth flows, *J. Differential Equations* **259** (2015) 4615–4633.

[7] A. B. Vasileva, V. F. Butuzov, L. V. Kalachev, *The Boundary Function Method for Singular Perturbation Problems*, SIAM 1995.

[8] P. Hartman, *Ordinary Differential Equations*, John Wiley & Sons, Inc. 1964.

[9] C. Chicone, *Ordinary Differential Equations with Applications*, Texts in Applied Mathematics 34, Springer 2006.

[10] M. Kunze, *Non-smooth Dynamical Systems*, Lecture Notes in Mathematics 1744, Springer 2000.

[11] P. Olejnik, J. Awrejcewicz, M. Fečkan, An approximation method for the numerical solution of planar discontinuous dynamical systems with stick-slip friction, *Appl. Math. Sci. (Ruse)* **8** (145) (2014) 7213–7238.

[12] J. Awrejcewicz, M. Fečkan, P. Olejnik, On continuous approximation of discontinuous systems, *Nonlinear Anal.* **62** (2005) 1317–1331.

[13] J. Awrejcewicz, J. Delfs, Dynamics of a self-excited stick-slip oscillator with two degrees of freedom, Part I: Investigation of equilibria, *Eur. J. Mech. A Solids* **9** (4) (1990) 269–282.

[14] J. Awrejcewicz, J. Delfs, Dynamics of a self-excited stick-slip oscillator with two degrees of freedom, Part II: Slip-stick, slip-slip, stick-slip transitions, periodic and chaotic orbits, *Eur. J. Mech. A Solids* **9** (5) (1990) 397–418.

[15] J. Awrejcewicz, P. Olejnik, Stick-slip dynamics of a two-degree-of-freedom system, *Internat. J. Bifur. Chaos Appl. Sci. Engrg.* **13** (4) (2003) 843–861.

[16] A. B. Vasileva, V. F. Butuzov, *Asymptotic Expansion of Solutions of Singularly Perturbed Equations*, Nauka 1973, in Russian.

[17] R. I. Leine, D. H. van Campen, B. L. van de Vrande, Bifurcations in nonlinear discontinuous systems, *Nonlinear Dynam.* **23** (2000) 105–164.

[18] K. Deimling, *Nonlinear Functional Analysis*, Springer-Verlag 1985.

[19] K. Deimling, *Multivalued Differential Equations*, Walter de Gruyter 1992.

[20] L. Cesari, *Asymptotic Behavior and Stability Problems in Ordinary Differential Equations*, Springer-Verlag 1959.

[21] M. Farkas, *Periodic Motions*, Springer-Verlag 1994.

[22] V. A. Yakubovich, V. M. Starzhinskii, *Linear Differential Equations with Periodic Coefficients and Its Applications*, NAUKA 1972, in Russian.

[23] M. U. Akhmet, On the smoothness of solutions of differential equations with a discontinuous right-hand side, *Ukrainian Math. J.* **45** (1993) 1785–1792.

[24] J. Guckenheimer, P. Holmes, *Nonlinear Oscillations, Dynamical Systems and Bifurcations of Vector Fields*, Springer-Verlag 1983.

[25] M. S. Berger, *Nonlinearity and Functional Analysis*, Academic Press 1977.

[26] S. G. Krantz, H. R. Parks, *The Implicit Function Theorem, History, Theory, and Applications*, Birkhäuser 2003.

[27] M. Fečkan, *Topological Degree Approach to Bifurcation Problems*, Springer Netherlands 2008.

[28] E. Gaines, J. Mawhin, *Coincidence Degree, and Nonlinear Differential Equations*, Springer-Verlag 1977.

[29] I. S. Gradshteyn, I. M. Ryzhik, *Table of Integrals, Series, and Products*, Academic Press 2007.

[30] D. F. Lawden, *Elliptic Functions and Applications*, Appl. Math. Sciences 80, Springer-Verlag 1989.

[31] J. K. Hale, H. Koçak, *Dynamics and Bifurcations*, Springer-Verlag 1991.

[32] M. Fečkan, M. Pospíšil, Discretization of dynamical systems with first integrals, *Discrete Cont. Dyn. Syst.* **33** (2013) 3543–3554.

[33] B. M. Garay, P. Kloeden, Discretization near compact invariant sets, *Random Comput. Dyn.* **5** (1997) 93–123.

[34] B. M. Garay, K. H. Lee, Attractors and invariant manifolds under discretization with variable stepsize, *Discrete Cont. Dyn. Syst.* **13** (2005) 827–841.

[35] E. Hairer, Ch. Lubich, G. Wanner, *Geometric Numerical Integration: Structure-Preserving Algorithms for Ordinary Differential Equations*, 2nd edition, Springer-Verlag 2006.

[36] A. M. Stuart, A. R. Humphries, *Dynamical Systems and Numerical Analysis*, Cambridge Univ. Press 1998.

APPENDIX A

A.1. Nonlinear functional analysis

A.1.1. Linear functional analysis

Let X be a Banach space with a norm $|\cdot|$. A sequence $\{x_n\}_{n\in\mathbb{N}} \subset X$ converges to $x_0 \in X$ if $|x_n - x_0| \to 0$ as $n \to \infty$, for short $x_n \to x_0$. A convex hull con S of a subset $S \subset X$ is the intersection of all convex subsets of X containing S.

Let X and Y be Banach spaces. The set of all linear bounded mappings $A \colon X \to Y$ is denoted by $L(X, Y)$, while we put $L(X) := L(X, X)$. The kernel (null space) $A^{-1}(0)$ and the range (image) $A(X)$ is denoted by $\mathcal{N}A$ and $\mathcal{R}A$, respectively.

In using the Lyapunov-Schmidt method, we first need the following Banach inverse mapping theorem.

Theorem A.1. *If $A \in L(X, Y)$ is surjective and injective then its inverse is linear, i.e. $A^{-1} \in L(Y, X)$.*

Then this lemma.

Lemma A.2. *Let $Z \subset X$ be a linear subspace with either $\dim Z < \infty$ or Z be closed with $\operatorname{codim} Z < \infty$. Then there is a bounded projection $P \colon X \to Z$. Note $\operatorname{codim} Z = \dim X/Z$ and X/Z is the factor space of X with respect to Z.*

If the norm $|\cdot|$ is generated by a scalar product $\langle\cdot,\cdot\rangle$, i.e. $|x| = \sqrt{\langle x, x\rangle}$ for any x, then X is a Hilbert space. Now we state a consequence of the Hahn-Banach theorem and the Riesz representation theorem.

Theorem A.3. *Let H be a Hilbert space with a scalar product $\langle\cdot,\cdot\rangle$. For any $x \in H$ and a closed subspace $M \subset H$ such that $x \notin M$, there exists $x^* \in H$ such that $\langle x^*, x\rangle = 1$ and $\langle x^*, z\rangle = 0$ for any $z \in M$.*

More details and proofs of these results can be found in [1, 2].

Poincaré-Andronov-Melnikov Analysis for Non-Smooth Systems.
http://dx.doi.org/10.1016/B978-0-12-804294-6.50019-9

A.1.2. Implicit function theorem

Let X, Y be Banach spaces and $\Omega \subset X$ be open. A map $F: \Omega \to Y$ is said to be Fréchet differentiable at $x_0 \in \Omega$ if there is a $DF(x_0) \in L(X, Y)$ such that

$$\lim_{h \to 0} \frac{|F(x_0 + h) - F(x_0) - DF(x_0)h|}{|h|} = 0.$$

If F is differentiable at each $x \in \Omega$ and $DF: \Omega \to L(X, Y)$ is continuous, then F is said to be continuously differentiable on Ω and we write $F \in C^1(\Omega, Y)$. Higher derivatives are defined in the usual way by induction. Similarly, partial derivatives are defined standardly [3, p. 46]. Now we state the implicit function theorem [4, p. 26].

Theorem A.4. *Let X, Y, Z be Banach spaces, $U \subset X$, $V \subset Y$ be open subsets and $(x_0, y_0) \in U \times V$. Consider $F \in C^1(U \times V, Z)$ such that $F(x_0, y_0) = 0$ and $D_x F(x_0, y_0)$: $X \to Z$ has a bounded inverse. Then there is a neighborhood $U_1 \times V_1 \subset U \times V$ of (x_0, y_0) and a function $f \in C^1(V_1, X)$ such that $f(y_0) = x_0$ and $F(x, y) = 0$ for $(x, y) \in U_1 \times V_1$ if and only if $x = f(y)$. Moreover, if $F \in C^k(U \times V, Z)$, $k \geq 1$ then $f \in C^k(V_1, X)$.*

We refer the reader to [5, 6] for more applications and generalizations of the implicit function theorem.

A.1.3. Lyapunov-Schmidt method

Now we recall the well-known Lyapunov-Schmidt method [7, 8] for solving locally nonlinear equations when the implicit function theorem fails. So let X, Y, Z be Banach spaces, $U \subset X$, $V \subset Y$ be open subsets and $(x_0, y_0) \in U \times V$. Consider $F \in C^1(U \times V, Z)$ such that $F(x_0, y_0) = 0$. If $D_x F(x_0, y_0): X \to Z$ has a bounded inverse then the implicit function theorem can be applied to solve

$$F(x, y) = 0 \tag{A.1}$$

near (x_0, y_0). So we suppose that $D_x F(x_0, y_0): X \to Z$ has no bounded inverse. In general this situation is difficult. The simplest case occurs when $D_x F(x_0, y_0): X \to Z$ is a Fredholm operator, i.e. $\dim \mathcal{N} D_x F(x_0, y_0) < \infty$, $\mathcal{R} D_x F(x_0, y_0)$ is closed in Z and $\operatorname{codim} \mathcal{R} D_x F(x_0, y_0) < \infty$. Then by Lemma A.2, there are bounded projections $P: X \to \mathcal{N} D_x F(x_0, y_0)$ and $Q: Z \to \mathcal{R} D_x F(x_0, y_0)$. Hence we split any $x \in X$ as $x = x_0 + u + v$ with $u \in \mathcal{R}(\mathbb{I} - P)$, $v \in \mathcal{R}P$, and decompose (A.1) as follows

$$H(u, v, y) := QF(x_0 + u + v, y) = 0, \tag{A.2}$$

$$(\mathbb{I} - Q)F(x_0 + u + v, y) = 0. \tag{A.3}$$

Observe that $D_uH(0,0,y_0) = D_xF(x_0,y_0)|_{\mathcal{R}(\mathbb{I}-P)} \to \mathcal{R}D_xF(x_0,y_0)$. So $D_uH(0,0,y_0)$ is injective and surjective. By the Banach inverse mapping Theorem A.1, it has a bounded inverse. Since $H(0,0,y_0) = 0$, the implicit function theorem can be applied to solve (A.2) in $u = u(v,y)$ with $u(0,y_0) = 0$. Inserting this solution to (A.3) we get the bifurcation equation

$$B(v,y) := (\mathbb{I} - Q)F(x_0 + u(v,y) + v, y) = 0. \tag{A.4}$$

Since $B(0,y_0) = (\mathbb{I} - Q)F(x_0,y_0) = 0$ and

$$D_vB(0,y_0) = (\mathbb{I} - Q)D_xF(x_0,y_0)(D_vu(0,y_0) + \mathbb{I}) = 0,$$

the function $B(v,y)$ has a higher singularity at $(0,y_0)$, so the implicit function theorem is not applicable, and the bifurcation theory must be used [4] (see also Theorem A.5).

A.1.4. Crandall-Rabinowitz type result

We present a Crandall-Rabinowitz type result (see also [3, 4] and [9, Theorem 4.1]) concerning the existence of a solution of a nonlinear equation having a manifold of fixed points at a certain value of a parameter.

Theorem A.5. *Let X, Y be Banach spaces and $F: X \times \mathbb{R} \to Y$ a C^2-map such that $F(x,0) = 0$ has a C^2-, d-dimensional manifold of solutions, $\mathcal{M} = \{x = \xi(\mu) \mid \mu \in \mathbb{R}^d\}$. Assume that for any μ in a neighborhood of $\mu = 0$ the linearization $L(\mu) = D_1F(\xi(\mu),0)$ has the null space $T_{\xi(\mu)}\mathcal{M} = \mathrm{span}\{D\xi(\mu)\}$. Assume further that $L(\mu)$ is Fredholm with index zero and let $\Pi(\mu): Y \to \mathcal{R}L(\mu)$ be a projection of Y onto the range of $L(\mu)$. Then if the Poincaré-Andronov-Melnikov function*

$$[\mathbb{I} - \Pi(\mu)]D_2F(\xi(\mu),0)$$

has a simple zero at $\mu = 0$, there exists $\bar{\varepsilon} > 0$ and a unique map $(-\bar{\varepsilon}, \bar{\varepsilon}) \ni \varepsilon \mapsto x(\varepsilon) \in X$ such that $F(x(\varepsilon), \varepsilon) = 0$. Moreover $D_1F(x(\varepsilon), \varepsilon)$ is an isomorphism for $\varepsilon \neq 0$.

Proof. The existence part is quite standard so we sketch it and give emphasis to the proof of invertibility of $D_1F(x(\varepsilon), \varepsilon)$ for $\varepsilon \neq 0$. Since $F(\xi(\mu), 0) = 0$, we get $L(\mu)D\xi(\mu) = 0$ and, differentiating time, $D_1^2F(\xi(\mu), 0)(D\xi(\mu), D\xi(\mu)) + L(\mu)D^2\xi(\mu) = 0$. As a consequence $D\xi(\mu) \in \mathcal{N}L(\mu)$ and

$$D_1^2F(\xi(\mu),0)(v,w) \in \mathcal{R}L(\mu) \text{ for any } v,w \in \mathcal{N}L(\mu).$$

Let $\Pi(\mu): Y \to \mathcal{R}L(\mu)$ be as in the statement of the theorem. We write $x = z + \xi(\mu)$, with $z \in \mathcal{N}L(\mu)^\perp$. Applying the implicit function theorem to the map $(z, \mu, \varepsilon) \mapsto \Pi(\mu)F(z + \xi(\mu), \varepsilon)$ we get the existence of a unique C^2-solution $z = z(\mu, \varepsilon) \in \mathcal{N}L(\mu)^\perp$

of the equation $\Pi(\mu)F(z + \xi(\mu), \varepsilon) = 0$. From uniqueness we also obtain

$$z(\mu, 0) = 0.$$

Next, differentiating the equality $\Pi(\mu)F(z(\mu, \varepsilon) + \xi(\mu), \varepsilon) = 0$ with respect to μ and to ε at $(\mu, 0)$ we get

$$\Pi(\mu)L(\mu)[z_\mu(\mu, 0) + D\xi(\mu)] = 0 \quad \Rightarrow \quad z_\mu(\mu, 0) \in NL(\mu),$$
$$\Pi(\mu)[L(\mu)z_\varepsilon(\mu, 0) + D_2 F(\xi(\mu), 0)] = 0 \quad \Rightarrow \quad L(\mu)z_\varepsilon(\mu, 0) = -\Pi(\mu)D_2 F(\xi(\mu), 0).$$

Next, for $\varepsilon \neq 0$, equation $[\mathbb{I} - \Pi(\mu)]F(z(\mu, \varepsilon) + \xi(\mu), \varepsilon) = 0$ is equivalent to $\varepsilon^{-1}[\mathbb{I} - \Pi(\mu)]F(z(\mu, \varepsilon) + \xi(\mu), \varepsilon) = 0$, but the left-hand side tends to $[\mathbb{I} - \Pi(\mu)]D_2 F(\xi(\mu), 0)$ for $\varepsilon \to 0$, which gives the Poincaré-Andronov-Melnikov condition. We conclude that, if the Poincaré-Andronov-Melnikov condition is satisfied, for $\varepsilon \neq 0$ (small) there exists a unique solution of equation $F(x, \varepsilon) = 0$, $x = x(\varepsilon) = z(\mu(\varepsilon), \varepsilon) + \xi(\mu(\varepsilon))$, with $\mu(0) = 0$.

Now we prove the invertibility of $D_1 F(x(\varepsilon), \varepsilon)$. Since $D_1 F(x(\varepsilon), \varepsilon)$ is Fredholm with index zero, it is enough to prove that equation $D_1 F(x(\varepsilon), \varepsilon)z = 0$ has, for $\varepsilon \neq 0$, the unique solution $z = 0$. Although $\mathcal{F}(z, \varepsilon) := D_1 F(x(\varepsilon), \varepsilon)z$ is only C^1 with respect to ε, it is linear in z. Thus we can still apply the existence and uniqueness argument given previously. Of course, $\mathcal{F}(z, 0)$ vanishes on the linear subspace $NL(0)$, and clearly $ND_1\mathcal{F}(z, 0) = NL(0)$. Next, $RD_1\mathcal{F}(z, 0) = RL(0)$. So, we can consider $M = NL(0)$ and $\Pi(\mu) = \Pi(0)$ in the context of the previous arguments. Thus, from the existence and uniqueness result it follows that $ND_1 F(x(\varepsilon), \varepsilon) = \{0\}$ if the following condition is satisfied:

$$z \in NL(0) \text{ and } [\mathbb{I} - \Pi(0)][D_1^2 F(0, 0)x'(0) + D_1 D_2 F(0, 0)]z = 0 \Rightarrow z = 0.$$

On account of $x(\varepsilon) = z(\mu(\varepsilon), \varepsilon) + \xi(\mu(\varepsilon))$, we are led to look at the solutions of

$$[\mathbb{I} - \Pi(0)][D_1^2 F(0, 0)(z_\mu(0, 0)\mu'(0) + z_\varepsilon(0, 0) + D\xi(0)\mu'(0)) + D_1 D_2 F(0, 0)]z = 0$$

with $z \in NL(0)$. From the previous remarks we get

$$D_1^2 F(0, 0)(z_\mu(0, 0)\mu'(0), z) \in RL(0), \quad D_1^2 F(0, 0)(D\xi(0)\mu'(0), z) \in RL(0)$$

for any $z \in NL(0)$, since $z_\mu(0, 0)\mu'(0), D\xi(0)\mu'(0) \in NL(0)$. So the claim to be proved is:

$$[\mathbb{I} - \Pi(0)]\left[D_1^2 F(0, 0)z_\varepsilon(0, 0) + D_1 D_2 F(0, 0)\right]D\xi(0) \neq 0,$$

where we have replaced z with $D\xi(0)$, since $NL(0) = \text{span}\{D\xi(0)\}$. Now, we differentiate the equality

$$L(\mu)z_\varepsilon(\mu, 0) = -\Pi(\mu)D_2 F(\xi(\mu), 0) = M(\mu) - D_2 F(\xi(\mu), 0)$$

with respect to μ at $\mu = 0$ to get

$$D_1^2 F(0,0)z_\varepsilon(0,0)D\xi(0) + L(0)z_{\varepsilon\varepsilon}(0,0) = DM(0) - D_1 D_2 F(0,0)D\xi(0).$$

Hence,

$$[\mathbb{I} - \Pi(0)]\left[D_1^2 F(0,0)z_\varepsilon(0,0) + D_1 D_2 F(0,0)\right]D\xi(0) = [\mathbb{I} - \Pi(0)]DM(0) \neq 0,$$

since from $[\mathbb{I} - \Pi(\mu)]M(\mu) = M(\mu)$ and $M(0) = 0$, we get $[\mathbb{I} - \Pi(0)]DM(0) = DM(0)$. The proof of Theorem A.5 is complete. □

A.1.5. Leray-Schauder degree

Let X be a Banach space and $\Omega \subset X$ be open and bounded. A continuous map $G \in C(\overline{\Omega}, X)$ is compact, if $G(\overline{\Omega})$ is compact in X. The set of all such maps is denoted by $K(\Omega)$. A triple (F, Ω, y) is admissible, if $F = \mathbb{I} - G$ for some $G \in K(\Omega)$ (so F is a compact perturbation of identity) and $y \in X$ with $y \notin F(\partial\Omega)$, where $\partial\Omega$ is the border of Ω. A mapping $F \in C([0,1] \times \overline{\Omega}, X)$ is an admissible homotopy, if $F(\lambda, \cdot) = \mathbb{I} - G(\lambda, \cdot)$ with $G \in C([0,1] \times \overline{\Omega}, X)$ compact, i.e. $G\left([0,1] \times \overline{\Omega}\right)$ is compact in X, along with $y \notin F([0,1] \times \partial\Omega)$. Now on these admissible triples (F, Ω, y), there is a $\mathbb{Z}$-defined function deg [3, p. 56].

Theorem A.6. *There is a unique mapping* deg *defined on the set of all admissible triples* (F, Ω, y) *determined by the following properties:*
i) *If* $\deg(F, \Omega, y) \neq 0$ *then there is an* $x \in \Omega$ *such that* $F(x) = y$.
ii) $\deg(\mathbb{I}, \Omega, y) = 1$ *for any* $y \in \Omega$.
iii) $\deg(F, \Omega, y) = \deg(F, \Omega_1, y) + \deg(F, \Omega_2, y)$ *whenever* $\Omega_{1,2}$ *are disjoint open subsets of* Ω *such that* $y \notin F\left(\overline{\Omega} \backslash (\Omega_1 \cup \Omega_2)\right)$.
iv) $\deg(F(\lambda, \cdot), \Omega, y)$ *is constant under an admissible homotopy* $F(\lambda, \cdot)$.

The number $\deg(F, \Omega, y)$ is called the Leray-Schauder degree of the map F. If $X = \mathbb{R}^n$ then $\deg(F, \Omega, y)$ is the classic Brouwer degree and F is just $F \in C(\overline{\Omega}, \mathbb{R}^n)$ with $y \notin F(\partial\Omega)$. If x_0 is an isolated zero of F in $\Omega \subset \mathbb{R}^n$, then $I(x_0) := \deg(F, \Omega_0, 0)$ is called the Brouwer index of F at x_0, where $x_0 \in \Omega_0 \subset \Omega$ is an open subset such that x_0 is the only zero point of F in Ω_0 (cf. [4, p. 69]). $I(x_0)$ is independent of such Ω_0. Note, if $y \in \mathbb{R}^n$ is a regular value of $F \in C^1(\overline{\Omega}, \mathbb{R}^n)$, i.e. $\det DF(x) \neq 0$ for any $x \in \Omega$ with $F(x) = y$, and $y \notin F(\partial\Omega)$, then $F^{-1}(y)$ is finite and $\deg(F, \Omega, y) = \sum_{x \in F^{-1}(y)} \operatorname{sgn} \det DF(x)$. In particular, if x_0 is a simple zero of $F(x)$, i.e. $F(x_0) = 0$ and $\det DF(x_0) \neq 0$, then $I(x_0) = \operatorname{sgn} \det DF(x_0) = \pm 1$.

A.2. Multivalued mappings

A.2.1. Upper semicontinuity

Let X, Y be Banach spaces and $\Omega \subset X$. By 2^Y we denote the family of all subsets of Y. Any mapping $F\colon \Omega \to 2^Y \setminus \{\emptyset\}$ is called multivalued or set-valued. For such mappings we define sets

$$\text{graph } F := \{(x,y) \in \Omega \times Y \mid x \in \Omega,\ y \in F(x)\}, \quad F(\Omega) := \bigcup_{x \in \Omega} F(x),$$

$$F^{-1}(A) := \{x \in \Omega \mid F(x) \cap A \neq \emptyset\} \quad \text{for} \quad A \subset Y.$$

Definition A.7. A multivalued mapping $F\colon \Omega \to 2^Y \setminus \{\emptyset\}$ is upper-semicontinuous, usc for short, if the set $F^{-1}(A)$ is closed in Ω for any closed $A \subset Y$.

This condition of usc is more transparent in terms of sequences: if $\{x_n\}_{n=1}^{\infty} \subset \Omega$, $A \subset Y$ is closed, $x_n \to x_0 \in \Omega$ and $F(x_n) \cap A \neq \emptyset$ for all $n \geq 1$, then also $F(x_0) \cap A \neq \emptyset$. The following result is a part of [1, Proposition 1.2.(b)].

Theorem A.8. *If* graph F *is closed and* $\overline{F(\Omega)}$ *is compact then* F *is usc. In particular, F is usc if it has a compact* graph F.

We refer the reader for more properties of usc mappings to [1, p. 3–11], [10] and [11].

A.2.2. Degree theory for set-valued maps

Let X be a Banach space and $\Omega \subset X$ be open and bounded. A triple (F, Ω, y) is admissible if $F = \mathbb{I} - G$ for some $G\colon \overline{\Omega} \to 2^X \setminus \{\emptyset\}$ which is usc with compact convex values and $\overline{G(\overline{\Omega})} \subset X$ is compact, and $y \in X$ with $y \notin F(\partial\Omega)$. Let M be the set of all admissible triples. Then it is possible to define (cf. [1, pp. 154–155]) a unique function $\deg\colon M \to \mathbb{Z}$ with the properties of Theorem A.6 with the evident differences that in i) $y \in F(x)$ is in place of $F(x) = y$ and the homotopy in iv) is compact usc with compact convex values. The number $\deg(F, \Omega, y)$ is the Leray-Schauder degree of the multivalued map F. We refer the reader for more topological methods for multivalued equations to the books [10, 11].

A.3. Singularly perturbed ODEs

A.3.1. Setting of the problem

In this section we recall the well-known result of Tichonov from [12, 13]. Let us consider the system of ODEs of the form

$$\kappa \dot{z} = F(z, y, t), \qquad \dot{y} = f(z, y, t), \tag{A.5}$$

where $F \in C^r(\Omega, \mathbb{R}^m)$, $f \in C^r(\Omega, \mathbb{R}^n)$ for $r \geq 1$ and an open subset $\Omega \subset \mathbb{R}^{m+n+1}$. Here we consider $\kappa > 0$ small. We also associate the initial value condition

$$z(0) = z_0, \qquad y(0) = y_0 \tag{A.6}$$

to (A.5). Setting $\kappa = 0$ in (A.5), we get the degenerate system

$$0 = F(\bar{z}, \bar{y}, t), \qquad \dot{\bar{y}} = f(\bar{z}, \bar{y}, t). \tag{A.7}$$

Since the limit equation (A.7) is not an ODE, (A.5) is called a singularly perturbed ODE. To solve (A.7), we shall suppose that $0 = F(\bar{z}, \bar{y}, t)$ is solvable for $\bar{z} = \varphi(\bar{y}, t)$. Plugging this solution into the ODE of (A.7), we arrive at the reduced ODE

$$\dot{\bar{y}} = f(\varphi(\bar{y}, t), \bar{y}, t), \qquad \bar{y}(0) = y_0. \tag{A.8}$$

Assuming the existence of solution $\bar{y}(t)$ of (A.8) on some interval $[0, T]$, we expect that (A.5) with (A.6) will have a solution $z(\kappa, t)$ and $y(\kappa, t)$ on $[0, T]$ close to $\bar{z}(t) = \varphi(\bar{y}(t), t)$ and $\bar{y}(t)$, respectively. The answer to this problem is given in the next subsection.

A.3.2. Tichonov theorem for singularly perturbed ODEs

We need the following assumptions:

I. F and f are continuous and Lipschitz with respect to z and y in Ω.

II. Equation $F(z, y, t) = 0$ is solvable in $z = \varphi(y, t)$ for any $(y, t) \in \overline{D}$ for a bounded open subset $D \subset \mathbb{R}^{n+1}$ such that

   1. $\varphi(y, t)$ is continuous on $\overline{D}$,

   2. $(\varphi(y, t), y, t) \in \Omega$ for any $(y, t) \in \overline{D}$,

   3. the root $z = \varphi(y, t)$ is isolated in $\overline{D}$, i.e. $\exists \eta > 0$ such that $F(z, y, t) \neq 0$ for any $z, (y, t) \in \overline{D}$ such that $0 < \|z - \varphi(y, t)\| < \eta$.

III. System (A.8) has a unique solution $\bar{y}(t)$ on the interval $[0, T]$ such that $(\bar{y}(t), t) \in D$ for any $t \in [0, T]$. Moreover, $f(\varphi(y, t), y, t)$ is Lipschitz in y on $\overline{D}$.

Now we consider the associated system

$$\frac{d\tilde{z}}{d\tau} = F(\tilde{z}, y, t), \qquad \tau \geq 0, \tag{A.9}$$

when y and t are considered as parameters. By II., $\tilde{z} = \varphi(y, t)$ is an isolated equilibrium of (A.9) for $(y, t) \in \overline{D}$. We suppose

IV. Equilibrium $\tilde{z} = \varphi(y, t)$ of (A.9) is asymptotically stable in the sense of Lyapunov uniformly with respect to $(y, t) \in \overline{D}$.

This means that for any $\varepsilon > 0$ there is $\bar{\delta}(\varepsilon)$ (uniformly for $(y, t) \in \overline{D}$) such that for $\|\tilde{z}(0) - \varphi(y, t)\| < \bar{\delta}(\varepsilon)$ it holds $\|\tilde{z}(\tau) - \varphi(y, t)\| < \varepsilon$ for $\tau \geq 0$ along with $\tilde{z}(\tau) \to \varphi(y, t)$ as $\tau \to \infty$.

Now we investigate (A.9) for $y = y_0$ and $t = 0$,

$$\frac{d\tilde{z}}{d\tau} = F(\tilde{z}, y_0, 0), \quad \tau \geq 0, \tag{A.10}$$

with the initial value condition

$$\tilde{z}(0) = z_0. \tag{A.11}$$

Since z_0 is not close, in general, to $\varphi(y_0, 0)$, we suppose

V. The solution $\tilde{z}(\tau)$ of (A.10) and (A.11) satisfies

    1. $\tilde{z}(\tau) \to \varphi(y_0, 0)$ as $\tau \to \infty$,

    2. $(\tilde{z}(\tau), y_0, 0) \in \Omega$ for $\tau \geq 0$.

Now we are ready to state the Tichonov theorem in its general form [12, 13].

Theorem A.9. *Under assumptions I.–V. there is a constant $\kappa_0 > 0$ such that for any $0 < \kappa \leq \kappa_0$ the solution $z(\kappa, t)$, $y(\kappa, t)$ of (A.5), (A.6) exists on $[0, T]$, it is unique and satisfies*

$$\lim_{\kappa \to 0^+} y(\kappa, t) = \bar{y}(t) \quad for\ 0 \leq t \leq T, \tag{A.12}$$

$$\lim_{\kappa \to 0^+} z(\kappa, t) = \bar{z}(t) = \varphi(\bar{y}(t), t) \quad for\ 0 < t \leq T. \tag{A.13}$$

By [12, 13], the limit is uniform in (A.12). To get a better asymptotic property than (A.13), we assume that F and f are at least C^2-smooth on Ω, and we strengthen assumption IV. as follows. Let $\bar{\lambda}_i(t)$, $i = 1, 2, \ldots, m$ be eigenvalues of the Jacobian matrix $D_z F(\bar{z}(t), \bar{y}(t), t)$. Moreover, we assume

IV'. $\max\{\sup_{t \in [0,T]} \Re \bar{\lambda}_i(t), i = 1, 2, \ldots, m\} < 0$.

Then it holds

$$z(\kappa, t) = \bar{z}(t) + \tilde{z}(t/\kappa) - \varphi(y_0, 0) + O(\kappa) \quad \text{uniformly on } [0, T]. \tag{A.14}$$

Furthermore, we get a higher-order asymptotic expansion of $z(\kappa, t)$ and $y(\kappa, t)$ by κ uniformly on $[0, T]$ when F and f are smoother. Moreover, when F and f depend smoothly also on κ and the other parameter $\mu \in \mathbb{R}^p$, then the previous asymptotic results remain also for derivatives with respect to μ.

A.4. Note on Lyapunov theorem for Hill's equation

This section is inspired by [14]. Let us consider the so-called Hill's equation

$$\ddot{z} + p(t)z = 0, \quad t \geq 0 \tag{A.15}$$

with T-periodic function $p \in C(\mathbb{R}, \mathbb{R})$. The next result is an improvement of [14, Theorem 2.5.3].

Lemma A.10. *If $p(t) < 0$ for all $t \in \mathbb{R}$, then equation (A.15) is asymptotically unstable with the characteristic multiplier greater than or equal to $e^{\sqrt{\mu}T}$ for $\mu := -\max_{t \in [0,T]} p(t) > 0$.*

Proof. By formula [14, (2.5.9)], the Lyapunov constant a of (A.15) is given by

$$a = 2 - T \int_0^T p(t)dt + \sum_{k=2}^{\infty} (-1)^k \int_0^T \int_0^{t_1} \dots \int_0^{t_{k-1}} (T - t_1 + t_k)$$
$$\times (t_1 - t_2) \dots (t_{k-1} - t_k) p(t_1)p(t_2) \dots p(t_k) dt_k \dots dt_2 dt_1.$$

Applying the estimation by μ, we get

$$a \geq 2 + \mu T^2 + \sum_{k=2}^{\infty} \mu^k \int_0^T \int_0^{t_1} \dots \int_0^{t_{k-1}} (T - t_1 + t_k)$$
$$\times (t_1 - t_2) \dots (t_{k-1} - t_k) dt_k \dots dt_2 dt_1 =: b.$$

That means that the Lyapunov constant of (A.15) is greater than or equal to the Lyapunov constant b of the equation

$$\ddot{z} - \mu z = 0, \quad t \geq 0. \tag{A.16}$$

Note that b can be explicitly calculated from the fundamental matrix solution of the system

$$\dot{z}_1 = z_2$$
$$\dot{z}_2 = \mu z_1$$

corresponding to equation (A.16). The fundamental solution is given by

$$\Phi(t) = \begin{pmatrix} \cosh \sqrt{\mu}t & \frac{\sinh \sqrt{\mu}t}{\sqrt{\mu}} \\ \sqrt{\mu} \sinh \sqrt{\mu}t & \cosh \sqrt{\mu}t \end{pmatrix}, \quad t \geq 0.$$

Then $b = \text{Tr}\, \Phi(T) = 2\cosh \sqrt{\mu}T \leq a$. Finally, the largest characteristic multiplier of

(A.15) satisfies

$$\frac{a + \sqrt{a^2 - 4}}{2} \geq \frac{b + \sqrt{b^2 - 4}}{2} = e^{\sqrt{\mu}T}.$$

The proof is finished.   □

A similar result is also derived in [15, p. 516], but we presented a proof sufficient for our purpose.

REFERENCE

[1] K. Deimling, *Multivalued Differential Equations*, Walter de Gruyter 1992.
[2] K. Yosida, *Functional Analysis*, Springer-Verlag 1965.
[3] K. Deimling, *Nonlinear Functional Analysis*, Springer-Verlag 1985.
[4] S. N. Chow, J. K. Hale, *Methods of Bifurcation Theory*, Texts in Applied Mathematics 34, Springer-Verlag 1982.
[5] M. S. Berger, *Nonlinearity and Functional Analysis*, Academic Press 1977.
[6] S. G. Krantz, H. R. Parks, *The Implicit Function Theorem, History, Theory, and Applications*, Birkhäuser 2003.
[7] C. Chicone, *Ordinary Differential Equations with Applications*, Texts in Applied Mathematics 34, Springer 2006.
[8] M. Medved', *Dynamické systémy*, Veda 1988, in Slovak.
[9] K. J. Palmer, Exponential dichotomies and transversal homoclinic points, *J. Differential Equations* **55** (1984) 225–256.
[10] J. Andres, L. Górniewicz, *Topological Fixed Point Principles for Boundary Value Problems*, Kluwer 2003.
[11] L. Górniewicz, *Topological Fixed Point Theory for Multivalued Mappings*, Springer-Verlag 1973.
[12] A. B. Vasileva, V. F. Butuzov, *Asymptotic Expansion of Solutions of Singularly Perturbed Equations*, Nauka 1973, in Russian.
[13] A. B. Vasileva, V. F. Butuzov, L. V. Kalachev, *The Boundary Function Method for Singular Perturbation Problems*, SIAM 1995.
[14] M. Farkas, *Periodic Motions*, Springer-Verlag 1994.
[15] V. A. Yakubovich, V. M. Starzhinskii, *Linear Differential Equations with Periodic Coefficients and Its Applications*, NAUKA 1972, in Russian.

BIBLIOGRAPHY

[1] V. Acary, O. Bonnefon, B. Brogliato, *Nonsmooth Modeling and Simulation for Switched Circuits*, Springer 2011.

[2] Z. Afsharnezhad, M. Karimi Amaleh, Continuation of the periodic orbits for the differential equation with discontinuous right hand side, *J. Dynam. Differential Equations* **23** (2011) 71–92.

[3] M. U. Akhmet, On the smoothness of solutions of differential equations with a discontinuous right-hand side, *Ukrainian Math. J.* **45** (1993) 1785–1792.

[4] M. U. Akhmet, Periodic solutions of strongly nonlinear systems with non classical right-side in the case of a family of generating solutions, *Ukrainian Math. J.* **45** (1993) 215–222.

[5] M. U. Akhmet, D. Arugaslan, Bifurcation of a non-smooth planar limit cycle from a vertex, *Nonlinear Anal.* **71** (2009) 2723–2733.

[6] M.U. Akhmet, C. Buyukadali, T. Ergenc, Periodic solutions of the hybrid system with small parameter, *Nonlinear Anal. Hybrid Syst.* **2** (2008) 532–543.

[7] J. Andres, L. Górniewicz, *Topological Fixed Point Principles for Boundary Value Problems*, Kluwer 2003.

[8] A. A. Andronov, A. A. Vitt, S. E. Khaikin, *Theory of Oscillators*, Pergamon Press 1966.

[9] J. Awrejcewicz, J. Delfs, Dynamics of a self-excited stick-slip oscillator with two degrees of freedom, Part I: Investigation of equilibria, *Eur. J. Mech. A Solids* **9** (4) (1990) 269–282.

[10] J. Awrejcewicz, J. Delfs, Dynamics of a self-excited stick-slip oscillator with two degrees of freedom, Part II: Slip-stick, slip-slip, stick-slip transitions, periodic and chaotic orbits, *Eur. J. Mech. A Solids* **9** (5) (1990) 397–418.

[11] J. Awrejcewicz, M. Fečkan, P. Olejnik, On continuous approximation of discontinuous systems, *Nonlinear Anal.* **62** (2005) 1317–1331.

[12] J. Awrejcewicz, M. M. Holicke, *Smooth and Nonsmooth High Dimensional Chaos and the Melnikov-Type Methods*, World Scientific Publishing Company 2007.

[13] J. Awrejcewicz, G. Kudra, Modeling, numerical analysis and application of triple physical pendulum with rigid limiters of motion, *Arch. Appl. Mech.* **74** (11) (2005) 746–753.

[14] J. Awrejcewicz, G. Kudra, C.-H. Lamarque, Dynamics investigation of three coupled rods with a horizontal barrier, *Meccanica* **38** (6) (2003) 687–698.

[15] J. Awrejcewicz, P. Olejnik, Stick-slip dynamics of a two-degree-of-freedom system, *Internat. J. Bifur. Chaos Appl. Sci. Engrg.* **13** (4) (2003) 843–861.

[16] D. Bainov, P. S. Simeonov, *Impulsive Differential Equations: Asymptotic Properties of the Solutions*, World Scientific Publishing Company 1995.

[17] F. Battelli, M. Fečkan, Homoclinic trajectories in discontinuous systems, *J. Dynam. Differential Equations* **20** (2008) 337–376.

[18] F. Battelli, M. Fečkan, Some remarks on the Melnikov function, *Electron. J. Differential Equations* **2002** (13) (2002) 1–29.

[19] F. Battelli, M. Fečkan, Fast-slow dynamical approximation of forced impact systems near periodic solutions, *Bound. Value Probl.* **2013** (71) (2013) 1–33.

[20] F. Battelli, M. Fečkan, On the chaotic behavior of non-flat billiards, *Commun. Nonlinear Sci. Numer. Simul.* **19** (2014) 1442–1464.

[21] A. M. Baxter, R. Umble, Periodic orbits of billiards on an equilateral triangle, *Amer. Math. Monthly* **115** (2008) 479–491.

[22] E. DiBenedetto, *Classical Mechanics: Theory and Mathematical Modeling*, Cornerstones, Birkhäuser 2011.

[23] M. di Bernardo, C. J. Budd, A. R. Champneys, P. Kowalczyk, *Piecewise-smooth Dynamical Systems: Theory and Applications*, Applied Mathematical Sciences 163, Springer-Verlag 2008.

[24] M. S. Berger, *Nonlinearity and Functional Analysis*, Academic Press 1977.

[25] M. Bonnin, F. Corinto, M. Gilli, Diliberto's theorem in higher dimension, *Internat. J. Bifur. Chaos*

    Appl. Sci. Engrg. **19** (2009) 629–637.
[26] B. Brogliato, *Nonsmooth Impact Mechanics*, Lecture Notes in Control and Information Sciences 220, Springer 1996.
[27] A. Buica, J. Llibre, Averaging methods for finding periodic orbits via Brouwer degree, *Bull. Sci. Math.* **128** (2004) 7–22.
[28] N. Chernov, R. Markarian, *Chaotic Billiards*, Mathematical Surveys and Monographs 127, Amer. Math. Soc. 2006.
[29] L. Cesari, *Asymptotic Behavior and Stability Problems in Ordinary Differential Equations*, Springer-Verlag 1959.
[30] C. Chicone, *Ordinary Differential Equations with Applications*, Texts in Applied Mathematics 34, Springer 2006.
[31] D. R. J. Chillingworth, Discontinuous geometry for an impact oscillator, *Dyn. Syst.* **17** (2002) 389–420.
[32] S. N. Chow, J. K. Hale, *Methods of Bifurcation Theory*, Texts in Applied Mathematics 34, Springer-Verlag 1982.
[33] L. O. Chua, M. Komuro, T. Matsumoto, The double scroll family, *IEEE Trans. Circuits Syst.* **33** (1986) 1073–1118.
[34] K. Deimling, *Nonlinear Functional Analysis*, Springer-Verlag 1985.
[35] K. Deimling, *Multivalued Differential Equations*, Walter de Gruyter 1992.
[36] N. Dilna, M. Fečkan, On the uniqueness, stability and hyperbolicity of symmetric and periodic solutions of weakly nonlinear ordinary differential equations, *Miskolc Math. Notes* **10** (1) (2009) 11–40.
[37] N. Dilna, M. Fečkan, About the uniqueness and stability of symmetric and periodic solutions of weakly nonlinear ordinary differential equations, *Dopovidi Nac. Acad. Nauk Ukraini (Reports of the National Academy of Sciences of Ukraine)* **5** (2009) 22–28.
[38] N. Dilna, M. Fečkan, Weakly non-linear and symmetric periodic systems at resonance, *Nonlinear Stud.* **16** (2009) 23–44.
[39] Z. Du, W. Zhang, Melnikov method for homoclinic bifurcation in nonlinear impact oscillators, *Comput. Math. Appl.* **50** (2005) 445–458.
[40] M. Farkas, *Periodic Motions*, Springer-Verlag 1994.
[41] M. Fečkan, *Topological Degree Approach to Bifurcation Problems*, Springer Netherlands 2008.
[42] M. Fečkan, R. Ma, B. Thompson, Weakly coupled oscillators and topological degree, *Bull. Sci. Math.* **131** (2007) 559–571.
[43] M. Fečkan, M. Pospíšil, On the bifurcation of periodic orbits in discontinuous sytems, *Commun. Math. Anal.* **8** (2010) 87–108.
[44] M. Fečkan, M. Pospíšil, Bifurcation from family of periodic orbits in discontinuous systems, *Differ. Equ. Dyn. Syst.* **20** (2012) 207–234.
[45] M. Fečkan, M. Pospíšil, Bifurcation from single periodic orbit in discontinuous autonomous systems, *Appl. Anal.* **92** (2013) 1085–1100.
[46] M. Fečkan, M. Pospíšil, Bifurcation of periodic orbits in periodically forced impact systems, *Math. Slovaca* **64** (2014) 101–118.
[47] M. Fečkan, M. Pospíšil, Bifurcation of sliding periodic orbits in periodically forced discontinuous systems, *Nonlinear Anal. Real World Appl.* **14** (2013) 150–162.
[48] M. Fečkan, M. Pospíšil, Discretization of dynamical systems with first integrals, *Discrete Cont. Dyn. Syst.* **33** (2013) 3543–3554.
[49] N. Fenichel, Geometric singular perturbation theory for ordinary differential equations, *J. Differential Equations* **31** (1979) 53–98.
[50] A. Fidlin, *Nonlinear Oscillations in Mechanical Engineering*, Springer 2006.
[51] A. F. Filippov, *Differential Equations with Discontinuous Righthand Sides*, Mathematics and Its Applications 18, Kluwer Academic 1988.
[52] U. Galvanetto, C. Knudsen, Event maps in a stick-slip system, *Nonlinear Dynam.* **13** (1997) 99–115.
[53] E. Gaines, J. Mawhin, *Coincidence Degree, and Nonlinear Differential Equations*, Springer-Verlag 1977.

[54] F. R. Gantmacher, *Applications of the Theory of Matrices*, Interscience 1959.

[55] B. M. Garay, P. Kloeden, Discretization near compact invariant sets, *Random Comput. Dyn.* **5** (1997) 93–123.

[56] B. M. Garay, K. H. Lee, Attractors and invariant manifolds under discretization with variable stepsize, *Discrete Cont. Dyn. Syst.* **13** (2005) 827–841.

[57] F. Giannakopoulos, K. Pliete, Planar systems of piecewise linear differential equations with a line of discontinuity, *Nonlinearity* **14** (2001) 1611–1632.

[58] M. Golubitsky, V. Guillemin, *Stable Mappings and Their Singularities*, Kluwer Academic Publishers 1999.

[59] L. Górniewicz, *Topological Fixed Point Theory for Multivalued Mappings*, Springer-Verlag 1973.

[60] I. S. Gradshteyn, I. M. Ryzhik, *Table of Integrals, Series, and Products*, Academic Press 2007.

[61] J. Guckenheimer, P. Holmes, *Nonlinear Oscillations, Dynamical Systems and Bifurcations of Vector Fields*, Springer-Verlag 1983.

[62] E. Gutkin, Billiard dynamics: a survey with the emphasis on open problems, *Regul. Chaotic Dyn.* **8** (2003) 1–13.

[63] E. Hairer, Ch. Lubich, G. Wanner, *Geometric Numerical Integration: Structure-Preserving Algorithms for Ordinary Differential Equations*, 2nd edition, Springer-Verlag 2006.

[64] A. Halanay, D. Wexler, *Qualitative Theory of Impulsive Systems*, Editura Academiei Republicii Socialiste Romania 1968.

[65] J. K. Hale, H. Koçak, *Dynamics and Bifurcations*, Springer-Verlag 1991.

[66] P. Hartman, *Ordinary Differential Equations*, John Wiley & Sons, Inc. 1964.

[67] A. Ivanov, Bifurcations in impact systems, *Chaos Solitons Fractals* **7** (1996) 1615–1634.

[68] C. Jones, *Geometric Singular Perturbation Theory*, C.I.M.E. Lectures, Montecatini Terme 1994, in: Lec. Notes Math., vol. 1609, Springer-Verlag 1995.

[69] H.-J. Jodl H. J. Korsch, T. Hartmann, *Chaos: A Program Collection for the PC*, 3rd edition, Springer-Verlag Berlin 2008.

[70] V. V. Kozlov, D. V. Treshhev, *Billiards: A Genetic Introduction to the Dynamics of Systems with Impacts*, Translations of Mathematical Monographs 89, American Mathematical Society 1991.

[71] A. Kovaleva, The Melnikov criterion of instability for random rocking dynamics of a rigid block with an attached secondary structure, *Nonlinear Anal. Real World Appl.* **11** (2010) 472–479.

[72] S. G. Krantz, H. R. Parks, *The Implicit Function Theorem, History, Theory, and Applications*, Birkhäuser 2003.

[73] P. Kukučka, Jumps of the fundamental solution matrix in discontinuous systems and applications, *Nonlinear Anal.* **66** (2007) 2529–2546.

[74] P. Kukučka, Melnikov method for discontinuous planar systems, *Nonlinear Anal.* **66** (2007) 2698–2719.

[75] M. Kunze, *Non-smooth Dynamical Systems*, Lecture Notes in Mathematics 1744, Springer 2000.

[76] M. Kunze, T. Küpper, *Non-smooth dynamical systems: an overview*, Springer 2001 pp. 431–452.

[77] M. Kunze, T. Küpper, Qualitative bifurcation analysis of a non-smooth friction-oscillator model, *Z. Angew. Math. Phys.* **48** (1997) 87–101.

[78] Yu. A. Kuznetsov, S. Rinaldi, A. Gragnani, One-parametric bifurcations in planar Filippov systems, *Internat. J. Bifur. Chaos Appl. Sci. Engrg.* **13** (2003) 2157–2188.

[79] V. Lakshmikantham, D. Bainov, P.S. Simeonov, *Theory of Impulsive Differential Equation*, World Scientific Publishing Company 1989.

[80] D. F. Lawden, *Elliptic Functions and Applications*, Appl. Math. Sciences 80, Springer-Verlag 1989.

[81] R. I. Leine, H. Nijmeijer, *Dynamics and Bifurcations of Non-smooth Mechanical Systems*, Lecture Notes in Applied and Computational Mechanics 18, Springer-Verlag 2004.

[82] R. I. Leine, D. H. van Campen, B. L. van de Vrande, Bifurcations in nonlinear discontinuous systems, *Nonlinear Dynam.* **23** (2000) 105–164.

[83] S. Lenci, G. Rega, Heteroclinic bifurcations and optimal control in the nonlinear rocking dynamics of generic and slender rigid blocks, *Internat. J. Bifur. Chaos Appl. Sci. Engrg.* **15** (2005) 1901–1918.

[84] J. Llibre, O. Makarenkov, Asymptotic stability of periodic solutions for nonsmooth differential equations with application to the nonsmooth van der Pol oscillator, *SIAM J. Math. Anal.* **40** (2009) 2478–2495.

[85] O. Makarenkov, F. Verhulst, Bifurcation of asymptotically stable periodic solutions in nearly impact oscillators (unpublished manuscript), preprint [arXiv:0909.4354v1].

[86] M. Medved', *Dynamické systémy*, Veda 1988, in Slovak.

[87] M. Medved', *Dynamické systémy*, Comenius University in Bratislava 2000, in Slovak.

[88] J. Moser, Regularization of Kepler's problem and the averaging method on a manifold, *Comm. Pure Appl. Math.* **23** (1970) 609–636.

[89] J. Murdock, C. Robinson, Qualitative dynamics from asymptotic expansions: local theory, *J. Differential Equations* **36** (3) (1980) 425–441.

[90] D. D. Novaes, M. R. Jeffrey, Regularization of hidden dynamics in piecewise smooth flows, *J. Differential Equations* **259** (2015) 4615–4633.

[91] P. Olejnik, J. Awrejcewicz, M. Fečkan, An approximation method for the numerical solution of planar discontinuous dynamical systems with stick-slip friction, *Appl. Math. Sci. (Ruse)* **8** (145) (2014) 7213–7238.

[92] K. J. Palmer, Exponential dichotomies and transversal homoclinic points, *J. Differential Equations* **55** (1984) 225–256.

[93] W. Rudin, *Real and Complex Analysis*, McGraw-Hill, Inc. 1974.

[94] T. Ruijgrok, Periodic orbits in triangular billiards, *Acta Phys. Polon. B* **22** (1991) 955–981.

[95] M. Rychlik, Periodic points of the billiard ball map in a convex domain, *J. Differential Geom.* **30** (1989) 191–205.

[96] A.M. Samoilenko, N.A. Perestyuk, *Impulsive Differential Equations*, World Scientific Publishing Company 1995.

[97] A. M. Stuart, A. R. Humphries, *Dynamical Systems and Numerical Analysis*, Cambridge Univ. Press 1998.

[98] S. Tabachnikov, *Geometry and Billiards*, Student Mathematical Library 30, American Mathematical Society 2005.

[99] A. M. Teixeira, P. R. Silva, Regularization and singular perturbation techniques for non-smooth systems, *Phys. D* **241** (2012) 1948–1955.

[100] A. B. Vasileva, V. F. Butuzov, *Asymptotic Expansion of Solutions of Singularly Perturbed Equations*, Nauka 1973, in Russian.

[101] A. B. Vasileva, V. F. Butuzov, L. V. Kalachev, *The Boundary Function Method for Singular Perturbation Problems*, SIAM 1995.

[102] W. Xu, J. Feng, H. Rong, Melnikov's method for a general nonlinear vibro-impact oscillator, *Nonlinear Anal.* **71** (2009) 418–426.

[103] V. A. Yakubovich, V. M. Starzhinskii, *Linear Differential Equations with Periodic Coefficients and Its Applications*, NAUKA 1972, in Russian.

[104] K. Yosida, *Functional Analysis*, Springer-Verlag 1965.

INDEX

Printed in the United States
By Bookmasters